Springer Series in
ADVANCED MICROELECTRONICS 24

Springer Series in
ADVANCED MICROELECTRONICS

The Springer Series in Advanced Microelectronics provides systematic information on all the topics relevant for the design, processing, and manufacturing of microelectronic devices. The books, each prepared by leading researchers or engineers in their fields, cover the basic and advanced aspects of topics such as wafer processing, materials, device design, device technologies, circuit design, VLSI implementation, and subsystem technology. The series forms a bridge between physics and engineering and the volumes will appeal to practicing engineers as well as research scientists.

18 **Microcontrollers in Practice**
By I. Susnea and M. Mitescu

19 **Gettering Defects in Semiconductors**
By V.A. Perevoschikov and V.D. Skoupov

20 **Low Power VCO Design in CMOS**
By M. Tiebout

21 **Continuous-Time Sigma-Delta A/D Conversion**
Fundamentals, Performance Limits and Robust Implementations
By M. Ortmanns and F. Gerfers

22 **Detection and Signal Processing**
Technical Realization
By W.J. Witteman

23 **Highly Sensitive Optical Receivers**
By K. Schneider and H.K. Zimmermann

24 **Bonding in Microsystem Technology**
By J.A. Dziuban

Volumes 1–17 are listed at the end of the book.

J.A. Dziuban

Bonding in Microsystem Technology

With 296 Figures

Springer

Dr. Jan A. Dziuban
Wrocław University of Technology
Faculty of Microsystems Electronics and Photonics
Wrocław, Poland
Email: ja.dziuban@pwr.wroc.pl

ISSN 1437-0387

ISBN-10 1-4020-4578-6 (HB)
ISBN-13 978-1-4020-4578-3 (HB)
ISBN-10 1-4020-4589-1 (e-book)
ISBN-13 978-1-4020-4589-9 (e-book)

Springer is a part of Springer Science+Business Media.

springer.com

Printed in The Netherlands

Camera-ready by the Author
Cover concept by eStudio Calmar Steinen using a background picture from Photo Studio "SONO".
Courtesy of Mr. Yukio Sono, 3-18-4 Uchi-Kanda, Chiyoda-ku, Tokyo
Cover design: *design & production* GmbH, Heidelberg

Printed on acid-free paper SPIN: 11544302 - 5 4 3 2 1 0

For my wife Małgorzata
and our children:
Hanna, Piotr and Tusia.

With love

The final version of this book has been prepared by the author during his "stage rouge" in the CNRS Institute FEMTO – St, Besançon, France.

I would like to thank my French friends for the warm atmosphere surrounding me during the stage. Special thanks to Michel de Labachelerie and Christophe Gorecki.

CONTENTS

ACKNOWLEDGMENTS

I would like to thank all my friends from the Faculty of Microsystem Electronics and Photonics of the Wrocław University of Technology, Wrocław, Poland. This is my mother university, where I teach students microsystem technology and develop technological procedures and devices.

Specially I would like to thank dr Irena Barycka, dr Anna Górecka-Drzazga, dr Rafał Walczak, dr Sylwester Bargiel, dr Lukasz Nieradko, Paweł Knapkiewicz and all my colleagues from the laboratory for their help. Special thanks to M.Sc. Ewa Bargiel for translation of the text from Polish. I would also like to thank my best friends from the Institute of Electron Technology of Warsaw (actually my second place of work): M.Sc. Jan Koszur, M.Sc. Paweł Kowalski, M.Sc. Jerzy Jazwiński and dr Zenon Gniazdowki, for their continuing cooperation.

LIST OF ACRONYMS AND SYMBOLS

ATE	*Anisotropic Electrochemical Etching*
bio-chip	*biochemical, biomedical chip*
CAGR	*Coefficient of Annual Growth Rate*
CHEM-FET	*Chemical Field-Electric-Transistor*
CMP	*Chemical-Mechanical-Polishing*
CSAC	*Chip-Scale Atomic Clock*
DRIE	*Deep Reactive Ion Etching*
EDP	*Ethylenodiamine/Pyrazine*
EMSi	*Etching Microwave Silicon*
ERDA	*Elastic Recoil Detection Analysis*
E2MSi	*Extended Etching Microwave*
FEA	*Field Emitting Array*
FED	*Field Emission Displays*
FSO	*Full Scale Output*
G-FEA	*Gated-Field Emitter Array*
IPA	*Isopropyl Alcohol*
lab-on-chip	*laboratory on-the-chip, type of μTAS*
LIGA	*Lithography Galvanotechnik Abformung*
LPCVD	*Low Pressure Chemical Vapor Deposition*
LTB	*Low Temperature Bonding*
MAP	*Manifold Absolute Pressure (Sensor)*
MEMS	*Micro-Electro-Mechanical-System(s)*
MEOMS	*Micro-Electro-Opto-Mechanical-System(s)*
MLE	*Molecular Layer Epitaxy*
MOLD	*method of fabrication of microdetails*
MST	*Micro-System-Technology*
NEG	*Non Evaporable Getter*
NHA	*Nitric Hydro fluoric Acid*
PECVD	*Plasma Enhanced CVD*
PERIE	*Plasma Enhanced RIE*
PMMA	*Polymethyl Methacrylate*
RIE	*Reactive Ion Etching*

SDB	***S**ilicon **D**irect **B**onding*
SFB	***S**ilicon **F**usion **B**onding*
SIMS	***S**econdary-**I**on **M**ass **S**pectroscopy*
SMA	***S**hape **M**emory **A**lloy*
SOG	***S**pin-**O**n-**G**lass*
SOI	***S**ilicon-**O**n-**I**nsulator*
STB	***S**ilicon **T**hermal **B**onding*
TEOS	***T**ri **E**thylo **O**xysilane*
TFFF	***T**hermal **F**ield-**F**low **F**ractionation*
TMAH	*Tetra-**M**ethyl-**A**mmonium-**H**ydoxide*
TTV	***T**otal **T**hickness **V**ariation*
VCSEL	*Verti**c**al **S**urface **E**mitting **L**aser*
M-MOSFET	***V**acuum **M**etal-**O**xide **F**ield **E**mitting **T**ransistor*
VSM	***V**ery **S**mall **M**achine(s)*
XPS	***X**-ray **P**hotoelectron **S**pectroscopy*
μTAS	*micro-**T**otal-**A**nalysis-**S**ystem(s)*

SYMBOLS IN CHAPTER 3

E	Young's modulus
E_a	activation energy
F	force
H	precursor height
I_e, I_c	electrical currents
K_o	constant
L	length
M	Mach's number
$N_{A,D}$	donors, acceptors concentration
P_{RF}	power of RF plasma excitation
S	etching selectivity
SF	etch process acceleration factor
T	absolute temperature in Kelvins
TCS	temperature coefficient of sensitivity
TCU_o	temperature coefficient of off-set voltage
U_o	off-set voltage
U_{DC}	self-polarization voltage
V	etch rate
V_E	electrical potential
V_{TR}	etch rate of silicon in RIE process
V_{hkl}	etch rate of (hkl) plane
V_1, V_2	electric potential
V_{100}	etch rate of (100) plane
V_{111}	etch rate of (111) plane
V_{110}, V_{221}, V_{231}	etch rate of (110), (221), (231) planes

$V_{100\mu W}$	etch rate of (100) for EMSi process
$V_{100temp}$	etch rate of (100) for thermally activated etching (only to compare to $V_{100\mu W}$)
W	window mask edge
W_{kr}	window mask edge ensuring self-stopping of etching
X_d	thickness of surface charge
a	edge of square pattern wet etched in silicon or lattice constant of silicon crystal $a = 0.543095$ nm
a_o	edge of square membrane wet etched in silicon
a_p	precursor mask dimension
b	boss edge
d	silicon wafer thickness
d_{100}	etch depth toward $\langle 100 \rangle$ direction
d_{111}	etch depth toward $\langle 111 \rangle$ direction
d_n	V-groove depth
f	frequency
h	tip height
hkl	Miller's indexes
k	Boltzman's constant
m	membrane thickness or distance between bottom of V-groove and silicon wafer surface
p	distance between bosses at a membrane
s	FEA grid distance
t	etching time or temperature in °C
u	underetching of pattern etched in (100) substrate
α	thermal coefficient of expansion
ΔL	elongation factor
Δx	distance between pattern edge etched in two steps procedure
ϕ	external diameter
10M, 7M,	concentration of etching solution in moles
v	Poisson's modulus
Θ	angle between (111) and (100) planes, equals 54.75°

SYMBOLS IN CHAPTER 4

A	area of sample
A_{void}	area of voids and bubbles or area of sample with voids and bubbles
C	capacitance of electrical capacitor
C_{void}	capacitance of sample with voids
C_{bp}	capacitance voids-free samples
E	Young's modulus
E_a	activation energy
E_p	electric field intensity

F	electrostatic clamping force
H	height (diameter) of dust
H_{test}	depth of test pattern
I	electrical current, bonding current
$I(t)$	bonding current as a function of time
I_{max}	maximal bonding current
J	bonding current density
J_{1max}, J_{2max}	maximal bonding current density for temperature T_1, T_2 in Kelvins
K	constant equaling 0.166 C
K_p	constant determining packing density of polishing particles
L	length of fracture propagation
L_{Test}	width of the test pattern
N	sodium ions concentration
N_a	boron concentration
$N_{composition}$	sodium ions concentration limited by composition of the glass
N_{theor}	theoretical sodium ions concentration
P	pressure, electrostatic pressure, surface density of dust particles
P_d	electrostatic pressure for glass-silicon bonding
P_i	inlet pressure
Q	charge
R	surface resistance of glass
R_{bp}	cross-resistance of void-free sample
R_{g1}, R_{g2}	heater and thermoresistor resistance
R_S	Hörtz's radius
R_{void}	cross-resistance of sample with bubbles
S	flat capacitor surface, depleted layer surface, substrate bow
T	absolute temperature
T_g	transformation point (temperature) of glass
U	electric voltage
U_{bond}	test bonding polarization
U_p	bonding polarization
a_o	proportion factor, membrane edge
a_{min}	pressure sensor chip edge
b	aluminum foil thickness
d	cathode-anode distance
d_o	distance between silicon and glass (air-gap)
d_{ox}	dielectric layer thickness
d_p	penetration depth
d_z	depleted layer thickness
d_{SiO_2}	SiO_2 layer thickness
d_1	thin-film dielectric layer thickness
d_2	depleted layer thickness

d_3	void (bubble) height
f	frequency
h	amplitude of wave, height of waviness
k	Boltzman's constant
k_{min}	minimal dimension of pressure sensor chip
k_{p}	factor
l	length of wave
m	membrane thickness
q	electron charge
r	dust radius, minimal width of frame of pressure sensor chip
r_{void}	void radius
t	temperature in °C, thickness of silicon wafer in razor test
u	(111) plane underetching
y	thickness of sharp edge in razor test
α_{g}	coefficient of thermal expansion of glass
α_{gZ}	coefficient of thermal expansion of glass in depleted layer
$\alpha_{\mathrm{g}\,20:450}$	coefficient of thermal expansion of glass at 20°C and 450°C
α_{m}	coefficient of thermal expansion of metal
α_{Si}	coefficient of thermal expansion of silicon
$\alpha_{\mathrm{Si}(230^{\circ}\mathrm{C})}$	coefficient of thermal expansion of silicon for 230°C
β	aperture of optical system
ε	dielectric constant, rectification constant
ε_{g}	dielectric constant of glass
ε_{o}	dielectric constant of vacuum
ε_{r}	dielectric constant of the sample
ε_1	dielectric constant of thin oxide layer
ε_2	dielectric constant of the depleted layer
ε_3	dielectric constant of the substances inside voids
ϕ	diameter of polishing powder particle, hole dimension
ϕ_{test}	test circles dimension
γ	surface bonding energy
γ_{GS}	surface tension at gas-solid state boundary
γ_{CS}	surface tension at solid state-liquid boundary
γ_{CG}	surface tension at liquid-gas boundary
σ_{R}	residual stress
λ	wavelength
μ	sodium ion mobility in hot glass
η	viscosity of glass
θ	wetting angle
σ	stress
σ_{y}	allowable stress corresponding to elastic deformation of given material
σ_{+}, σ_{-}	hydrodynamic conductance
ξ_{g}	glass sample unit elongation

ξ_{Si}	silicon sample unit elongation
Δx	resolution

Chapter 1

INTRODUCTION

The first procedures of silicon microengineering were developed in the 1960s and applied to the fabrication of three-dimensional, silicon, electromechanical pressure sensors.

In the following years a number of deep silicon micromachining techniques have been developed and used in mass scale fabrication of three-dimensional silicon, miniaturized sensors and actuators, utilizing the excellent mechanical properties of this material. These sensors and actuators were called micromechanical devices; the new section of microelectronics was called micromechanics integrated on silicon or simply silicon micromechanics.

Further developments in silicon micromechanics resulted in fabrication of the complex micro-electro-mechanical devices that contained many micromechanical components and parts, working together as a system.

In the 1990s the complexity of micromechanical devices became increased. Continuous development of micromachining techniques, introduction of new materials and growing integration of micromechanical devices and digital/analog circuits led toward the new type of sophisticated highly developed solutions, recognized as microsystems. Microsystem technology was born.

The microsystem is an electromechanical micro-device, which consists of integrated, micromechanical and microelectronic components co-working as a system. Micromechanical components of a microsystem are three-dimensional (3-D) and must be fabricated by means of special 3-D microengineering procedures, based on microelectronic technology; most often deep wet etching. Apart from deep micromachining based on wet etching techniques, several other microengineering procedures are applied in microsystem technology; they are of microelectronic origin: surface silicon micromachining, LIGA (Litography – Galvanotechnik – Abformung), stereolithographic surface deposition (stereo CVD) or developments from mechanical machining: precise micromachining, electroerosion, powder blasting, laser machining, molding etc.

It is difficult to estimate how all of the above-mentioned techniques and procedures will be used in microsystem technology in the future, but it is common knowledge that the development of silicon and non-silicon microsystems is currently very intensive.

Chapter 2

SOME REMARKS ON MICROSYSTEM SYSTEMATICS AND DEVELOPMENT

The terms "micromachine technology" and "microsystems" or "microsystem technology" (in German *mikrosystemtechnik*) are interchangeably applied. Also the terms "micromechanics integrated" [1] and "microengineering", meaning the same in popular usage, are often cited [1–12].

The term "micromachine technology" derives from "machining", followed by "micromachining", that is precise machining. The term "microsystem technology" derives from "micro-electronic-system" [3, 4]. Both terms, "micromachine technology" and "microsystem technology" describe electromechanical precision objects, in which at least one of the characteristic dimensions is in the range of micrometers (Fig. 2.1) [8].

In spite of many years of development, the nomenclature concerning microsystems and microsystem technology – the names most often used in Europe –

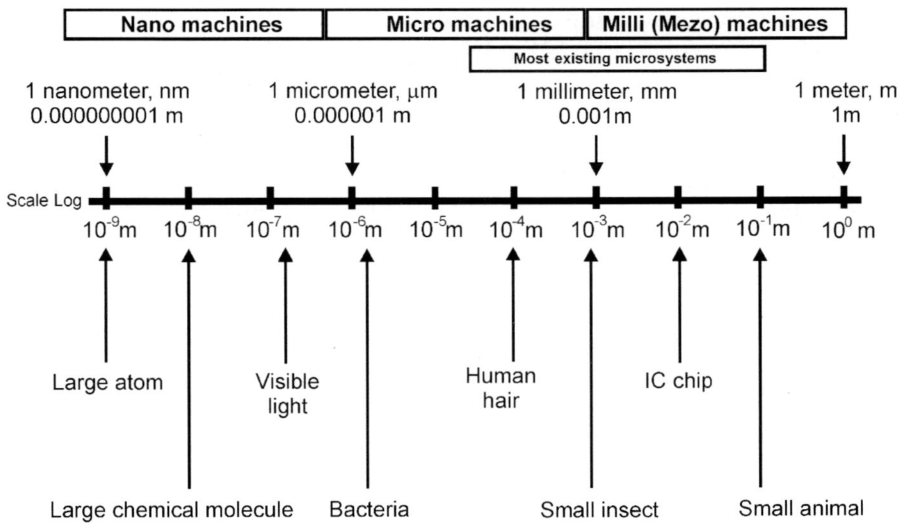

Fig. 2.1. Very small machines and microsystems.

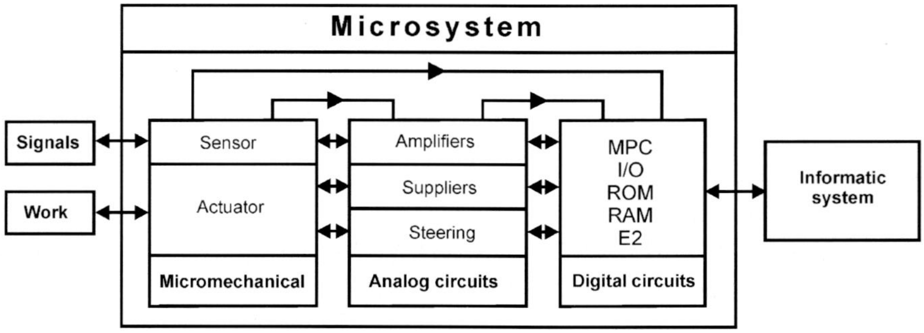

Fig. 2.2. Block diagram of microsystem in a full configuration.

remains vague. Generally, it appears that the use of all the terms mentioned above is somehow accidental, depending on the main field of research and application, development path, etc. For example, in the USA microsystem is equivalent to MEMS (**M**icro-**E**lectro-**M**echanical-**S**ystems), the name emphasizes the microelectronic origin of the device. In Japan microsystems are recognized as micromachines, the mechanical nature of the device is stressed*.

So, the terms used most frequently for the field discussed in this volume are: microsystem technology, MEMS technique, and micromachine technology. Unfortunately, exactly defining the key terms seems to be a difficult task. Definition and systematization applied here, in this book indicate basic or most characteristic functions of microsystems.

Thus, microsystem technology (MST), i.e. the whole discipline, is concerned with design, fabrication, research and development of microsystems, that is small devices (which have at least one of the characteristic dimensions in the range of micrometers), consisting of sensors, actuators and electronic circuits, co-working as a system (Fig. 2.2). A microsystem is a coherent system of sensing and conversion of physical and chemical quantities into electrical signals and, what is more, a microsystem is able to actuate (to move). The fabrication of microsystems requires methods of microengineering technology, micromachining and assembling, methods adapted from microelectronic technology.

According to the classification accepted in this book, microsystems would be divided into the following groups: MEMS, MEOMS, μTAS and micromachines (VSMs – **V**ery **S**mall **M**achines).

MEMS is defined as a silicon or silicon/glass (the most often used materials) integrated electromechanical device with mechanical and electronic elements as well as software joined together to form a system. MEOMS (**M**icro-**E**lectro-**O**pto-**M**echanical-**S**ystems) (MOMS, MOEMS, OMEMS) is an integrated optical, electromechanical device, usually made of silicon and glass, although metals and plastics are often used. The MEOMS can possess the capability of optical

* It is possible that this situation is a result of the huge interest in new technologies of the precision machine industry, including the toy industry of Japan.

detection of physical and chemical parameters, and may be designed as a "pure" micro-optical device or an opto-mechanical instrument.

μTAS (micro-**T**otal-**A**nalysis-**S**ystem) (miutas, microtas) is an integrated, miniaturized device, in which chemical or biochemical processes or analysis can be carried out. Some microsystems of this type are produced in the form of biochips (biochemical chip, biomedical chip) or lab-on-chips (laboratory-on-the-chip). The importance of μTAS's, which are a new group of microsystems, has grown significantly during the past decade. Definitions and systematization of μTAS's are not yet completed. They are made of silicon or silicon and glass, and other types of materials such as photosensitive glasses or plastics are used widely.

Micromachine (VSM – **V**ery **S**mall **M**achine) is a miniaturized, moving mechanism, produced with the use of microelectronic and micromechanical technologies, applying photolithography procedures or methods of fabrication taken from traditional precise mechanics. The most complex micromachines are the industrial microrobots. In general, VSM micromachines are made of various materials, among which silicon is rather marginal. Microsystems are becoming more and more sophisticated. The number of mechanical and electronic parts integrated in a single device can reach millions. For the majority of known microsystems the mechanical/electronic integration equals 100/100, but in the projection matrixes of DMD multimedia projectors, produced by Texas Instruments Company, the integration $10^6/10^6$ has been obtained (Fig. 2.3).

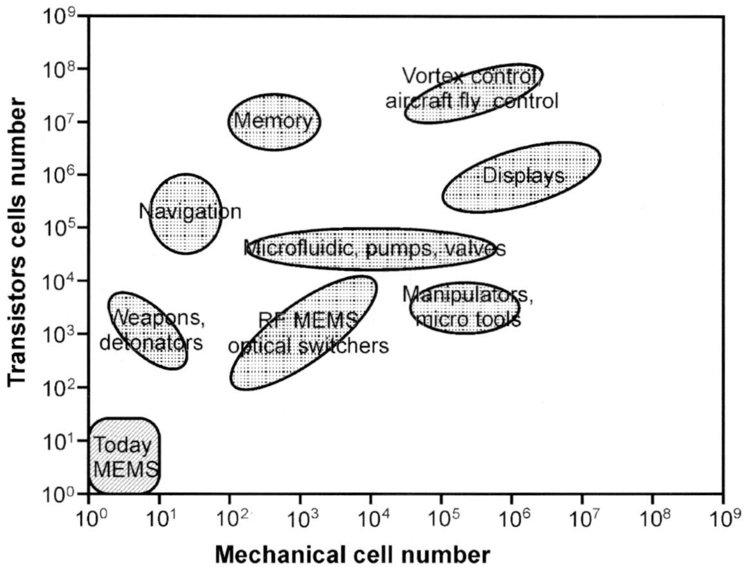

Fig. 2.3. The complexity of microsystems – some chosen examples.

Examples and field of application of microsystems

The list of examples of microsystems is shown in Table 2.1 on the basis of recognized worldwide scientific papers, catalogs and notes of producers as well as of economic reports and Internet sources. The author has decided to present the table, although in the Internet Age these kinds of lists have lost their significance. However, following the sources considered to be models (Agnell *et al.* in 1983 [1], Greenwood in 1988 [4], Bubley in 1995 [8], Fluitman in 1996 [10], Nexus report [12]), in which this type of analysis has been carried out, selected examples describing the field of microsystem technology are compiled.

The microsystem industry, often called the M^3 industry (Microelectromechanical systems & Microsystem technology & Micromachines), has becomes an increasingly important branch of the high technology industry. Its development is characterized by high CAGR (**C**oefficient of **A**nnual **G**rowth **R**ate), which has amounted to over 21% for the past few years. Many European, US and Japanese academic, research and industrial centers, including huge producers of house-keeping facilities, automobiles, electro-equipment, etc. are working with microsystem technology. Particular regions and countries concentrate on the development of these domains of microsystem technology, which are recognized to be crucial for the industry. Hence, micromachines are of great significance in Japan, while in the USA the most important products are MEMS, MEOMS and genetic research devices, and in Germany: microsystems for biomedicine and motorization (Fig. 2.4) [13].

The characteristic of the development of microsystem research in the world (data for Japan, the USA and Europe) is a high participation of private capital in research and a large number of research and teaching centers [14, 15].

Market for Microsystems

According to market data the most significant group of microsystems with established market position are heads of ink-jet printers and different micromechanical sensors and actuators (Table 2.2), while the most rapidly growing group of microsystems include: medicine feeders, integrated laboratories, hard disks as well as various sensors and actuators (Table 2.3) [8, 12, 13]. As early as 1996 the market value of microsystems was comparable with the value which the microprocessor market reached in 1999 (about 13 billion USD). According to the prognosis from 1999, the microsystem market was supposed to increase to 40 billion USD in 2002. The data from the end of 2000 (Expo 2000 in Hannover) showed that the market was already bigger and growing rapidly [16]. It is estimated, that in 2000 there were about 100 M^3 enterprises in the world with over 1 million USD of capital each [17]. Some of these were founded by huge corporations looking for credible deliverers of key components for their final products. As an example we quote the company Xros ($\sim$3 billion

Table 2.1. Microsystems and the key components for microsystems

EMS	MEOMS	μTAS, bio-chip Lab-on-chip	VSM Micromachines
• pressure sensors • acceleration sensors • flowmeters • gyroscopes • tonometers • electromagnetic radiation sensors and bolometers • prosthesis of senses (ear, eye) • various biomedical sensors • devices for non-invasive surgery, colonoscopes, endoscopes, tools, biomedical robots • regenerators of nerves • neurological probes • microdialyzers • electron tunneling microscopes • atomic forces microscopes • electric and electromagnetic microengines • turbine microgenerators of electric power • hard disk – new types of memory • ink-jet printers • fuel (fluids) atomizers • electronic components: filters, transformers, capacitors • RF components: switchers, antennas, groups of antennas, detectors, couplers and resonators, cathodes	• mirrors and adaptive reflectors • static lens • dynamic lens • modulators of light beams • switchers • scanners • projectors • diffractive grating and photolithographic masks • interferometers, • monochromators • spectrophotometers • optical detectors • pressure sensors • acceleration sensors • SNOM	• micro- pico- and nano-reactors • mixers, filters, separators • micropumps, valves • fluid and medicine feeders, • gas and liquid chromatographs • biosensors • pH-sensors • bioanalyzers • DNA analyzer • cell manipulators • insulin dosers • intelligent pills	• screws, springs, gear wheels • gear boxes • microengines, rotational, linear, electrical, electromagnetic • steam machines • grippers • micro automobiles milli and micro flying machines • intelligent industrial microrobots • intelligent "dust" • gyroscopes • surgery tools • "walking" diagnostic and repairing devices • flowmeters with turbines • micro explosive charge • rocket, jet-propelled engines • submarines

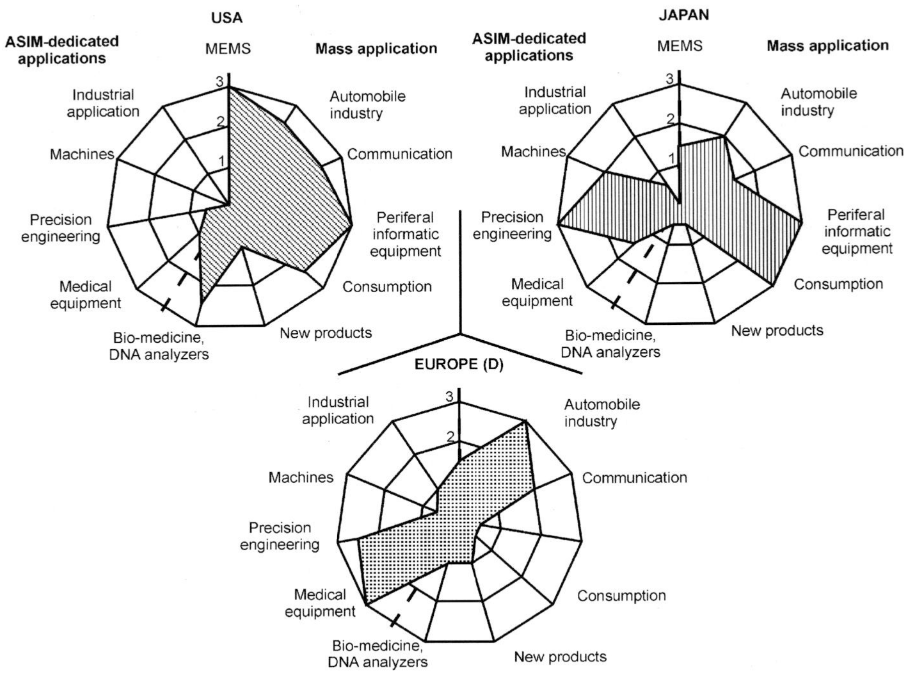

Fig. 2.4. Microsystem technology in national configurations.

Table 2.2. World microsystem market, its structure and range in 1996–2002; products well established on the market (prognosis from 1999) [12, 13]

Product name	1996 Pieces million	1996 USD million	2002 Pieces million	2002 USD million
Heads of hard disks	530	4500	1500	12 000
Heads of ink-jet printers	100	4400	500	10 000
Heart stimulators	0.2	1000	0.8	3700
Devices for *in vitro* diagnosis	700	450	4000	2800
Non-hearing aid devices	4	1150	7	2000
Pressure sensors	115	600	309	1300
Acceleration sensors	24	240	90	430
Chemical sensors	100	300	400	800
Infrared cameras	0.01	220	0.4	800
Gyroscopes	6	150	30	360
Magneto resistive sensors	15	20	60	60
Microspectrometers	0.0006	3	0.150	40
Total		13 033		34 290

Table 2.3. World microsystem market: newly developed microsystems in 1996–2002 (prognosis from 1999) [12, 13]

Product name	1996		2002		Expected growth	
	Pieces million	USD million	Pieces million	USD million	In number	In value
Drug delivery/dosing systems	1	10	100	1000	10 ×	10 ×
Optical switches	1	50	40	1000	50 ×	25 ×
Lab-on-chip (DNA, HPLC, etc.)	0	0	100	1000	—	—
Magneto-optical hard disks	0.01	1	100	500	10 000 ×	500 ×
DMD picture projectors	0.1	10	1	300	10 ×	30 ×
Coils on chip	20	10	600	100	30 ×	10 ×
Microrelays	—	0.1	50	100	—	1000 ×
Micromotors	0.1	5	2	80	20 ×	16 ×
Inclination meters (level lines)	1	10	20	70	20 ×	7 ×
Injection dishes	10	10	30	30	3 ×	3 ×
Anticollision sensors	0.01	0.5	2	20	200 ×	100 ×
Electronic noises	0.001	0.1	0.05	5	50 ×	50 ×
Total		106.7		4205		> 40 ×

USD), established by the concern Nortel, or Intellisense (750 billion USD), founded by the enterprise Corning. Other firms were bought, for instance Nova Sensors by the concern Lucas, or Micro Technology Instruments, bought by Hewlett-Packard (Agilent). In the 2000–2003 period a particularly big growth of M^3 stock companies, operating in the field of biotechnology, was observed.

It is expected that the biggest market of M^3 products *(killer applications)* in the near future will be predominated by the components for portable communication devices, microwave electronic circuits, remote-controlled sensors (so-called RF M^3), heads of hard disks and analytical μTAS [17].

In the coming years microsystem technology will continue to exert a growing influence on the development of almost all fields of technology [17–20]. This is confirmed by research and marketing reports, the opinions of politicians as well as financial and scientific authorities and the over twenty times growth in the numbers of patents in the field of microsystem technology in the 1989–1999 period (694 patents in 1999 [17]).

For the moment, the most important microsystems are made of monocrystalline silicon and/or silicon sealed with glass in a form of multi-layer stack. Micromechanical silicon three-dimensional (3-D) parts of microsystems; membranes, grooves, beams, cavities, etc. (Fig. 2.5) are deeply micromachined, most often wet/dry etched (Fig. 2.6). Glass can be mechanically machined or isotropically wet etched (Fig. 2.7). A stack of layers is fabricated by use of a special sealing technique, sometimes called micromechanical bonding or, in short, bonding.

The most widely used bonding processes are direct (fusion) bonding and anodic bonding (Fig. 2.8). Direct bonding is a thermally activated process – the final formation of a bond needs annealing of the bonded stack of wafers at

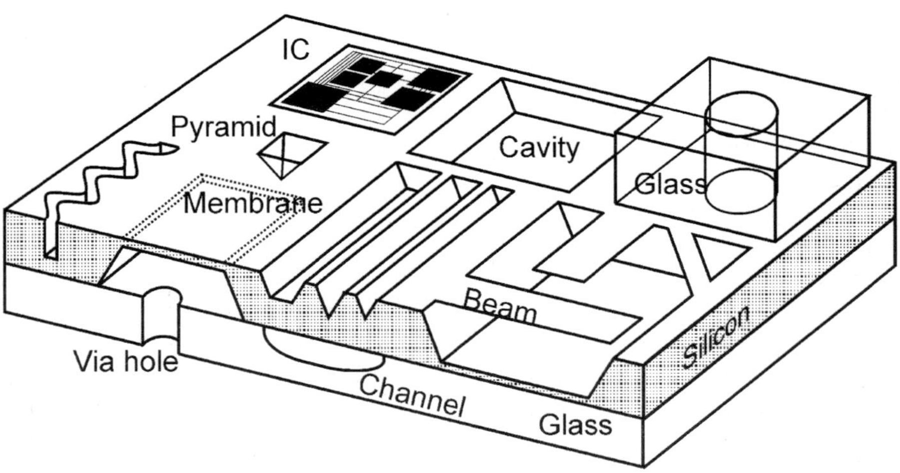

Fig. 2.5. Silicon–glass microsystem – schematic view.

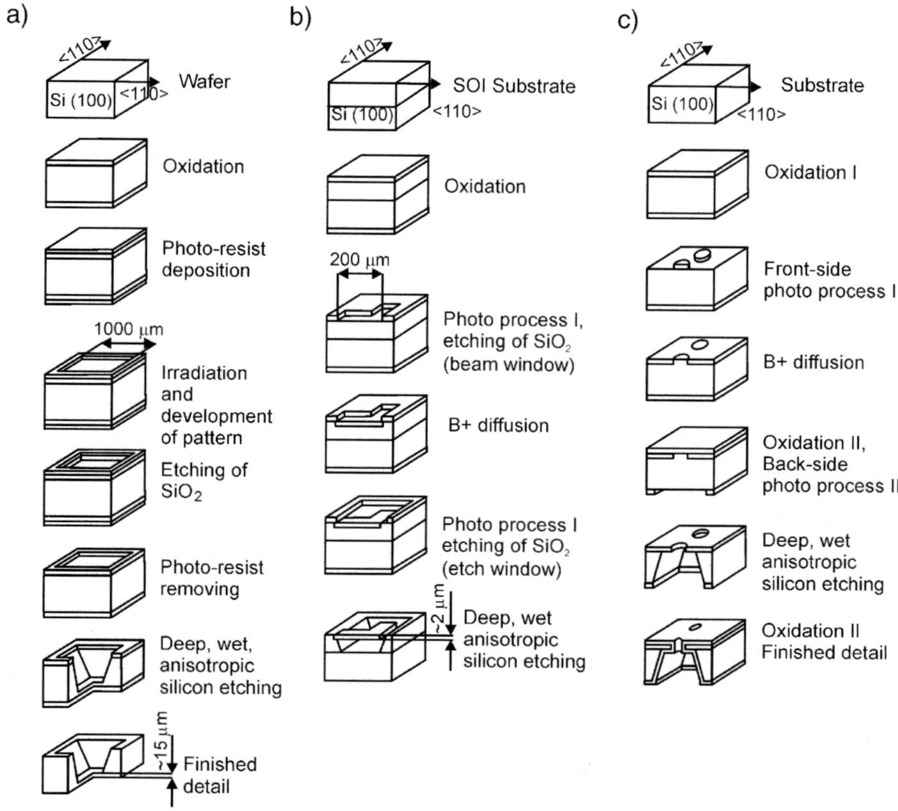

Fig. 2.6. Deep, wet micromachining of silicon – examples of procedures: a) membrane fabrication, b) beam fabrication, c) fabrication of holes. For clarity single chips are presented.

medium or high temperature. Anodic bonding is performed at medium temperature under electric field excitation. Other bonding methods, such as low-melting glass bonding, eutectic bonding, and foil bonding, are used occasionally. Deep micromachining and bonding are technologically linked very closely together. Thus, the description of bonding should be introduced by a short course on deep wet anisotropic machining of silicon, the most important micromachining method used in microsystem technology.

The main content of this book is a description of bonding and its applications in microsystem technology. The experimental material presented has been mainly based on experiments conducted by the author and his co-workers. The literature sources have been collected from the main sources but, because microsystem technology is expanding rapidly, the author would like to apologize to readers for any inconvenience caused by insufficient literature sources given in this book. Parts of experimental results on anodic bonding have not been published. The bonding procedures, which are described in this book can be utilized in day-to-day technological practice.

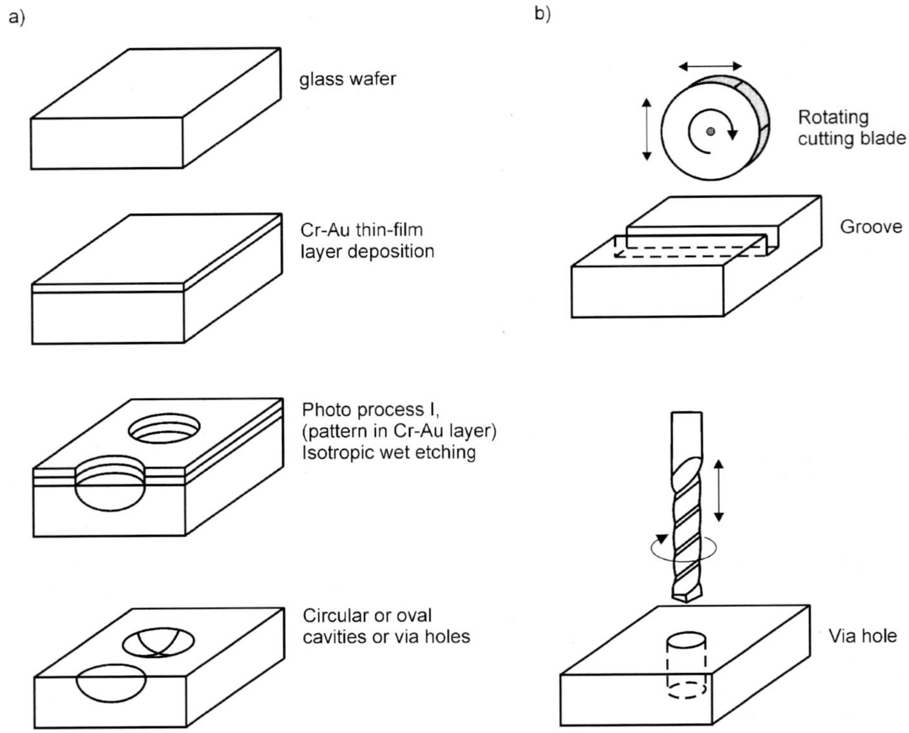

Fig. 2.7. Glass machining: etching, sewing, drilling.

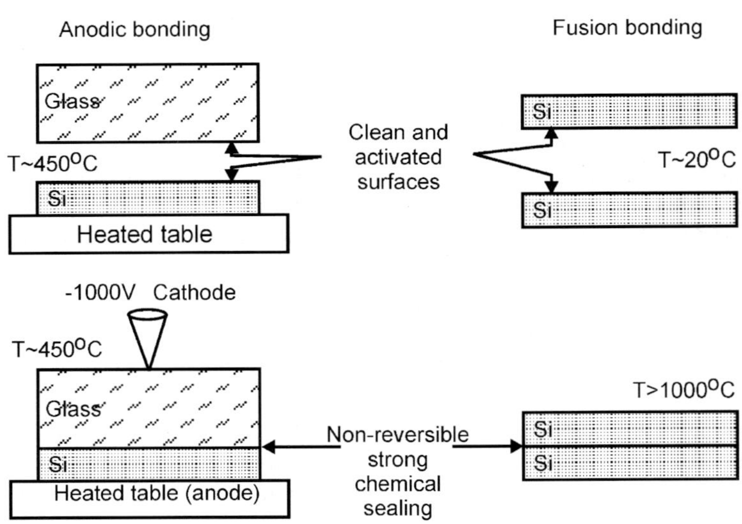

Fig. 2.8. Direct and anodic bonding – two main sealing procedures used in microsystem technology.

The narration starts from the systematic of microsystems and state-of-the-art of microsystem technique, followed by a deep wet anisotropic silicon micromachining presentation (Chapter 3). Next, bonding in microsystem technology is described (Chapter 4), followed by a short summary (Chapter 5).

LITERATURE

[1] J. B. Agnell, S.C. Terry, P.W. Barth, Silicon micromechanical devices, *Sc. Am.*, 44, 1983, 44–54.
[2] J. Bryzek, K. Petersen, W. Mc. Culley, Micromachines on the march, *IEEE Spectrum*, 5, 1984, 20–31.
[3] L. Csepregi, Micromechanics: a silicon microfabrication technology, *Microelectronics* 3, 1985, 221–234.
[4] J.C. Greenwood, Silicon in mechanical sensors, *J. Physics*, 21, 1988, 1114–1128.
[5] K. Gabriel, J. Jarvis, W. Trimmer, *Small machines, large opportunities*, A report of the Workshop on Micromechanical System Research, 1987–1988, AT&T Bell Laboratories Report, 1–31.
[6] S. Middelhoek, S.A. Audet, *Silicon Sensors*, Academic Press, New York, 1989.
[7] J. Bryzek, K. Petersen, J. Mallon, Jr., L. Christel, F. Pourahmedi, *Silicon Sensors and Microstructures*, Nova Sensor, Fremont, CA, USA, 1991.
[8] J. Bubley, *Micromachines: Applications Markets and Trends*, A Financial Times Management Report, published by Pearson Professional Ltd, London UK, 1995.
[9] K. Petersen, From microsensors to microinstrumentation, *Sensors and Actuators A*, 56, 1996, 143–149.
[10] J. Fluitman, Microsystem technology objectives, *Sensors and Actuators A*, 56, 1996, 151–156.
[11] R. Wechsung, J.C. Eloy, *Market analysis for microsystems*, an internal report of the NEXUS TASK FORCE, Proceed. Eurosensors XI, Warsaw, Poland, 1997, Vol. 2, 519–626.
[12] NEXUS analysis of microsystems, 1996–2002, Nexus Office, Fraunhofer ISiT, Dillenburger Strasse 53, D-14199 Berlin.
[13] M. Schueremann, V. Huentrug, R. Bierhals, Economic potentials and miniaturization from the industrial viewpoint, *MST News*, 2, 1999, 34–36.
[14] K.D. Wise, J.M. Giachimo, H. Guckel, G.B. Hocker, S.C. Jacobsen, R.S. Muller, *Micromechanical systems in Japan*, Jap. Techn. Eval. Center, The Panel Report, www.itri.loyola.edu/mems.
[15] *JTEC Panel Report on Microelectromechanical Systems in Japan*, NSF, ARPA, Dept. of Commerce of USA, AFSC USA, 1997, www.itri.loyola.edu/mems.
[16] Statement of the Federal Minister of Education and Research of Federal Republic of Germany, World Congress of Microsystems, MicroTECH 2000, September 25–27, Hannover Expo 2000, Germany.
[17] R.H. Grace, P. Salomon, Microsystems/MEMS/Micromachines – on the move from technology to business, *MST News*, 5, 2001, 4–8.
[18] J. Bryzek, Impact of MEMS technology on society, *Sensors and Actuators A*, 56, 1996, 1–9.
[19] K. Petersen, From sensors to microinstruments, *Sensors and Actuators A*, 56, 1996, 143–149.
[20] H. Christ, German MST Program News, *MST News*, 3, 1999, 36–37 H.

Chapter 3

DEEP, THREE-DIMENSIONAL SILICON MICROMACHINING

Three-dimensional (3-D) silicon structures (also called silicon micromechanical structures, micromechanical constructions or just micromechanical structures) are commonly produced by means of wet deep micromachining of silicon substrates. Structures are truly three-dimensional; they are micromachined "through" a substrate.

The description of wet deep micromachining, presented in this chapter, begins with determination of the mechanical properties of silicon and characterization of silicon substrates used for fabrication of micromechanical structures and basic features of wet anisotropic etching of silicon (focused on etching in KOH). Next, fabrication procedures of basic silicon microconstructions, and examples of its application in microsystems, are described.

3.1. MICROMECHANICAL SUBSTRATES AND MECHANICAL PROPERTIES OF SILICON

Micromechanical substrates

Silicon (discovered in 1824 by J.J. Berzelius) is a brightly glittering, shiny, metallic, hard, light material. It can be ground and polished. Silicon crystallizes in a FCC (**F**ace **C**entred **C**ubic) crystal structure, forming regular crystals of Fd3 space group symmetry (two Bravais F type lattices moved $1/4a$, and $\sqrt{3}/4a$ in the $\langle 111 \rangle$ direction, where a is a lattice constant equal to $a = 0.543095$ nm. The position of three basic crystallographic planes (100), (111) and (110) and the unit cell of the silicon crystal is shown in Fig. 3.1. Angles formed between individual crystallographic planes can be expressed by:

$$\cos \phi = \frac{h_1 h_2 + k_1 k_2 + l_1 l_2}{\sqrt{(h_1^2 + k_1^2 + l_1^2)(h_2^2 + h_2^2 + l_2^2)}}, \tag{3.1}$$

where: h_1, k_1, l_1, h_2, k_2, l_2 = Miller indexes for a given plane.

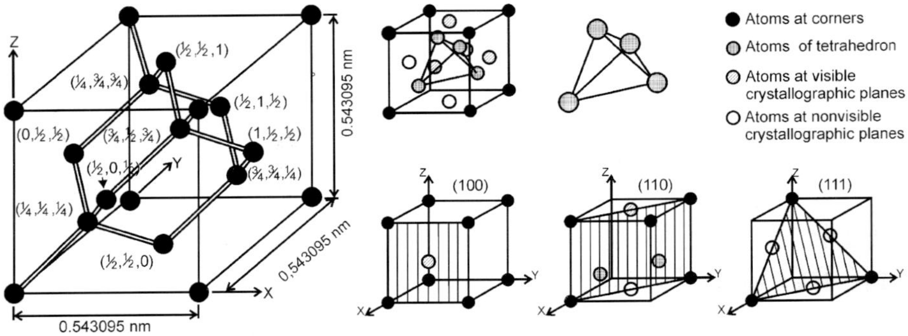

Fig. 3.1. The structure of the silicon crystal.

For example, to specify the angle between planes (100) and (111), Miller's indexes h_1, k_1, l_1, h_2, k_2, l_2 in equation 3.1 should be: 1,0,0 and 1,1,1.

The angles between three basic planes (100), (111) and (110) are given in Table 3.1.

The position of planes and, first of all, the angle between them, can be specified from Wulf's stereographic projections of a crystal. The stereographic projection of a silicon crystal onto the three crystallographic planes (100), (110) and (111) is presented in Fig. 3.2. These data are sufficient for the analysis of geometry (shapes) of simple patterns deeply wet anistropically etched in silicon. As an example, let us analyze geometrical properties characterizing the (100) plane, according to the stereographic projection of a silicon crystal on plane (100). The angle between four (111) planes, represented in Fig. 3.2 by large black dots, is 90°, the angle between the (111) plane and the (100) plane (smaller black dots) is 54°44′ (54.74°). Symmetry of (111) planes on the (100) plane is based on the shape of the square. Continuing, the symmetry of (111) planes on the (110) plane is rhomboidal, and on the (111) plane is triangular.

Mechanical properties of silicon

Monocrystalline silicon is a material with almost perfect, highly anisotropic, mechanical properties [1–8]. Its tear resistance factor along the ⟨111⟩ direction is two times higher than the factor of constructional steel [6]. Monocrystalline silicon cracks after exceeding the maximum permissible mechanical load (Fig. 3.3) without the sample "flow" which is typical for metals*. Silicon is an almost ideal mechanical construction material [3], because it is stronger than steel, light (like aluminum), has a proportion between mechanical strength,

* In ref. [9] the method of plastic permanent strain of silicon microstructures is described, as well as the permanent strain of a silicon substrate (heated to 800°C) by means of the method of applying a controlled pressure in such a way that material deformation does not exceed 0.1 mm/min. This means that silicon "flows" mechanically at a higher temperature. This phenomenon is of no significance in silicon microsystems, because their working temperature does not exceed a few hundred degrees centigrade.

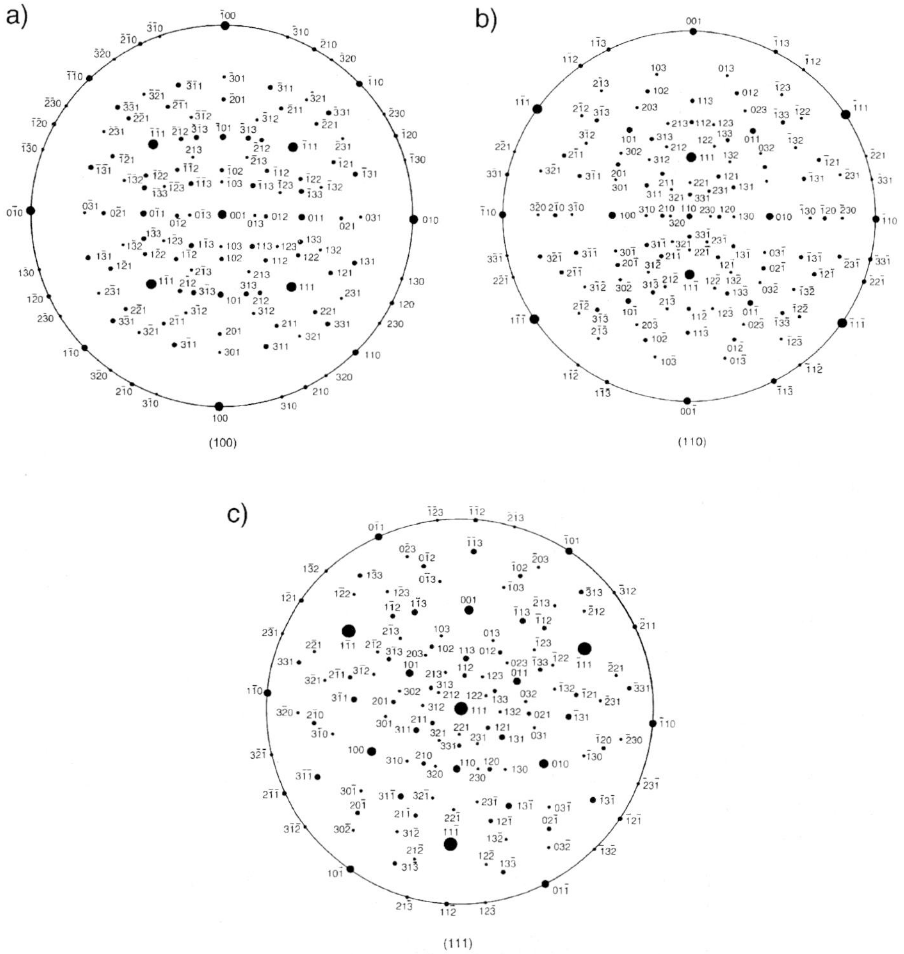

Fig. 3.2. Wulf's stereographic projections of the silicon crystal: a) onto plane (100), b) onto plane (110), c) onto plane (111).

Table 3.1. Angles between basic planes in silicon crystal

Miller indexes $(h_1k_1l_1)$	Miller indexes $(h_2k_2l_2)$	Angles	
	100	0°	90°
100	110	45°	90°
	111	54°44′	
110	110	0°	60°
	111	35°16′	90°
111	111	0°	70°

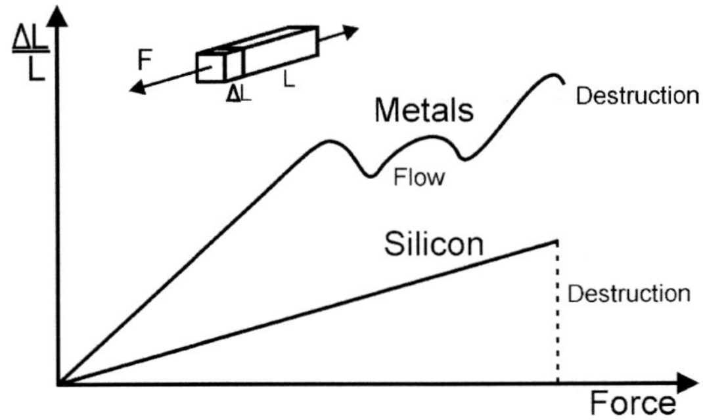

Fig. 3.3. Strain–force curves: an example of the relative elongation of a silicon and metal sample under the influence of tensile force.

rigidity and specific gravity much higher than that for steel, with negligible mechanical hysteresis. Mechanical and electric parameters of silicon are very stable in time and are technologically repeatable.

The excellent mechanical properties of silicon, researched among others in publications [10]–[14], led to stable and continuing work on miniature micromechanical constructions. The Q-factor of vibrating micromechanical constructions easily reaches 10^4, the frequency of vibration lies in the range of 100 kHz or more. The mechanical properties of silicon are well illustrated in Fig. 3.4, where a thin (15 μm), strongly deflected silicon membrane is presented. The planar dimensions of the membrane are: 5000 μm × 5000 μm, while its deflection equals over 450 μm. The beam 800 μm long and 150 μm wide is deflected over 500 μm down from its neutral position by the tip of a metal needle probe; such

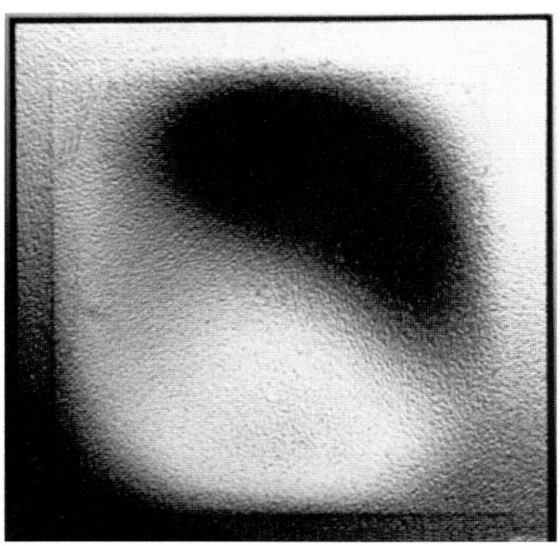

Fig. 3.4. Deflected silicon membrane.

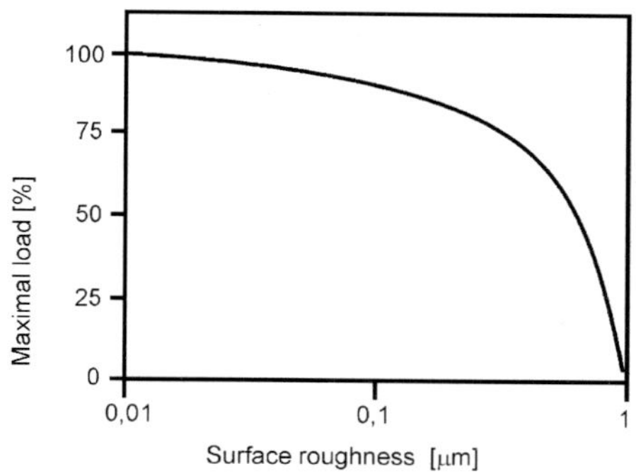

Fig. 3.5. Strength of microconstruction versus surface roughness [1].

a membrane or beam can be deflected many times (10^{14} or more) without any change in its parameters.

The very high mechanical quality of silicon microconstructions, and the strength of silicon microsystems and micromechanisms, is a direct consequence of small dimensions and smooth surface of the devices, characterized by the small number of defects (Fig. 3.5) [1]. Usually, technologically induced steps are below 0.1 μm; the smoothness of the top layer is in order of a few nanometers.

Silicon micromechanical structures utilize, first of all, the interactions of forces aligned to main crystallographic directions ⟨100⟩, ⟨110⟩ and ⟨111⟩. Young's modulus E and Poisson's ratio v for these directions are given in Table 3.2.

Many authors have discussed the influence of doping of monocrystalline silicon on the mechanical properties of this material [2, 15–17]. No influence has been observed in weakly doped material ($N < 10^{16}\,\mathrm{cm}^{-3}$). However, for boron concentration over $N_A \geq 7 \times 10^{19}\,\mathrm{cm}^{-3}$, tensile stresses appear (amounting to about 60 MPa), Young's modulus in the ⟨110⟩ direction becomes smaller, about 108 ± 20 GPa [18]; this is approximately 40% smaller than measured for the non-doped silicon. On the other hand, when the boron concentration is higher than $10^{20}\,\mathrm{cm}^{-3}$, compressive stresses are noticed [19]. The compensation of stresses induced by heavy boron doping can be obtained by the simultaneous diffusion of germanium [17].

Other materials – although more rare than silicon – are also used in microsystem technology. This group of materials includes, among others, SiC, Si_3N_4, SiO_2, diamond-like layers and various metals. The basic mechanical parameters

Table 3.2. Coefficients E and v of silicon for the most important crystallographic directions, $T = 300$ K [4]

Crystallographic direction	E [GPa]	v [GPa]	$(dE/dt)/5E$ [K^{-1}]
⟨100⟩	129.5	79.0	−63
⟨110⟩	168.0	61.7	−80.3
⟨111⟩	186.5	57.5	−45.6

Table 3.3. Selected mechanical parameters of some materials utilized in microsystem technology [3, 7]

Materia	Young's modulus E [GPa]	Tearing resistance [GPa]	Thermal expansion α [10^{-6}/K]	Knoop's hardness [GPa]	Density [mg/m^3]
Diamond	103.5	53	1.2	7 (9.85)	3.5
SiC	700	21	$3.3 \div 5.2$	2.48	3.2
Al_2O_3	530	15.4	5.4	2.06	4.0
Si_3N_4	385	14	0.8	3.486	3.1
SiO_2	730	8.4	$0.5 \div 0.6$	$0.47 \div 1.2$	2.5
$Si_{(111)}$	~ 190	7	2.33	$0.83 \div 0.85$	2.3
Steel (max.)	$195 \div 210$	$2 \div 3.5$	16	do 1.5	7.9
W	410	4	4.5	0.4	19.3
Mo	343	2.1	5.39	0.27	10.3

of the mentioned materials (with silicon as comparison) are presented in Table 3.3.

Micromechanical substrates

Monocrystalline silicon in the form of silicon substrates, that is flat, usually double-side polished wafers 1″ to 8″ in diameter and a few hundred micrometers thick, has been applied in the planar technology of integrated microelectronics for many years.

Specific aspects of batch fabrication of silicon microsystems can be achieved by use of specially fabricated silicon micromechanical substrates (wafers). In order to fabricate the substrates Czochralski's method is used most often. Material obtained by this method is known as CZ Si. In this method a small monocrystalline silicon nucleation crystal is affixed to the front of the slowly turning mandrel. The crystallographic orientation of the flat frontal surface of the nucleation crystal is aligned to one of the chosen crystallographic planes, (100) or (110). Then the nucleation crystal frontal surface is brought into contact with the liquid silicon. The mandrel is then retracted (pulled out) very slowly, rotating all the time. The monocrystal grows and its diameter enlarges. The described process is slow and results in the formation of a relatively regular cylinder a few hundred millimeters in diameter and a few hundred centimeters long (2–3 m!), and weighing over a few dozen kilograms (over 50 kg).* After cooling down, the block of material is subjected to complicated mechanical and chemical machining, in order to divide it into thin wafers with a polished surface, with crystallographic orientation corresponding to the orientation of the nucleation crystal (Fig. 3.6).

* This weight is maintained during silicon monocrystal fabrication through a narrow (2 mm) silicon seed neck. Tensile stress generated in the neck amounts to over 12 MPa [1]. However, the material does not crack.

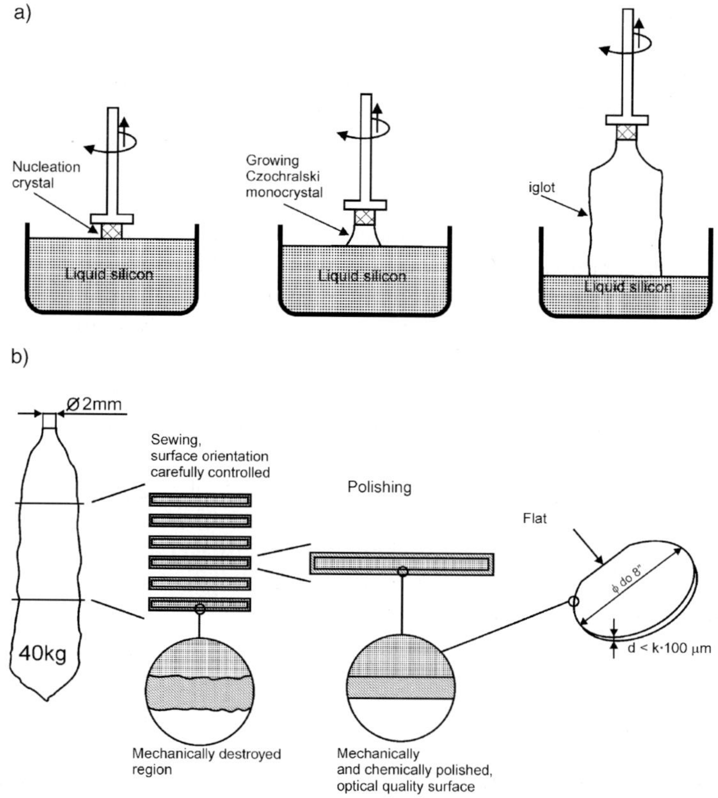

Fig. 3.6. Silicon substrate manufacture: a) Czochralski's method, b) wafer fabrication.

As mentioned above, the surface of micromechanical substrates is usually aligned to crystallographic planes (100) or (110). These substrates are often called substrates (wafers) (100) or (111). To mark the type of conductivity and crystallographic orientation of the surface of substrates, these so-called flats (coding marks) are produced. The flats (usually two) are made by suitably grinding off the edges of the wafers. Flats correspond to the chosen crystallographic direction, seen from a bird's eye view on the substrate (Fig. 3.7).

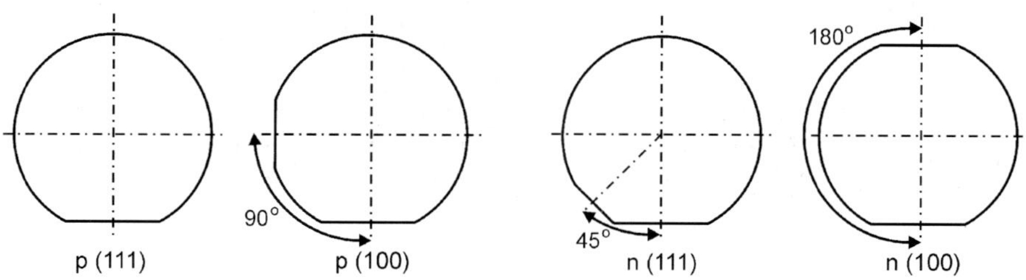

Fig. 3.7. Flats of substrates (100), (111).

For instance, for *p*-type substrates (100) the main, longer flat is aligned to crystallographic direction ⟨110⟩, while the auxiliary, shorter flat is fabricated perpendicular to the main one. Flats must be precisely aligned to the desired crystallographic direction as all of the photolithographic masks used in a microsystem fabrication process have to be aligned to them. This is very important, because – as will be shown in subsequent parts of this book – the geometry of micromechanical structures depends on the orientation-dependent wet etching process. Precise alignment of masks then limits the accuracy of micromachining of the structures defined by (100) and (111) planes (Fig. 3.8).

Silicon substrates which are used in microsystem technology are usually smaller than those used in the VLSI microelectronic industry; most often they are 3″ to 5″ inches. The substrates need to meet the following requirements: the shape and dimensions of substrates, especially their flat-parallelism, need to be kept in a strictly determined deviation range, total thickness variation (TTV) should not exceed a few micrometers, the surface of substrates on both sides should be optically smooth, and the disorientation of substrate surface from required crystallographic plane (wafer off-orientation) should not exceed 0.5°. The flats, that allow fabrication of the anisotropically wet etched micromechanical patterns aligned precisely to a chosen crystallographic direction, have to be fabricated with an accuracy of no less than 1° (the smaller the better) angular misalignment to the specified crystallographic direction – usually ⟨110⟩ at the (100) oriented wafer.

The typical resistivity of silicon equals a few Ωcm. Usually *n*-type, phosphorus doped substrates (100) are used; *p*-type substrates (100) with a thin *n*-type epitaxial layer are also used quite often. The thickness of this layer ranges from a few to a dozen or so micrometers. Quality parameters of the epitaxial layers cannot differ significantly from the standards common for microelectronics processes. For special applications SOI (silicon-on-insulator) substrates with a

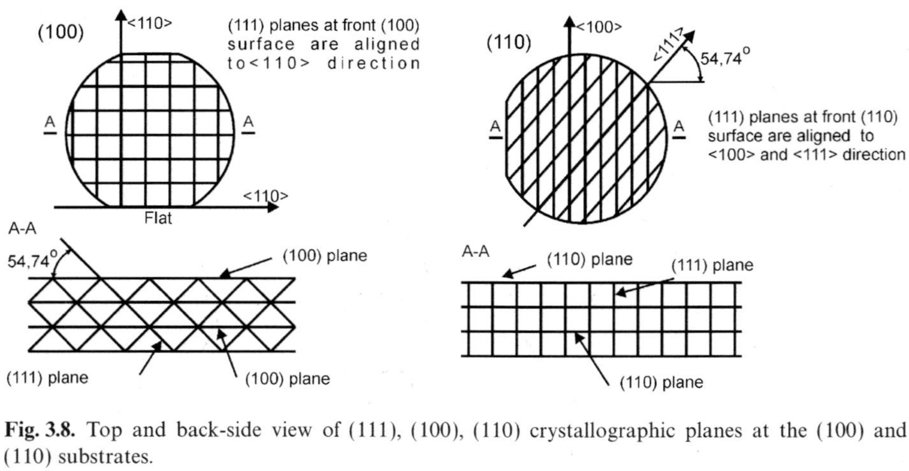

Fig. 3.8. Top and back-side view of (111), (100), (110) crystallographic planes at the (100) and (110) substrates.

Table 3.4. The micromechanical silicon substrates parameters

Parameter	Desired value of parameter
Diameter ϕ (most often used)	76, 100 $\pm$ 0.5 mm (3 inches, 4 inches)
Flatness	$< \pm 1.5$ μm
Bow and warp	< 20 μm
Surface parallelism	$< \pm 2.5$ μm
Orientation (mostly)	(100)
Off-orientation	$< 0.5°$
Flats (for (100))	$\langle 110 \rangle \pm 0.5°$
Thickness	360 to 450 μm
TIR	3 inches: ~ 10 μm > 4 inches: ~ 10 μm
TTV	3 inches: < 10 μm > 4 inches: < 20 μm
Dislocation density	< 100 cm
Particle density (front surface)	max. 5

buried silicon dioxide layer are utilized. This will be discussed in later sections of the book.

The example of data specification of the micromechanical silicon substrates is shown in the following table (Table 3.4).

3.2. WET ANISOTROPIC ETCHING OF SILICON

Deep anisotropic etching of monocrytalline silicon was applied for the first time for the precise machining of thin silicon membranes of pressure sensors in the 1960s. Further studies focused on the finding of a precise and repeatable method of machining of simple micromechanical structures, insulation islands for integrated circuits and beam leads [1, 2, 20–23]. After calculating fundamental process parameters [4, 5, 17, 24–26], silicon etching in alkaline solutions became the basic technology of deep, three-dimensional silicon micromachining of microsystems. Selected examples of wet micromachined microsystems are listed below:

spectrum analyzers
bolometers
gas and liquid chromatographs
pressure sensors
laser beam deflectors
feeders of fluids and medicines
jets of ink-jet printers
microfluidic elements
car carburettors
jets of thermal printers
microlamps
gas and liquid micropumps
electric microswitches
microreactors
microsatellites – constituent elements
electric, steam, jet-propelled microengines
tunnel and atomic forces microscopes
microvalves
optical modulators
flowmeters

"intelligent" mirrors and lens (auto focus)
surgical lancets
photolithographic masks
membranes and sieves for microchemistry
vibration, angle and position meters
mixers for microchemistry
micro safety devices
microdialyzers
casting and injection micromolds
diaphragms for light collimation
nerve regenerators – prosthesis of senses
pressure regulators
mechanical, electric and optical resonators
rocket/jet engines
spectrophotometers
optical waveguide couplers
turbines and dynamos.

The technology of deep wet etching of silicon for microsystems is being developed continuously. Technological progress in this field is well exemplified in Fig. 3.9, where two structures of silicon pressure sensors are presented: one sensor from the sixties with ultrasound drilled membrane and one sensor from the 1990s with a membrane anisotropically etched using the wet method. The second structure is more than 238 times smaller in volume than the first one. Reduction in proportion (similar in silicon consumption) means reducing the material costs of sensor production, so that a proportional reduction in sensor price may be expected.

Wet anisotropic etching of silicon is characterized by many parameters. The most important are:

- anisotropy of the process*, particularly etch rates of (100), (110) and (111) planes, and its proportion

a) b)

Fig. 3.9. Structures of piezoresistive silicon pressure sensors; an example of progress in deep silicon micromachining; membrane-side view: a) late 1960's; chip dimensions: 10 mm × 10 mm × 4 mm, membrane diameter $\phi = 8$ mm, minimum membrane thickness $d > 150$ μm; damage to membrane edges caused by ultrasound drilling is visible, b) 1990s; chip's dimensions: 2.1 mm × 2.1 mm × 0.38 mm, planar dimensions of membrane 1032 μm × 1032 μm, thickness 10 μm. Photographs are not to the same scale.

* Much credit for the discovery of anisotropic properties of crystals should be given to René-Just Haüy, a canon from the Notre-Dame Cathedral of Paris in about 1802 (Traitè de Crystalographie, 1822).

- selectivity of etching versus silicon oxide and nitride masks,
- factor of under-etching of the convex edges of patterns, being a result of high etch rates of high index (*hkl*) planes (221), (321) etc.
- quality of etched surfaces; smoothness, flatness
- temperature of the process
- industrial safety of the process, costs, environmental influences.

Etch rate V_{hkl} of crystallographic plane (*hkl*) can be expressed by

$$V_{hkl} = \frac{d_{hkl}}{t}, \tag{3.2}$$

where: d_{hkl} = depth of etched cavity measured perpendicular to the plane (*hkl*), t = the time of etching.

Hence:

$$V_{100} = \frac{d_{100}}{t}, \qquad V_{111} = \frac{d_{111}}{t}, \tag{3.3}$$

where: V_{100}, V_{111} = etch rate of planes (100) or (111), d_{100}, d_{111} = depth of etched cavity in crystallographic direction ⟨100⟩ or ⟨111⟩. The parameters V_{110}, V_{221}, V_{231}, that is etch rates of planes (110), (221), (231) and others, can be obtained by a similar equation.

The anisotropy factor of etching is usually determined by the ratio of the etch rate V_{100} to the etch rate V_{111}; it is sometimes helpful to use an anisotropy factor determined by the ratio of etch rates V_{110} to V_{111}.

The selectivity of etching (often abbreviated simply to selectivity) is usually described by the ratio of silicon etch rate V_{100} to etch rate of a mask, defined as a change in thickness of a mask divided by the time of etching.

The shape and dimensions of 3-D structures, etched in silicon by means of wet anisotropic etching, depend on four main factors:

- crystallographic orientation of substrate front surface,
- shape of mask and its position in relation to the main crystallographic directions,
- anisotropy of etching
- etch rate of plane (100).

The two last factors are related to the type of etching solution, its concentration, the temperature of the process and mixing intensity*.

A technically acceptable, good process should be very anisotropic (ratio of V_{100} to V_{111} over 100), fast and selective. Most often etch rates form the relation:

* Many authors believe that mixing does not influence the etching process. In our own research it was observed that, although this rule is correct for strong solutions, it does not apply to weak solutions, in which rheology and geometry of etched shapes depend on mixing conditions. Thus, our own results presented later in this book were attained in etching processes in which the mixing of solutions was performed in accordance with the rule "it will not harm, but it can help".

$V_{100} > V_{110} > V_{111}$, the ratio $V_{100}:V_{111}$ equals 100, etch rate V_{100} amounts to about 1 μm/min.

So, wet anisotropic etching of silicon can be used for etching a limited number of three-dimensional patterns deeply etched in (100) and (110) substrates (Fig. 3.10). The shape of the etched pattern is a function of parameters of the etch process, the shape of the window made in a mask, and – last but not least – of the alignment of the window and crystallographic directions valid at the particular chosen orientation of the wafer. Thus, this type of etching is named orientation-dependant-etching (ODE).

Regardless of the shape of etch window, after a sufficient time of wet anisotropic etching, the patterns etched at (100) wafers may become square or rectangular. The etched cavity will then be formed by four (111) planes, angled between them 90°, and 54.74° to the substrate surface (see stereographic projection of (111) planes onto (100) plane on Fig. 3.2). This results from the features of wet anisotropic etching of silicon, in which plane (111) is always etched the most slowly. This is a very important rule in the etching of micromechanical structures on substrates (100).

The cavity deeply etched on the substrate (110), obtained by use of mask windows of any shape, after a sufficiently long etching time, will also be limited by four walls (111), forming on the substrate surface a rhombus with corner

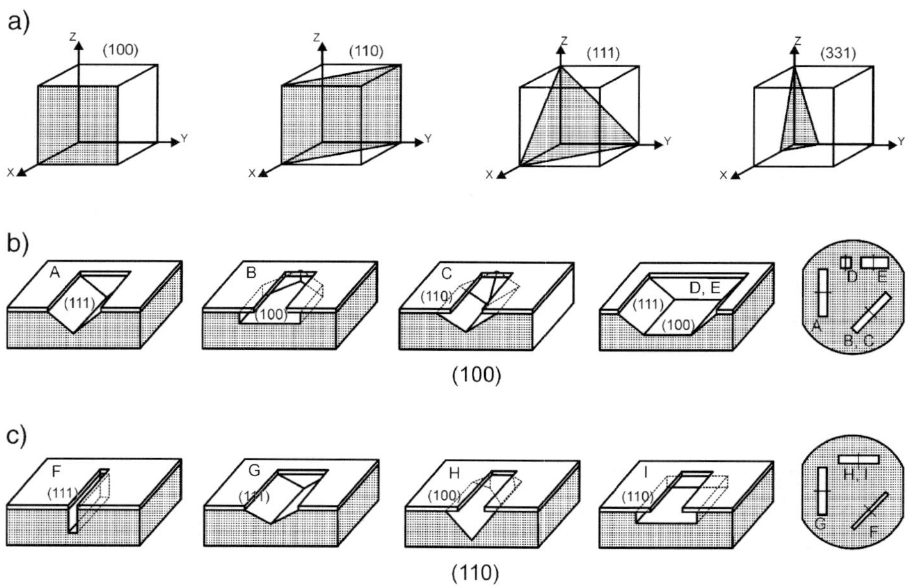

Fig. 3.10. Patterns wet anisotropically etched in silicon: a) crystallographic planes in the cubic crystal cell, b) possible cavities at substrate (100), c) at substrate (110); lay-out of the planes limiting etching as well as position of mask are marked. The picture perspective distorts the shape of cross-sections.

angle 70.5°. Side (111) walls will be perpendicular to the (110) oriented bottom of the cavity.

Usually, wet micromachining procedures are "tailored" according to the technological specification of microsystem fabrication by proper adjustment of the mask shape, a kind of wafer orientation, and choice of the proper parameters for etching.

3.2.1. Etching solutions, chemical reactions, etching models

The solutions that anisotropically etch silicon are alkali hydroxide aqueous solutions. Chronologically, the oldest solution of organic base, used for the fabrication of a three-dimensional silicon structure, was a water solution of hydrazine with an additive of pyrocatechol, introduced in 1962 in publication [27] and researched in the following years (e.g. [28–31]). In spite of many advantageous features of this solution, especially its high selectivity to SiO_2 and inertness to thin-film metallic connections of integrated circuits, strongly carcinogenic and explosive properties of this solution stopped its wider application in silicon micromechanics.

Solutions based on ethylene-diamine and pyrocatechol with pyrazine (EDP) have been applied in many processes [32–37]. EDP shows very good compatibility to CMOS technology. Particularly significant is a very good selectivity of this solution to SiO_2 (2000:1) and aluminum (few hundreds:1). Unfortunately, dehydrated EDP is explosive, solutions based on EDP are carcinogenic, they decompose when exposed to atmospheric oxygen and can form explosive mixtures.

A water solution of tetramethylammonium hydroxide TMAH [38–43] is also used for silicon etching. The solution shows very high selectivity against LPCVD silicon nitrides and thermal silicon dioxide (few thousands:1) and remains harmless to health [41]. Because this hydroxide is widely applied in integrated technology as a positive resist developer, the etch process is CMOS compatible.

Anisotropic silicon etching in inorganic alkali aqueous solutions has been known since the 1960s [44–75]. The water solution of potassium hydroxide (abbreviated to KOH) is usually used [45–56] pure or with organic additions. Usually isopropyl alcohol – IPA – is added. Other inorganic alkali solutions, applied more rarely, are sodium hydroxide NaOH [62] or ammonium hydroxide NH_4OH [57–60]. The RbOH, LiOH, CsOH solutions are not used technically, although some laboratory studies have been reported [59–60].

The widest application in industrial processes of silicon microsystem deep micromachining has found aqueous solutions of KOH, EDP and TMAH (Table 3.5).

The chemical reaction of anisotropic etching of silicon has been the subject of scientific discussion for many years. Finne and Klein [28] first proposed a model of silicon etching in KOH in which hydroxyl ions OH^- and water reacted

Table 3.5. Alkaline solutions for silicon etching used in micromechanics

Name	Chemical formula	Comments
Hydrazine and pyrocatechol	$N_2H_2 + C_6H_4(OH)_2 + H_2O$	First used (1962)
EDP (ethylene-diamine, pyrocatechol, pyrazine)	$NH_2(CH_2)_2NH_2 + C_6H_4(OH)_2 + C_4H_4N_2 + H_2O$	Carcinogenic, explodes, CMOS-compatible.
TMAH (tetramethylammonium hydroxide)	$N(CH_3)_4OH + H_2O$	Positive resist developer, does not etch silicon dioxides or nitrides
Potassium or sodium hydroxide	$KOH + H_2O$ $NaOH + H_2O$	Applied on a mass scale Used occasionally

on the silicon surface:

$$Si + 2(OH)^- + 4H_2O \rightarrow Si(OH)_6^- + 2H_2, \tag{3.4}$$

and they suggested the next step of chemical reaction valid for KOH and EDP in case of pyrocatehol presence:

$$Si(OH)_6^- + 3C_6H_4(OH)_2 \rightarrow Si(O_2C_6H_4)_3^- + 6H_2O. \tag{3.5}$$

Pyrocatechol has to accelerate significantly the reaction of silicon etching by increasing the solubility of reaction products.

Palik, and co-authors [56], observed that the silicon rate depends on the concentration of H_2O/OH^- complexes and that the reaction goes as follows:

$$Si + 2H_2O + 2(OH)^- \rightarrow Si(OH_2)O_2^- + 2H_2. \tag{3.6}$$

The mechanism of etching proposed by Finne and Klein [28] and Palik [56] assumes that only OH^- groups take part in the chemical reaction. Dziuban [63] has shown that silicon can be anisotropically etched in pure water excited by microwave irradiation, which strongly supports this assumption. It has been stated unambiguously [56] that transport of an electron from an OH^- group to surface silicon bonds is a fundamental condition of the chemical reaction. Raley and co-workers [35] widened this conception and assumed transport of four electrons injected to a conduction band during the initial reaction of silicon oxidation.

Price [46] related the anisotropy of etching to a number of chemical bonds existing on the crystallographic plane, which is etched away. The most dense plane (111) has to be etched most slowly, and plane (100) most quickly [46]. However, the phenomenon of a very high anisotropy of etching of silicon etched in 55% KOH:water solution, cannot be convincingly explained on the basis of the above-mentioned theory for planes (110) and (111). The proportion of bonds on (100) and (111) planes is as 1:2, while anisotropy V_{110}/V_{111} reaches 500:1. Kendall [49] says that the low etch rate of plane (111) results from its higher susceptibility to oxidation. The plane, immediately after the silicon draught in

Table 3.6. Activation energy E_a and constant V_0 for planes (100), (110) [51]

	(100) Si		(110) Si	
Solution type	E_a (eV)	V_0 (μm/h)	E_a (eV)	V_0 (μm/h)
EDP type S (*slow*)	0.40	9.33×10^6	0.33	1.16×10^6
20% KOH	0.57	1.23×10^{10}	0.59	3.17×10^{10}
20% KOH	0.62	4.08×10^{10}	0.58	4.28×10^9
23% KOH	0.60	2.69×10^{10}	0.62	8.98×10^{10}
27% KOH	0.60	3.61×10^{10}	0.57	1.50×10^{10}
32% KOH	0.57	9.28×10^9	0.62	9.88×10^{10}
34% KOH	0.61	3.10×10^{10}	0.60	3.66×10^{10}
42% KOH	0.57	1.60×10^{10}	0.60	4.37×10^{10}
24% NaOH	0.65	1.59×10^{11}	0.68	7.00×10^{11}
10% LiOH	0.60	3.12×10^{10}	0.62	8.03×10^{10}

etching solution, is probably passivated with a chemically inert layer of oxide, which reduces the chemical reaction rate.

In papers [41] and [74] the anisotropy of etching of monocrystalline silicon has been connected with the activation energy of the etching process, as well as with the density and geometry of bonds of the silicon atom of each of the etched crystallographic planes. Simulations illustrating anisotropy of silicon etching analogous to the model of anisotropic crystal growth observed in MBE epitaxy are presented in paper [53] and [75].

Seidel with co-authors [51], similarly to Palik's models [56, 73], have assumed that only hydroxide ions take part in the chemical reaction of silicon etching in aqueous alkali solutions, while the etch rate is dependent on the transport of electrons from the broken silicon bonds. The electrons move from the fundamental band to the conduction band through surface states, located in the forbidden zone. Seidel and co-workers have made an assumption, in their model, that the silicon etch rate V is given by Arhenius' law (equation 3.7), where the activation energy E_a is attributed to surface states energy, which depends on the type of crystallographic plane:

$$V = V_0 \cdot \exp\left(\frac{-E_a}{kT}\right), \tag{3.7}$$

where: $V_0 = \text{constant}$.

The activation energy E_a of the etch process for planes (100) and (110) calculated for several types of solutions is given in Table 3.6. Etch rates calculated according to equation 3.7 are in good agreement with experimental data.

3.2.2. Etching in KOH

A water solution of potassium hydroxide with additives, the mixture anisotropically etching silicon, is most often used in silicon microsystem technology. This is a result of many advantageous features of this solution: it is cheap and easy accessible, stable and inert in air in a wide range of temperatures, in the anhydrous state it can be stored for many years without losing its quality, it is relatively not very toxic, sufficiently selective to SiO_2 and very highly selective to Si_3N_4 masks, and inert to glass technological apparatus in which the etching process is most often carried out.

The major disadvantage of KOH is the fact that its temperature and concentration exert a great influence on the morphology of etched surfaces, etch rates and anisotropy. It may be assumed that temperature mainly influences the etch rate and anisotropy factor, while concentration affects the anisotropy and, to a smaller extent, etch rates. Moreover, water solution KOH etches silicon slowly, but this is a disadvantage common to all known alkaline solutions*.

Etching of silicon substrates is performed in a glass apparatus (Fig. 3.11) under atmospheric pressure. The solution is heated to the required temperature,

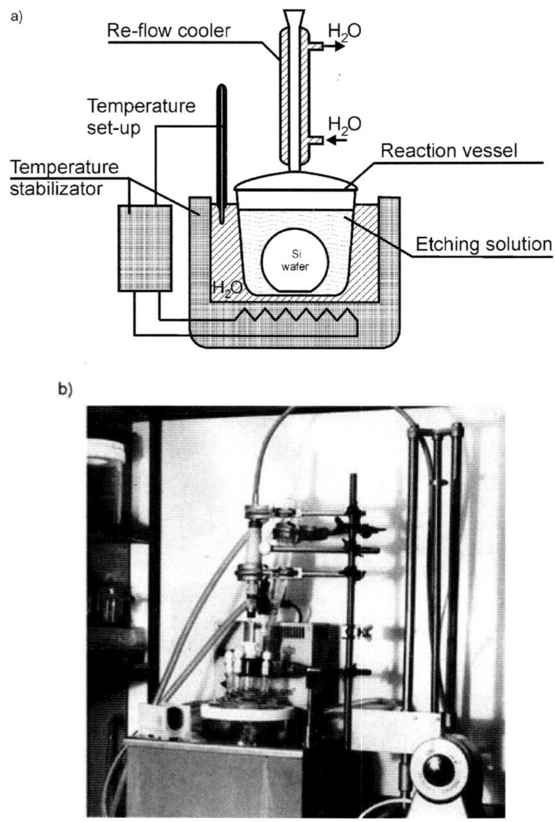

Fig. 3.11. Stand for silicon etching: a) scheme, b) its appearance.

* This feature applies to classic procedures, in which the solution is heated thermally. The introduction of microwave excitation accelerates etching significantly. This type of etching is described in section 3.2.4.

usually +80°C, which is stabilized ±0.5°C during many hours of etching. The reaction vessel is equipped with a re-flow water cooler to stabilize the composition of the solution during prolonged etching. In the batch etch process silicon substrates are situated in a Teflon® chuck (holder), parts of wafers or small silicon details must be tightly positioned in the dedicated holders. The stand shown in Fig. 3.11 makes possible batch etching of 50 pieces of 3″–4″ substrates and contains about 5 liters of an etchant.

3.2.2.1. Basic properties of the process

Etch rates and anisotropy

Etch rates and anisotropy of a process are experimentally specified using three methods:

- from the change of a pattern of planar circular concave or protrusion patterns etched in the silicon substrate,
- by etching of a model sphere and observation of the geometry of a three-dimensional figure [59, 66],
- from the wagon wheel test; observation of the surface pattern of the shape of a wheel with a large number of spokes (hence the name of the test) [51].

The wagon wheel test undoubtedly has the widest practical application. In this method the shape of the spokes of the wheel is etched so that a shallow under-etching of the dielectric mask can be observed, proportionally to the silicon etch rate, which depends on crystallographic direction. Because the wagon wheel shape has 360° symmetry the pattern appearing on the etched (100) oriented surface of (100) wafer has a quadruple symmetry and reflects precisely the relation between distribution of etch rate and crystallographic direction (Fig. 3.12). The etch rate graphs, prepared as described above for silicon substrates (100) and (110) (as explained, most often applied in silicon microsystem technology), obtained for a 1:1 water KOH solution, are presented in Fig. 3.13a [51]. A similar graph, attained for a T-type water solution EDP of composition: H_2O 470 ml, ethylene-diamine 1000 ml, pyrocatechol 176 g (Fig. 3.13b) [51] has been added for comparison.

The etch rate of silicon in KOH depends on temperature and on the concentration of solution. Etch rate V of planes (100) and (110) for water KOH solutions with concentration from 10% to 60% can be calculated using empirical relation [51]:

$$V = K_0 [H_2O]^4 [KOH]^{1/4} e^{-E_a/kT}, \tag{3.8}$$

where for plane (100): $K_0 = 2480\ \mu m/h \cdot (mole/liter)^{-4.25}$ and $E_a = 0.595$ eV, while for plane (110): $K_0 = 4500\ \mu m/h \cdot (mole/liter)^{-4.25}$ and $E_a = 0.60$ eV.

The experimental etch rates of silicon in KOH are usually lower than the one theoretically estimated from relation 3.8. This is shown in Table 3.7, where results of calculation for the temperature range from 70°C to 90°C based on relation 3.8 and experimentally obtained results are presented. Etch rate characteristics as a function of the concentration of KOH solution at 80°C demonstrate

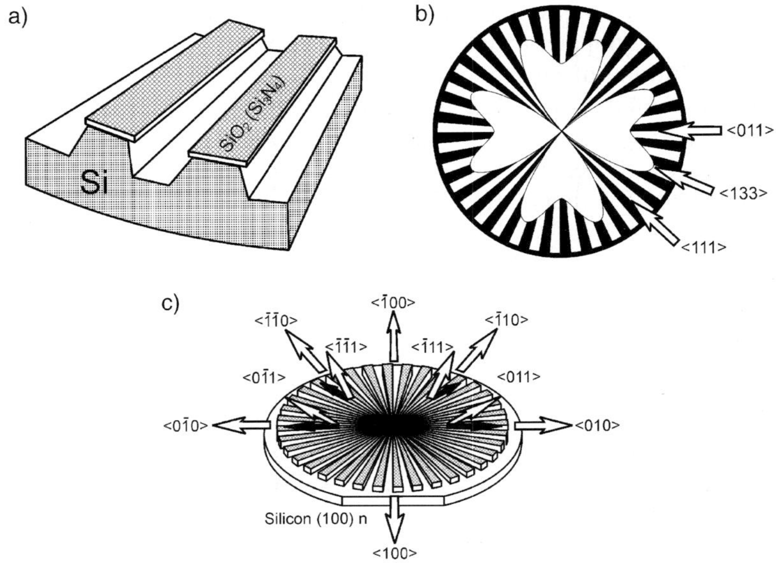

Fig. 3.12. Wagon wheel test: a) cross-section of the test mask sector with the outline of the cavity, b) wagon wheel pattern at the (100) wafer, c) the pattern and crystallographic directions at the (100) wafer.

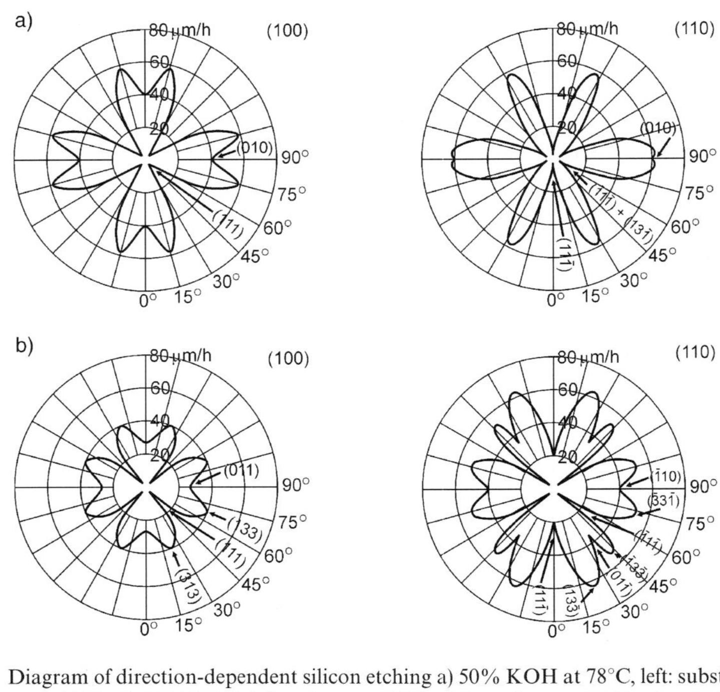

Fig. 3.13. Diagram of direction-dependent silicon etching a) 50% KOH at 78°C, left: substrate (100), right: substrate (110), b) 95% EDP, left: substrate (100), right: substrate (110), after [51].

Table 3.7. Etch rates of silicon planes (100) and (110), in µm/h for various KOH concentrations and different temperature of the etchant: according to formula (3.8) (f) and calculated in our own experimental works (e)

Plane	(100)						(110)					
Temperature [°C]	70		80		90		70		80		90	
3M KOH	46	24	82	32.4	140	49.8	71	—	126	—	216	—
5M KOH	—	25.8	—	40.8	—	82.8	—	72.6	—	126	—	210
7.5M KOH	49	29.4	86	53.4	148	100.8	75	82.2	133	141	229	222
10M KOH	45	30	79	50	135	87	68	79.8	121	132	209	220
Method of obtaining data	f	e	f	e	f	e	f	e	f	e	f	e

the maximum value for medium concentration of the etchant, 10–20% (3–5 moles). The highest V_{100} etch rate at 70°C is noted for strong solution, 30–35% (8 moles), at 80°C, the highest V_{100} etch rate is obtained in weaker solution, 20% (4 moles) (Fig. 3.14).

The numerical data describing etch rates V_{hkl} of silicon differ in various sources. For example, according to paper [51] the etch rate V_{100} equals 80 μm/h, in [42] the same etch rate amounts to 50 μm/h. This kind of discrepancy between the data results from the large influence of the type of etched substrate; apparatus configuration (mixing) and impurities in solution on etch rates. In addition the observed solution "fatigue", coming from its saturation with reaction products, also influences the process. All of these factors influence the repeatability of etching and make the process technologically difficult.

Curves of silicon etch rate in log $V - 1/T$ are straight lines (Fig. 3.15) in good agreement to Arhenius's law. This shows that the etch process is thermally activated, despite the discussion on its chemical or electrical nature.

The concentration of KOH exerts a crucial influence on the quality of etched surfaces. Smooth surfaces are obtained in strong solutions (40%, 10 moles or more), above 70°C. Below 70°C etched surfaces become rough (Fig. 3.16).

As the concentration of KOH decreases, various etch errors, unevenness and

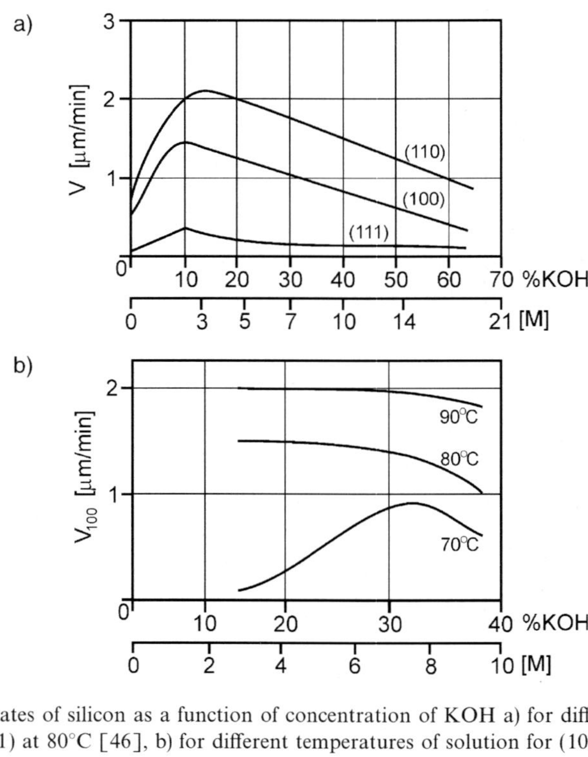

Fig. 3.14. Etch rates of silicon as a function of concentration of KOH a) for different orientations (110), (100), (111) at 80°C [46], b) for different temperatures of solution for (100) plane [43].

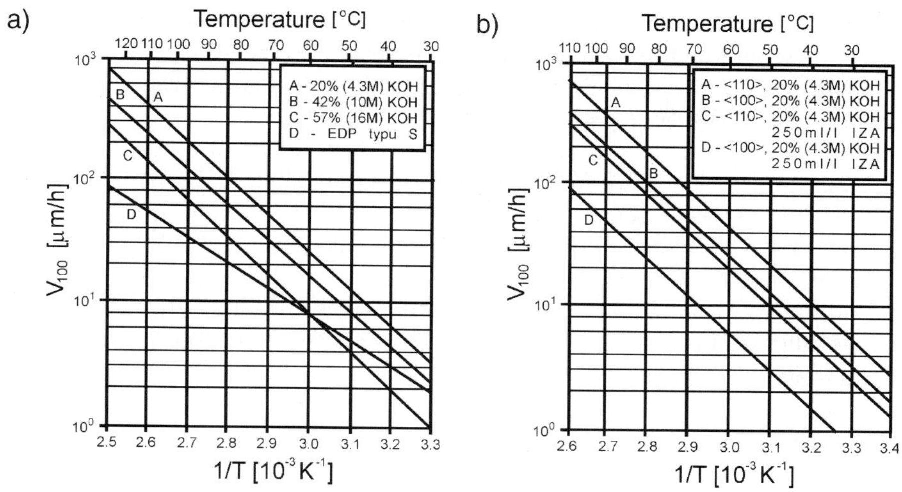

Fig. 3.15. Etch rates of silicon V in KOH as a function of temperature and concentration of solution: a) V_{100} for solution with different concentration, b) V_{100} and V_{110} in 20% pure KOH or with additive isopropyl alcohol, after [51].

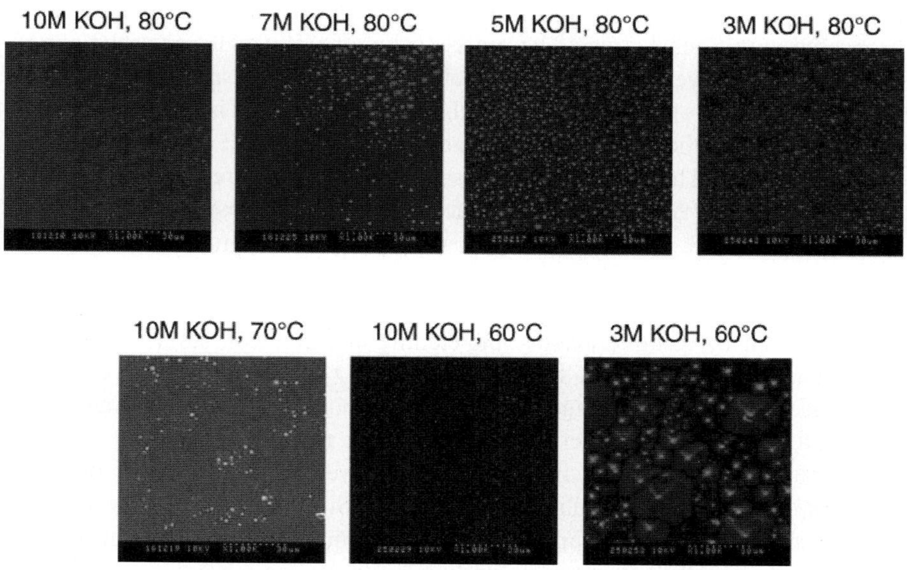

Fig. 3.16. (100) surface at the bottom of a cavity deeply etched in pure KOH, obtained for various etch process parameters. Note that, for concentration below 10M and temperature below 80°C, the surface becomes rough, for 3M KOH at 60°C the (100) surface is covered with hillocks.

hillocks appear, even in hot etchant. Thus the process engineer, selecting conditions of etching to form three-dimensional structures, needs to reconcile two opposite tendencies. A high KOH concentration (>7M KOH) and a high temperature of process ($\geq 80°C$) lead to:

- good basic anisotropy of etching, that is a high value of the ratio $V_{100}:V_{111}$,
- smooth surface,
- very serious undercutting of high *hkl* index planes,
- low selectivity against SiO_2 masks,
- pinhole-like strong local penetration of mask, resulting in formation of residual unintended, pyramidal, concave shallow cavities at the front surface of (100) wafers.

Low concentration of KOH at high temperature causes:

- smaller ratio of $V_{100}:V_{111}$,
- poor quality of surfaces,
- smaller undercutting of planes with higher *hkl* indexes,
- better selectivity against SiO_2 masks.

Lowering of temperature worsens the quality of surfaces, reduces etch rate, worsens anisotropy but increases selectivity. Selectivity of etching against a dielectric mask is one of the most important parameters of this process. Selectivity (S) of silicon etching is expressed by a ratio of etch rate of silicon to etch rate of mask in a direction perpendicular to the surface of substrate which is covered with a mask layer. The precise value of S is usually obtained experimentally and is given as a proportion of the etch rate V_{100} and the etch rate of the thin-film layer of silicon dioxide and/or silicon nitride. The value of parameter S depends on the method of fabrication, type of etchant and parameters of the etch process. Thermal oxides and silicon nitrides ensure required masking of silicon surfaces from the undesirable impact of the most often used water solutions of KOH. Selectivity is not a simple proportional function of the thickness of a mask layer. Etching solutions easily penetrate small defects – pinholes – that exist in a thicker thermal oxide layer of a mask. As an effect, defects caused by residual anisotropic etching of the silicon surface occur, and their density increases with the increase of SiO_2 mask thickness. Complete avoidance of this disadvantageous effect is not possible, although annealing of SiO_2 in a reducing atmosphere may decrease the number of defects.

Pinhole penetration of a mask is not observed for SiO_2–Si_3N_4 bi-layers. Thermal oxide protects the silicon surface, while Si_3N_4 "inundates" the pinholes because silicon nitride is characterized by an almost complete chemical resistance to alkaline. Silicon nitride thin-film layer can be employed as a single-layer mask. Selectivity S noted for Si_3N_4 may exceed the value 10 000:1.

Selectivity S is strongly dependent on temperature and KOH concentration (Fig. 3.17) [51].

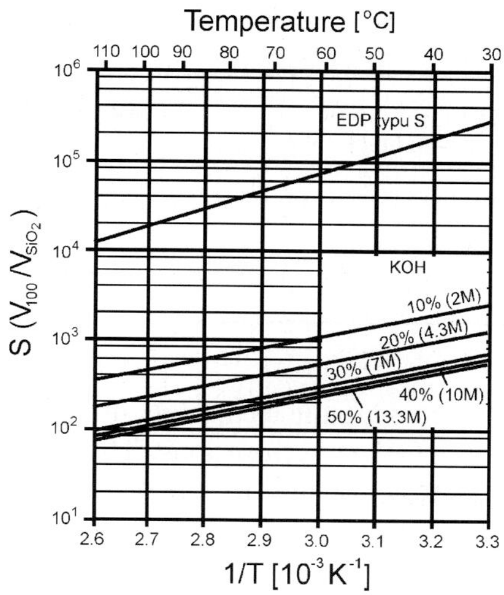

Fig. 3.17. Selectivity (S) of silicon (100) etching as a function of temperature, for different KOH concentrations, curve for EDP-S given for comparison, thermal silicon dioxide used as a mask [51].

Additives

Deep silicon etching in weak KOH solutions in water or/and in temperatures below 70°C requires the use of additives. In silicon micromechanics the additive mainly applied is isopropyl alcohol (IPA). Other additives are used more rarely [64, 65], e.g. poly-cyclic aromatic compounds, which catalyze chemical reactions of etching away of silicon, accelerate etching and even change the ratio from $V_{110} > V_{100}$ to $V_{100} > V_{110}$ (Fig. 3.18), as well as a small addition of As_2O_3 (1–2%) reported in [76], improving etched surface smoothness.

A low concentration of IPA in etching solution ensures a significant reduction of high index planes etch rates. Moreover, IPA additive smooths the etched

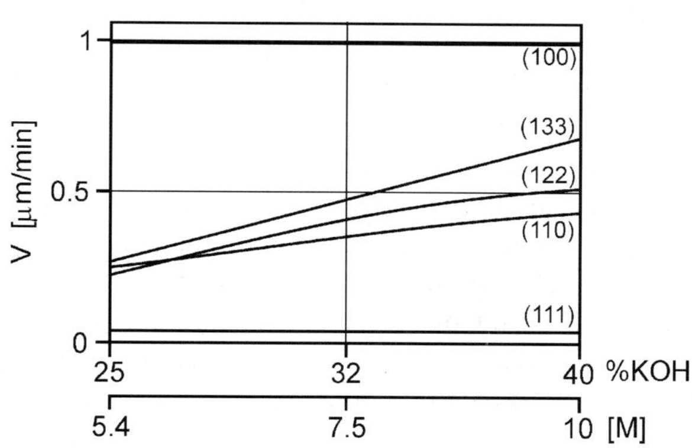

Fig. 3.18. Etch rates of planes (*hkl*) as a function of concentration of water KOH solution with addition of isopropanol at a temperature of 80°C [77].

Table 3.8. IPA solubility in water KOH solutions, at 80°C [77]

KOH concentration [mole/liter]	IPA solubility [% of weight]
3	12
5	5
7.5	2.4
10	1.8

surface (especially (100)), moving the technically useful concentration of etching solution towards lower KOH concentrations. The best effect is obtained for weak KOH solutions because solubility of IPA decreases with the increase of KOH concentration in a solution (Table 3.8).

In paper [51] two mechanisms of interaction between IPA and anisotropic silicon etching in KOH are given:

- addition of isopropyl alcohol modifies the percentage of contents of H_2O/OH^- complexes without changing the solution reaction, thereby influencing the mechanism of electron transport and etch rates of crystallographic planes with high hkl indicators,
- isopropyl alcohol covers silicon with a thin-film layer, making H^+ penetration difficult, increasing the resistance of, particularly, plane (110) to etching effects, on this plane a strong channeling effect of protons into silicon occurs, which decreases the force of silicon–silicon bond and facilitates breaking of the bonds during the multi-step process of silicon etching away.

Interpretations described above, quoted from [51], seem to apply to all etching alkaline solutions with addition of IPA used in micromechanical technology.

3.2.2.2. Stop-diffusion

The etch rate of anisotropic wet etching of weakly doped silicon ($N_{A,D} < 7 \cdot 10^{19}$ cm^{-3}) does not depend on the concentration and type of impurity. This is confirmed constantly in everyday experiments. Detailed research carried out for all alkaline solutions which etch anisotropically silicon has revealed the existence of the phenomenon of etch stop-effects (stopping of etching, etch-stop, stop-diffusion) on heavily doped areas, both of p^+ and n^+ type. Etch-stop effects have been noticed for boron, phosphorus, germanium, carbon, nitrogen [35, 78–84]. It was confirmed that the etch rate of boron-doped silicon decreases significantly, about 100-fold, for the concentration of a dopant from $7 \cdot 10^{19}$ cm^{-3}, to the degeneration of material (10^{22} cm^{-3}) (Fig. 3.19) and can be observed for various types of etchant (Fig. 3.20). The phenomenon of etch-stop of silicon wet etching on a heavily doped layer was described for the first time in 1969 by Greenwood [32], who mistakenly interpreted the

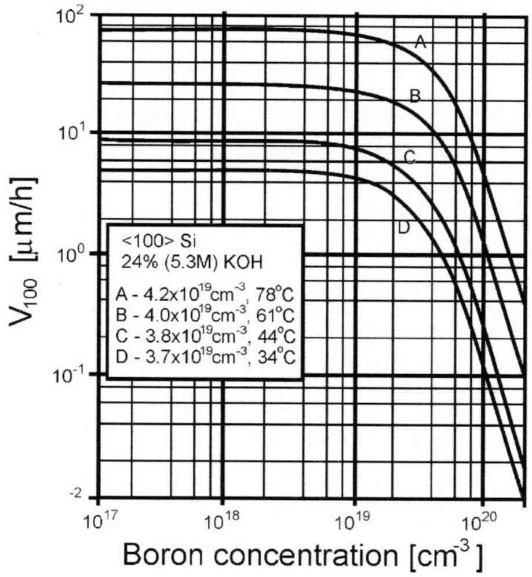

Fig. 3.19. Etch rate V_{100} as a function of boron concentration for KOH at different temperatures [81].

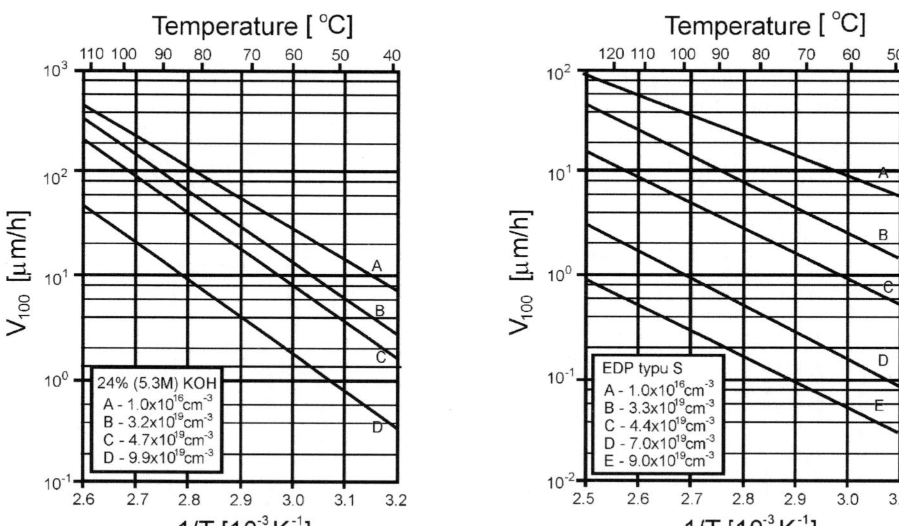

Fig. 3.20. Etch rate V_{100} as a function of $1/T$ for different doping levels. Etching solution 24% KOH (on the left) and EDP S-type (on the right) [81].

observed etch stop-effects, in ethylene-diamine and catechol solutions, as new, unknown properties of the *p-n* junction. Boughs [34] interpreted properly etch stop-effects as a result of the heavy boron doping of silicon, not attributed to *p-n* junction properties.

3.2.3. Electrochemical etching

Electrochemical etching of silicon was developed at the beginning of the integration of semiconductor devices as an auxiliary technique of silicon wafer thinning, fabrication of insulating islands and beam-leads connections [85–89]. Such an etching process of silicon, developed in the 1980s [31, 84, 90–101], has also been researched, to date, from the point of view of its application to silicon micromechanics [93, 95–97, 102–104]. The essence of electrochemical etching is that it makes possible repeatable control of the etch rate of silicon by attaching external potentials to a silicon and an etching solution – an electrolyte. Of course, the phenomenon mentioned above is connected with different limitations, related to the chemical reaction, occurring during the anisotropic etching of silicon substrates in aqueous alkaline solutions. Anisotropic electrochemical etching (ATE) of silicon can be performed in a two-, three- or four-electrode configuration of electrical circuits (Fig. 3.21), in all types of alkali etchant*. Most frequently the three-electrode circuit configuration is used. One of the electrodes works as a so-called current electrode, the second electrode (reference, potential) stabilizes the potential of an etchant. The third electrode (usually thin-film, metallic) is deposited on the back-side face of the etched substrate. During etching, an electric current flows in the circuit; through electrolyte, silicon substrates and external measurement/supply instruments.

Time-dependent graphs, called voltmetric curves or anodic polarization curves, obtained for *n*- and *p*-type silicon, are shown in Fig. 3.22.

The highest etch rate of silicon is noted for negative polarization of the etched substrates versus an etchant, in open circuit (OCP), for current $I = 0$. In the

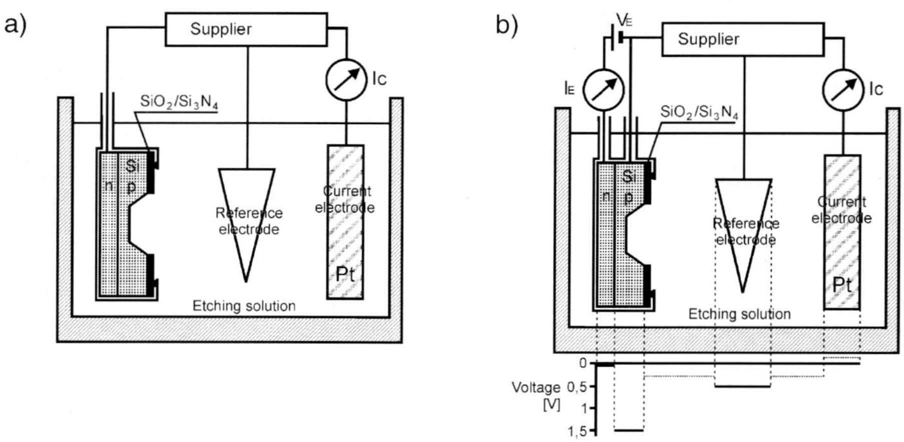

Fig. 3.21. Anisotropic electrochemical etching of silicon ATE, a) three-electrode configuration, b) four-electrode configuration (distribution of potential has been given).

* Isotropic, electrochemical, selective etching in acid solutions, e.g. $HF–H_2O$, can also be carried out in similar assemblies.

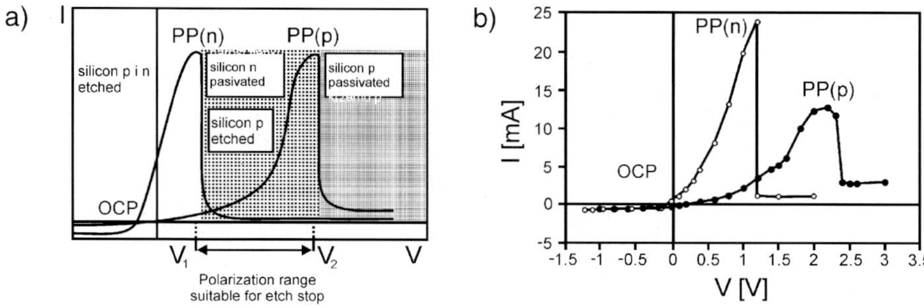

Fig. 3.22. Theoretical (a) and experimental (b) voltmetric curve $I = f(V)$ for *n*- and *p*-type silicon [97].

controlled etch-stop mode current flows in the circuit, the polarization is kept in the range in which *p*-type material is etched, while *n*-type material remains non-etched. Figure 3.22 shows this situation. Etching of *p*-type material is obtained by maintaining the potential of etching solution in relation to *n*-type silicon in the range determined by V_1 and V_2, where V_1 and V_2 potentials, for which $I \rightarrow 0$, are called the passivation potentials PP, by a suitable supply of the reference electrode. The etch process proceeds through *p*-type material, when the front of etching reaches the *n*-type silicon the etching practically stops. This moment can be observed as a jump in the current.

The mechanism of the potential-induced stop-etch phenomenon is a subject of scientific discussions. According to Seidel's model [51, 100. 101], for anodic polarization of etched silicon, the accumulation of holes occurs on a silicon/electrolyte boundary. Holes recombine with electrons needed to provide etching and stop the chemical reaction. Palik and others [92] tied the etch-stop to mechanical stress induced at the buried *p-n* junction.

According to Puers and co-workers [102] the mechanism of etch-stop of electrochemical etching for anodic polarization can be explained by blocking of the transport of hydroxide groups which occurs in the thin layer of oxide appearing at the etched silicon surface. Despite discussions on the mechanisms of electrochemical etching, this method has found wide application in micro-system technology.

The two-electrode configuration, in which the stabilization of the potential of etching solution in relation to the etched silicon wafer is not possible, is sufficient to attain the etch-stop effect, but the etch rate may fluctuate. The four-electrode configuration makes possible the additional *p-n* back-bias, which improves the quality of etched 3-D shapes, by the elimination of local holes that stop the etching. These holes are generated at local defects existing in thin oxides. This is especially important in the technology of thin silicon membranes for submicron photolithography masks [103]. The three-electrode configuration with a calomel reference electrode Ag/AgCl, polarizing etching solution, and a platinum current electrode is applied. Etched substrates have to be assembled in the holder that protects the back-side electric metallic thin-film layer (the contact to the *n*-type layer) and the boundary of the *p-n* junction. For the two-electrode configuration the *n*-type area should be strongly anodically polarized in relation to solution (for $t = 60°C$ and KOH, $V_{1n} = -1.080$ V). For the three-electrode configuration the *p*-type area needs to be anodically polarized in relation to solution ($V_p < V_{2p}$, where $V_{2p} \approx -1.04$ V). For the four-electrode configuration the *p*-type area is polarized using potential close to OCP (about -1.5 V), the additional voltage V_E polarizes the *p*-type area by $V_{1n} = -1.080$ V, kept over $+0.4$ V in relation to the *n*-type area. As mentioned earlier, in all of the electrode configurations, complete removal of the *p*-type layer can be observed as a temporary increase, a jump, of an electrical current, measured in the external circuit.

The optimum values of potentials PP V_{1n} and V_{2p}, as well as of the OCP potential for an open circuit, that is for a complete electrochemical etching, depend on the type and concentration of etching solutions and on the intensity of the light illuminating etched surface of a silicon substrate (Fig. 3.23) [102].

Electrochemical etching is employed first of all in the micromachining of very thin, large-area silicon membranes. The disadvantage of this method is the use of the protecting holder. The holder may induce mechanical stresses in an etched substrate, which may break large-area thin membranes. What is more, the use of many holders and electrical connections, connected to all substrates etched in a batch-type process complicates etching equipment. This drawback

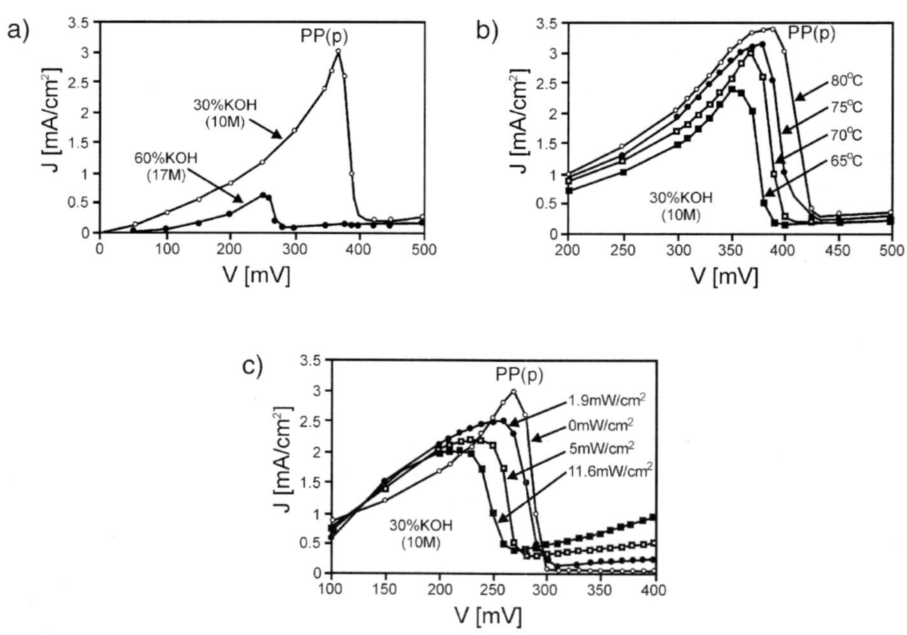

Fig. 3.23. Voltmetric curves obtained for a KOH solution – an example of electrochemical etching; a) the influence of concentration of solution, b) the influence of temperature, c) the influence of illumination [102].

does not occur in the electrochemical non-contact etching methods of silicon in water solution of TMAH, using an inner galvanic gold-silicon-TMAH potential source [104]. In this technique the *p*-layer is anodically polarized by a thin-film electrode deposited onto silicon immersed in the etchant.

3.2.4. Fast wet etching

So far, according to the state-of-art on wet anisotropic silicon etching, forming three-dimensional silicon microconstructions applied in microsystem technology, the process has been performed in a hot etchant – mainly an aqueous solution of KOH – heated in so-called water baths. The temperature of an etchant, matching technological requirements, is usually kept below 90°C. In such conditions the etch rate of plane (100), the most important from the point of view of microsystem technology, ranges – according to the literature [46, 51, 81] – from 0.8 μm/min to about 1.2 μm/min.

Acceleration of silicon wet anisotropic etching can be obtained by increasing the solution temperature, but at the expense of worsening of the anisotropy of the process. Because the silicon wet anisotropic etching process is activated thermally, consistent with Arhenius' law, increasing the temperature by 10 degrees centigrade results in a two times faster etching of silicon. However, at boiling point the etch rate V_{100} is still quite low, and for KOH solutions does not exceed 2 μm/min (according to the author's work), or 5 μm/min (according to [81]), while the anisotropy factor V_{100}/V_{111} is below 10. In addition, the selectivity of etching in relation to oxide masks becomes very pure. The etch rate V_{100}, about 10 μm/min at 130°C, was theoretically estimated in paper [51]. It was also stated there that the anisotropy factor V_{100}/V_{111} might be so low that such a process could not be useful in deep silicon micromachining.

Etching of silicon can be accelerated by an increase in the pressure of a gaseous atmosphere in a hermetic reaction vessel filled with etching solution. An experimentally determined etch rate V_{100} in 10M KOH, at a temperature of 80°C, in hydrogen under a pressure of 4 MPa, equals about 5.5 μm/min [105].

Therefore, according to the technology condition described above, increasing the temperature of etching solutions to boiling point, as well as increasing the pressure in reaction vessels, lead to a quite small acceleration of wet anisotropic silicon etching. The usefulness of a process accelerated in such ways for 3-D silicon structure micromachining is doubtful.

A small etch rate V_{100} is one of the most disadvantageous features of wet anisotropic silicon etching; it causes fabrication of 3-D micromechanical structures to last for a long time. For instance, fabrication of a silicon membrane a few dozen micrometers thick, in a typical micromechanical silicon substrate a few hundred micrometers thick, lasts typically for six hours or longer. Special procedures of fabrication of micromechanical pressure sensors need longer etching, lasting over twelve hours. For many reasons etching for so long is technically difficult, wearisome and, in mass production, economically ineffective. That is why the development of a fast, short, technologically useful wet silicon etching, apart from its scientific significance, is extremely important for microsystem technology.

3.2.4.1. *EMSi etching*

The phenomenon of fast wet anisotropic etching of silicon in solutions exposed to microwaves (EMSi – **E**tching **M**icrowave **Si**licon) was discovered by Dziuban [63, 106] and tested by him and collaborators [107–109]. In this etching method the vessel (reactor) filled with etching solution is located with substrates in a microwave resonator (Fig. 3.24) and then exposed to microwaves of a few GHz (2.45 GHz). Microwaves heat the solution to the desired temperature and simultaneously activate it, which significantly accelerates etching. The etching process can be performed in a closed reactor or in an open reactor at barometric pressure. The basic features of the EMSi process are shown in Table 3.9. The most important parameters of the process, such as temperature of solution, pressure in reactor and microwave power, has to be controlled independently.

In a closed reactor etching proceeds very quickly. For a pressure of 3 MPa, in 3M KOH, an etch rate V_{100} may reach 100 μm/min. The anisotropy ratio V_{100}/V_{111} stays no smaller than 10. However, such a quick etching is not repeatable. Etching in 10M KOH at a temperature of 105°C shows especially beneficial technological properties: medium etching rate $V_{100} = 35$ μm/min, anisotropy ratio V_{100}/V_{111} about 20, and very smooth surfaces of deeply etched micromechanical patterns (Fig. 3.25).

The essential feature of EMSi etching in an open reactor is that anisotropic (V_{100}/V_{111} equals about 30), selective, deep silicon etching can be easily obtained in weak KOH solutions at temperatures below 70°C (Fig. 3.26). What is more, under these conditions, not applied until now, very smooth surfaces of deeply etched patterns can be obtained (Fig. 3.27). Such etching procedures had never

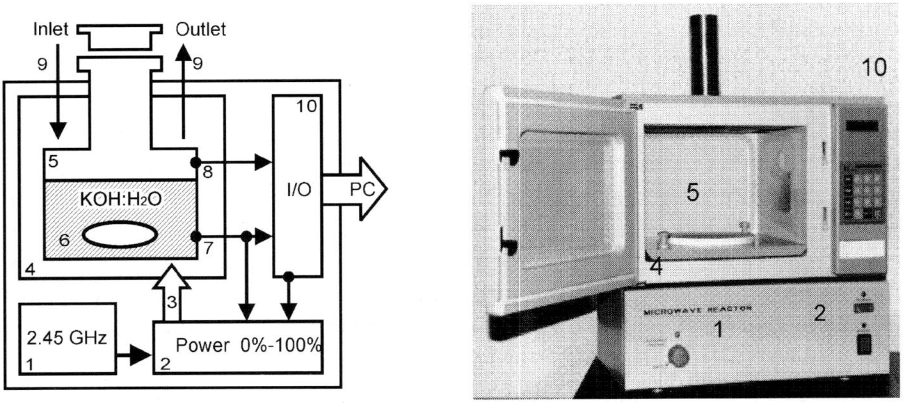

Fig. 3.24. Scheme and appearance of apparatus for fast silicon etching by the EMSi method: 1, microwave generator; 2, power attenuator; 3, feeder; 4, resonator; 5, reaction vessel; 6, etched wafer; 7, temperature meter; 8, pressure meter; 9, inlet and outlet of cooling fluid; 10, I/O interface.

Table 3.9. Comparison of etching methods

Standard thermal procedure	EMSi
Low etch rate $V_{100} \sim 1$ μm/min for 80°C	Very high ($V_{100} > 40$ μm/min for pressurized process) or high ($V_{100} \sim 10$ μm/min) etch rate
Etching rate dependent exponentially on the temperature of solution	Etching rate scarcely dependent on the temperature of solution
Surfaces influenced by hillocks below 7M KOH or/and below 80°C	Smooth surfaces from 3M to 10M KOH, even at 60°C
High anisotropy (V_{100}/V_{111}) (10M KOH $\sim 40 \div 50$) increases in stronger solutions	Good anisotropy (V_{100}/V_{111}) $\sim 30 \div 20$, increases in weaker solutions
Long-lasting process (hours)	Process of short duration (minutes)
Etching at 60°C is very slow, not used in practice	Etching at 60°C technologically useful ($V_{100} \sim 5$ μm/min)
Etching in 3M KOH is useless	Etching in 3M KOH technologically useful
Mask: SiO_2, Si_3N_4	Mask: SiO_2, SiO_2/Si_3N_4 or only Si_3N_4
Selectivity Si:$Si_3N_4 \sim k \cdot 10^4$	Selectivity Si:$Si_3N_4 > 10^4$ Si:$SiO_2 \sim 200$

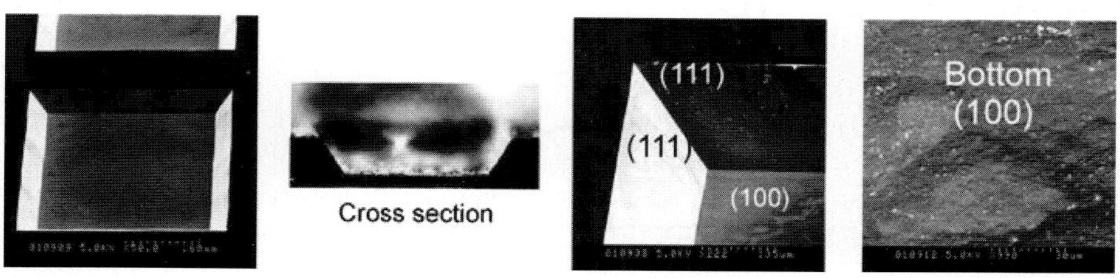

Fig. 3.25. A deep pattern etched very quicly in 10M KOH at $t = 105$°C, pressure $P = 2.7$ MPa, average etch rate V_{100} is 35 μm/min.

been taken into consideration in three-dimensional forming of micromechanical structures before EMSi etching was discovered. This is because, apart from the very slow etching of silicon and hillocks formation onto etched surfaces, they are not smooth in cold and weak KOH solutions (Fig. 3.28).

The minimal thickness of a wet, thermal silicon dioxide layer, sufficient for a cross-wafer EMSi etching of patterns, equals 1.6 μm for 380 μm-thick silicon wafers. Selectivity of EMSi etching against the typical high-temperature CVD deposited layers of silicon nitride, equals 1:10 000.

Many various 3-D micromechanical structures have been etched by means of the EMSi method [108, 109] (Fig. 3.29). The EMSi process seems to be a good alternative to the standard, thermally activated process. However, highly complicated and sophisticated apparatus, in which all of process parameters

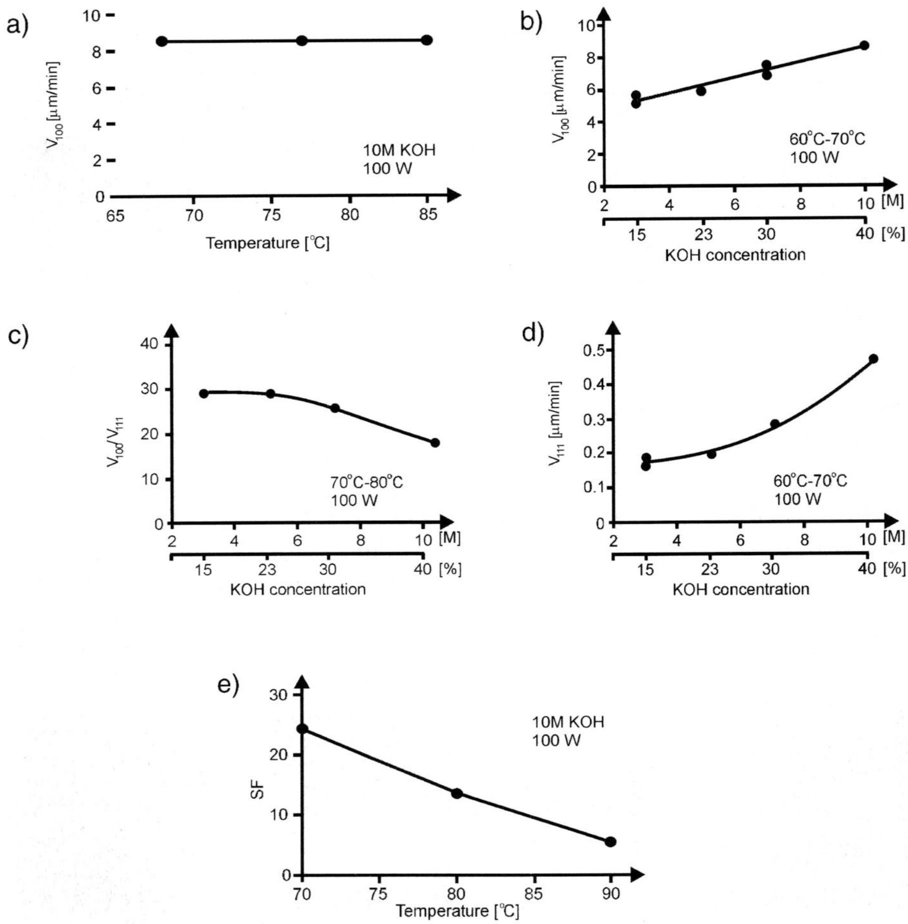

Fig. 3.26. Characterization of EMSi etching in an open reactor: a) V_{100} as a function of temperature for 10M KOH, b) acceleration factor of SF etching as a function of temperature, c) V_{100} versus concentration of the solution, d) V_{111} versus concentration of the solution, e) V_{100}/V_{111} versus concentration of the solution.

determining quality of deep silicon micromachining have to be controlled, limit the technological usefulness of EMSi.

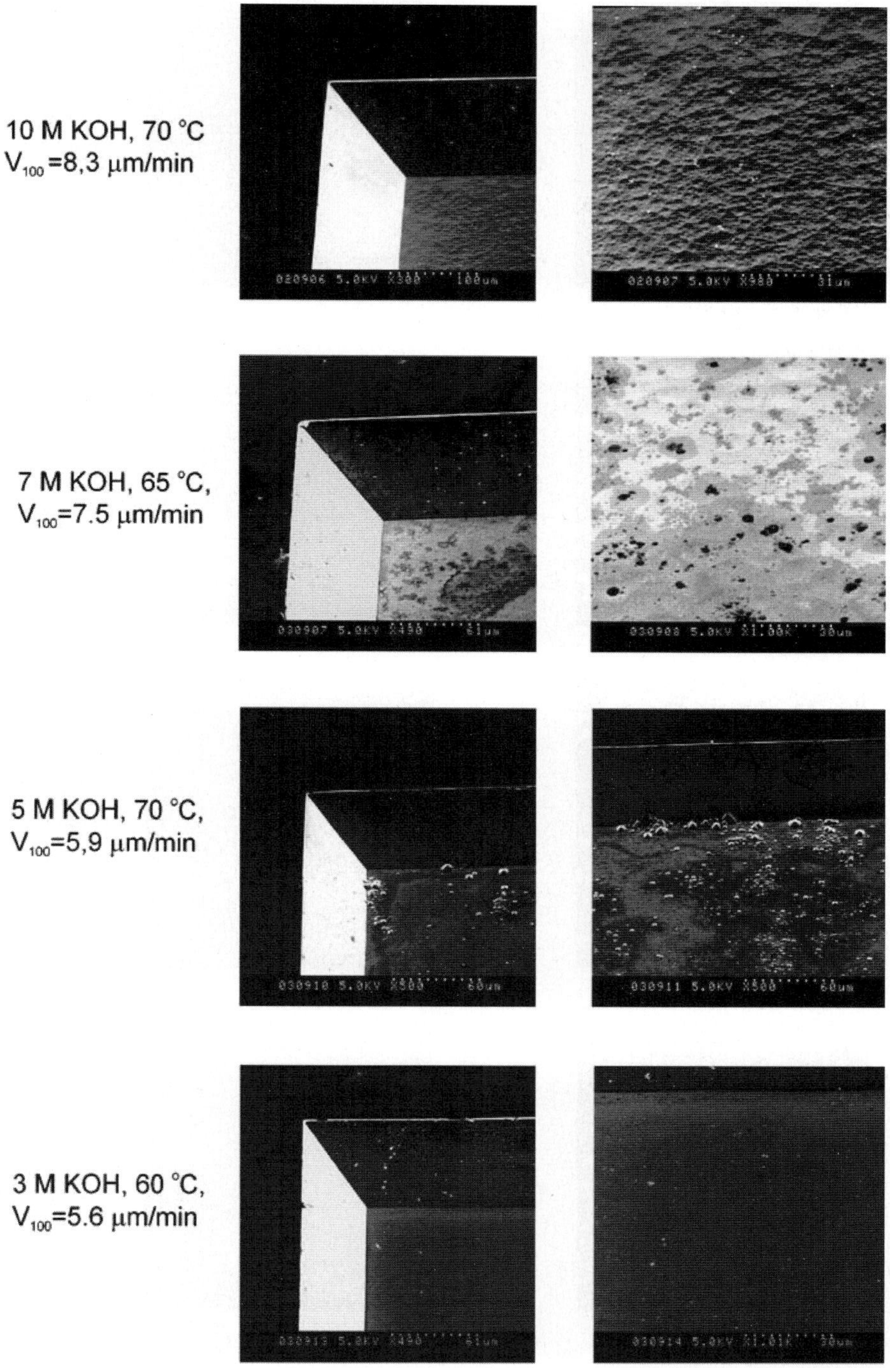

Fig. 3.27. EMSi etching – deeply etched patterns for varying process parameters.

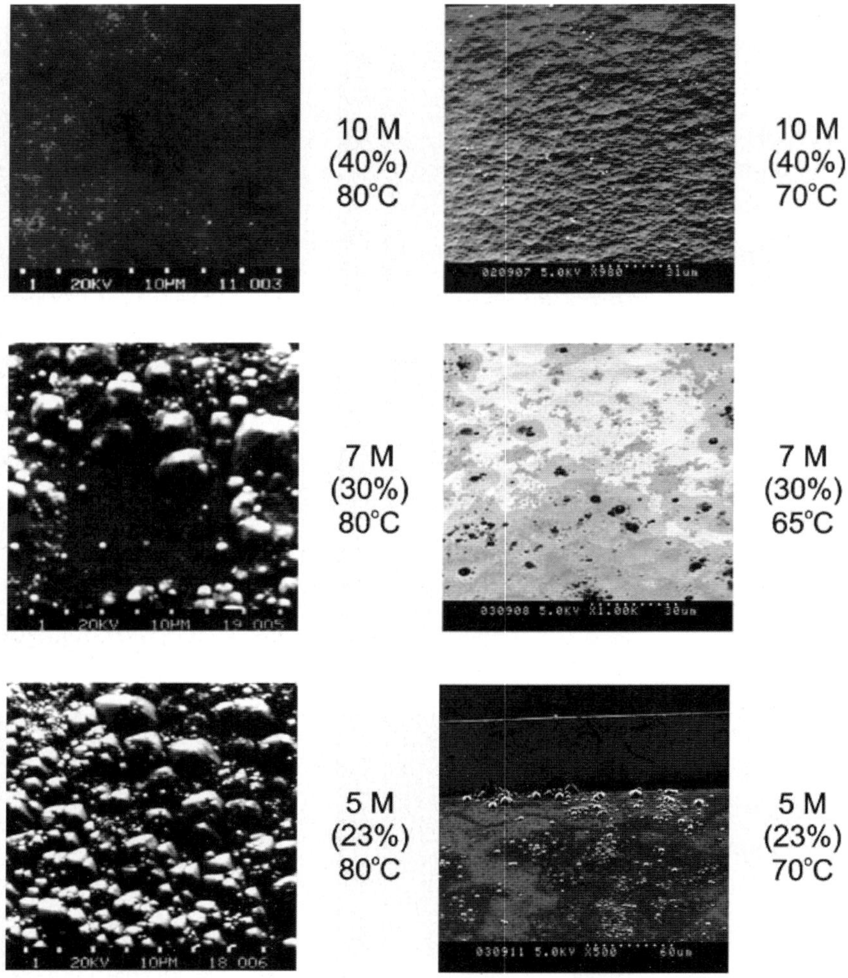

Fig. 3.28. Comparison of thermal and EMSi etching: patterns deeply etched in KOH with different concentration, on the left – thermal etching, on the right – EMSi etching.

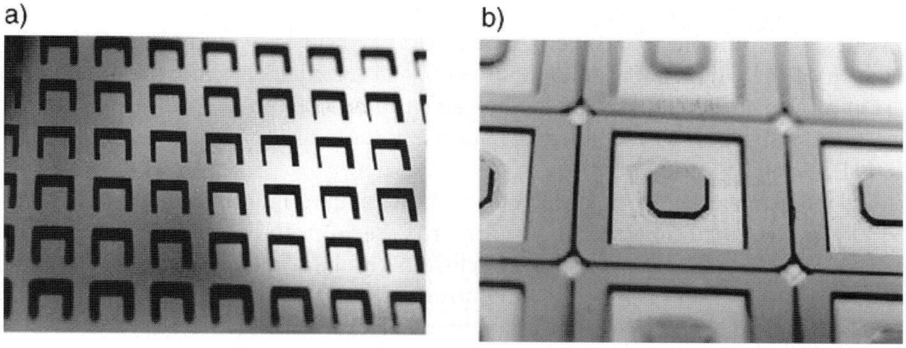

Fig. 3.29. Micromechanical structures etched by means of the EMSi method: a) membranes of piezoresistive pressure sensors, b) membranes of capacitive pressure sensors, the average etching rate equaled $V_{100} = 8.5$ μm/min for 7M KOH at 80°C.

The nature of EMSi

The acceleration of etching is an effect of change of water properties and, resulting from this, an increase in the reactivity of alkali aqueous solutions. As a proof of this phenomenon take the anisotropic etching of deep patterns in deionized water irradiated by microwave [63, 106]. In the original experiment, a 1500 μm × 1500 μm pattern was etched in the *n*-type silicon, through windows made in an LPCVD silicon nitride 100 nm-thick mask, under elevated pressure, approximately 4 MPa. First, a 10 mm × 10 mm silicon sample was positioned inside the tightly closed Teflon® vessel filled with deionized water. The vessel was positioned inside a microwave resonator supplied with 2.54 GHz, 100 W microwave radiation. Deionized water was warmed to 183°C. Next, microwave power was pulsed, in order to stabilize the pressure inside the vessel (Fig. 3.30a), after 15 minutes the microwave irradiation was ended and the solution was

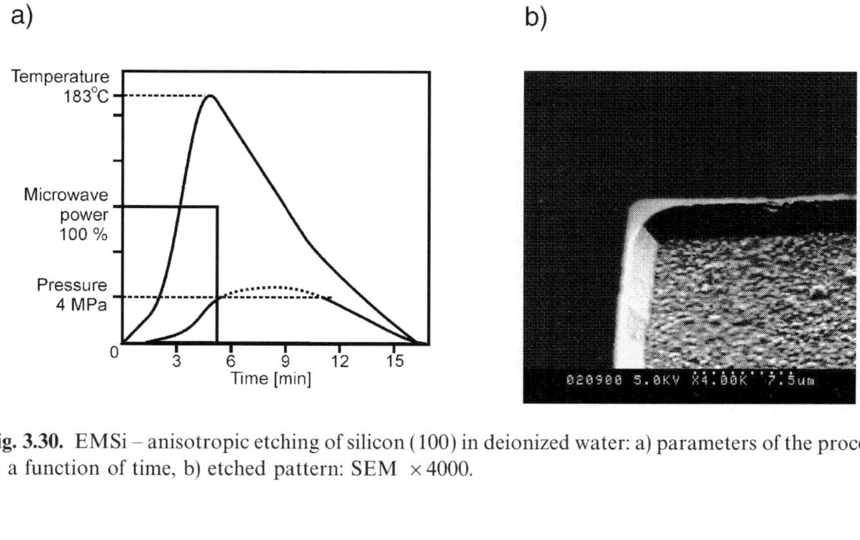

Fig. 3.30. EMSi – anisotropic etching of silicon (100) in deionized water: a) parameters of the process as a function of time, b) etched pattern: SEM ×4000.

cooled down. As a result of such treatment, anisotropically etched, 6 μm-deep cavities were obtained. The average etch rate V_{100} equaled 0.2 μm/min. Sidewalls of each of the cavities were formed by easily observable crystallographic planes (111) angled at 90° (Fig. 3.30b). The (100) surface of the bottom of the cavity was not smooth, and resembled the surface attained in thermally activated 3M KOH at 70°C.

This spectacular experimental result shows unambiguously that microwave irradiation leads to the generation of hydroxyl groups in deionized water, which anisotropically etch silicon. This conclusion results indirectly from the course of chemical reaction of silicon etching proposed by Finne and Klein [28], Palik [56], and Seidel [51, 81]. They documented that only hydroxyl groups take part in the reaction of silicon etching in alkali etchants formed on the basis of water. From the other viewpoint, anisotropic etching of silicon obtained in DI water confirms the mechanism of etching proposed by the above-cited authors.

3.2.4.2. E2MSi etching

Experimental works on EMSi etching have shown that the increased reactivity of etching alkali solution exposed to microwave remains for a dozen or so seconds after the exposure (Fig. 3.31). This is a surprising result. It should be concluded that effects of microwave irradiation are "stored", "memorized" by a solution for quite a long time, which appears to be a phenomenon not known in science until now.

This phenomenon has been applied in a new variant of microwave-enhanced silicon etching, called E2MSi (**E**xtended **E**tching **M**icrowave **Si**licon) [110], in which the microwave irradiation (excitation) of KOH and etching of silicon substrates have been separated in time and space. In the simplest configuration of the apparatus for E2MSi etching, a KOH aqueous solution, exposed in a microwave resonator, flows gravitationally to a reactor with etched substrates, or is driven by a pump (Fig. 3.32). Time from exposure to etching can be adjusted by selecting the time of flow of solution through a pipe-connecting

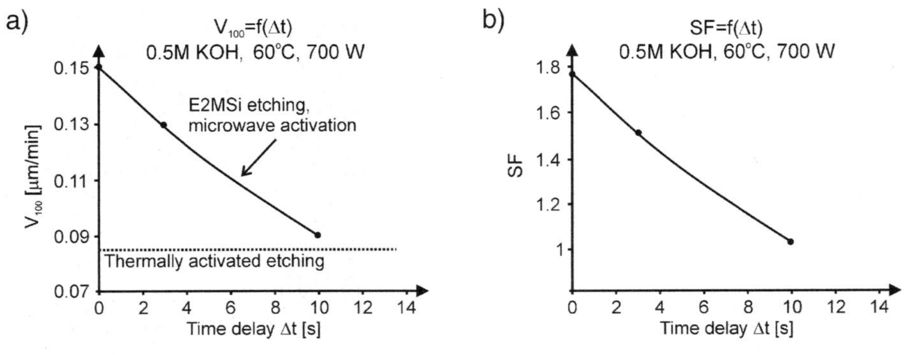

Fig. 3.31. Etch rate V_{100} (a) and acceleration factor SF (b) as a function of time period from the exposure to EMSi etching in 0.5M KOH, microwave power 700 W.

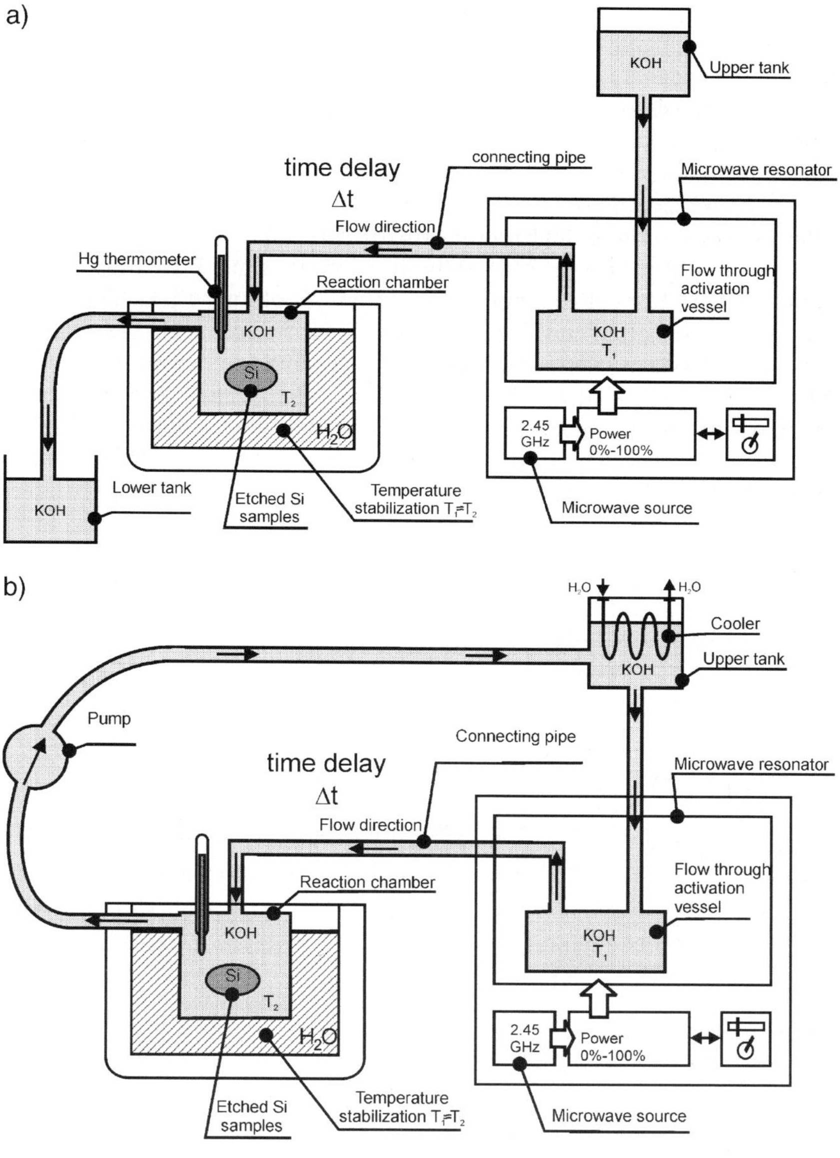

Fig. 3.32. Scheme of an apparatus configuration for E2MSi etching; a) gravitational flow, b) circulating flow.

resonator with a reactor. The reactor is located in a thermally stabilized chamber, in such a way that the temperature of the etching solution does not vary. The temperature of solution exposed in a microwave resonator and in a reactor has to be kept at the same level. The resonator has to be equipped with a microwave field mixer.

The basic features of the E2MSi process have been determined by etching of *n*-type and *p*-type silicon wafers in 0.5M KOH: DI water solution, by CVD nitride masks, at etchant temperature varying from 60°C to 70°C, for a three-second delay between exposure and etching. In such conditions the etch rate V_{100} is several times higher than noted for thermal etching and is only slightly dependent on the temperature of the solution. Acceleration of etching is proportional to microwave power level. The highest acceleration is obtained in cold solutions, irradiated by a high-power microwave (Fig. 3.33).

The main features of E2MSi etching are very similar to those of EMSi etching. Temperature-dependent characteristics of both processes have a similar course. Etch rates of silicon are many times higher than etch rates noted for thermally activated processes. However, the dependence of features of E2MSi etching on microwave power is more distinct. The quality of etched surfaces depends mainly on microwave power level irradiating a solution – "pumping activation". The smooth (100) and (111) surfaces may be obtained in cold solutions, exposed to a sufficiently high-power microwave (Fig. 3.34).

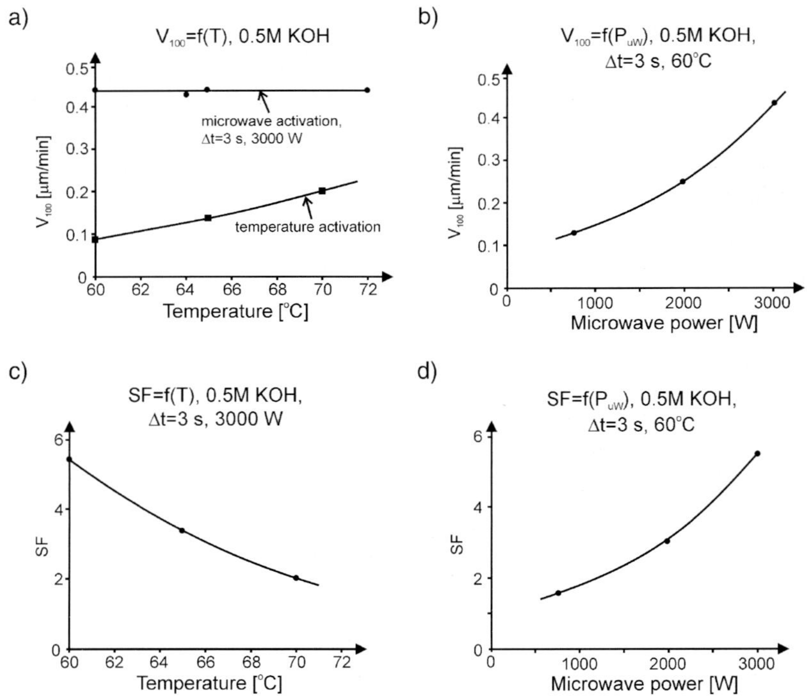

Fig. 3.33. Etch rate V_{100} and acceleration factor SF etching – E2MSi process, time from exposure 3 s, 0.5M KOH: a) V_{100} as a function of temperature for 3000 W of microwave power; a curve characterizing thermal etching carried out in a reactor in a non-irradiated solution is given for comparison, b) V_{100} as a function of microwave power at temperature of solution 60°C, c) SF as a function of temperature for 3000 W of microwave power, d) SF as a function of microwave power at a solution temperature of 60°C.

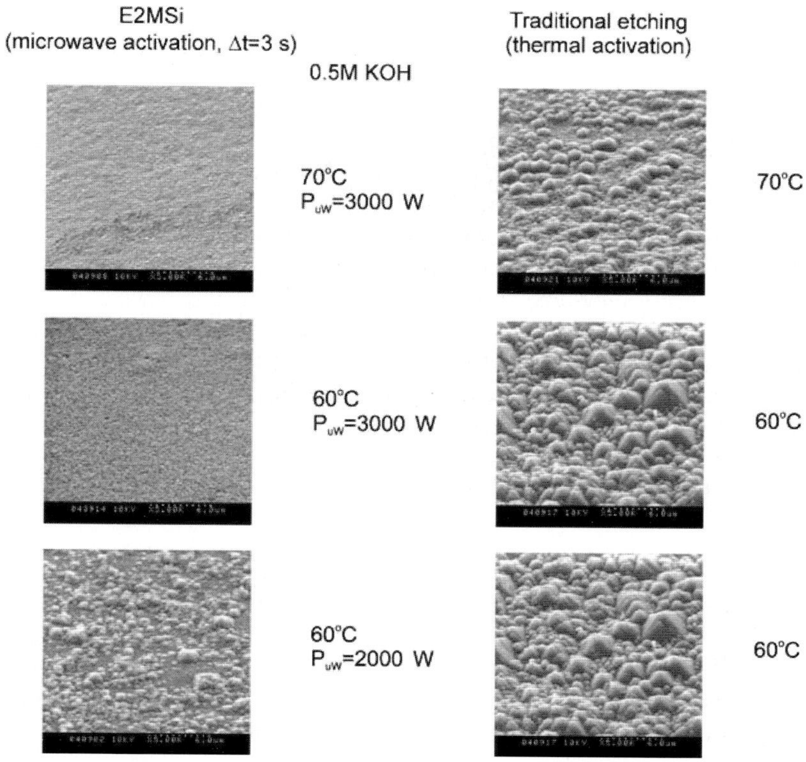

Fig. 3.34. Bottom surface of deeply etched patterns – E2MSi (left) and thermally activated (right) etching, 0.5M KOH. Note that in the conventional etching process the (100) surface remains roughened at 60°C and 70°C.

The technological application of E2MSi etching can be much wider than the application of EMSi etching. E2MSi etching preserves the advantageous features of the EMSi process; first of all a high etch rate with maintenance of a very good anisotropy in weak and cold alkaline solutions. The separation of exposure and etching greatly simplifies the construction of a microwave resonator. The reactivity of etching solutions can be changed by independent control of microwave power, exposure time and flow rate (time of presence) of liquid in a microwave resonator.

E2MSi etching ensures a smoothness of etched surface (100) in very low concentration solutions. These features of E2MSi etching can be useful in wide-area silicon substrate fabrication, for removal of a defective (after machining and polishing) surface layer.

3.2.5. Isotropic etching

It is a well-known fact that solid-state monocrystalline silicon easily dissolves in a mixture of concentrated nitric acid (HNO_3) and hydrofluoric acid (HF).

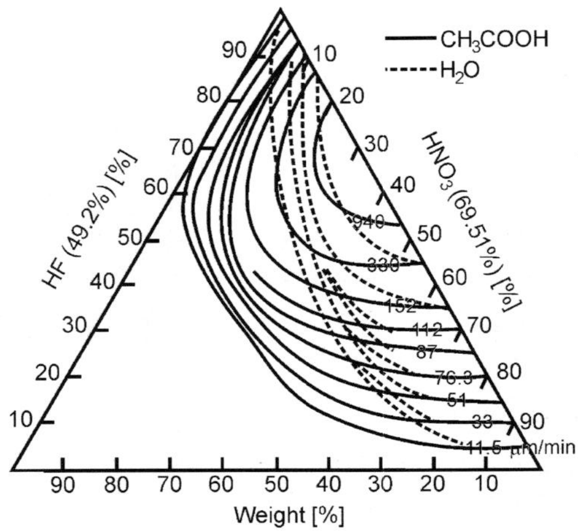

Fig. 3.35. Iso-etch curves [112].

Water or acetic acid (CH_3COOH) is used as a diluent. This mixture is called NH or NHA, and was well recognized in the 1960's and 1970's by Schwartz and Robins in a series of articles [111–114] and Bochenschuetz and co-workers in [115]. The so-called iso-etch curves of this etchant are shown in Fig. 3.35. Some examples of compositions and etch rate values, are shown in Table 3.10.

The overall reaction of NHA with silicon is as follows:

$$Si + HNO_3 + 6HF = H_2SiF_6 + HNO_2 + H_2O + H_2 \quad (3.9)$$

The reaction needs holes, which are injected in the valence band of silicon. Holes break bonds of silicon atoms; material oxidizes. Hydrofluoric acid dissolves silicon oxide.

Holes are produced in the specific reaction:

$$HNO_3 + H_2O + HNO_2 = 2HNO_2 + OH^- + 2h^+ \quad (3.10)$$

The reaction of silicon oxidation is:

$$Si^{++++} + 4OH^- = SiO_2 + H_2 \quad (3.11)$$

Table 3.10. Examples of compositions and etch rates of NHA etchant ($HF{:}HNO_3{:}CH_3COOH$), a mixture that isotropically etches silicon. T = temperature, V = etch rate

Composition of etching solution	T [°C]	V [nm/min]
$HF{:}HNO_3{:}CH_3COOH$ 50:3:8	30	$2 \cdot 10^5$
$HF{:}HNO_3{:}CH_3COOH$ 1:3:8	20	$0.7–3 \cdot 10^3$
$HF{:}HNO_3$ 9:91	30	$6 \cdot 10^3$

Isotropic etching of silicon is faster than anisotropic etching. The highest etch rate is noted for a HF:HNO_3 ratio of 2:1, adding of water or acetic acid slows down the process. In a HNO_3:HF 2:1 mixture of concentrated acids the etch rate exceeds dozen of micrometers per minute. Commonly used etching solutions have proportions of HF:HNO_3:CH_3COOH 1:3:8 and an etch rate circa 3 μm/min. The etch rate of NHA depends on the dopant concentration in silicon – as observed for anisotropic etching – but NHA etching slows down in silicon doped below 10^{17} /cm^3. The etch rate is reduced by one hundred. This effect did not find any application in deep silicon micromachining. The very disadvantageous feature of NHA etching is the strong influence of self-heating of the solution during longer, deeper etching and the weak homogeneity of this process in the wafer scale. Homogeneous etching is obtained at (111) oriented wafers.

Nieradko [116] has recently verified the literature data on NH and NHA etching. Four configurations of constituents of polishing etchants have been tested: NHA/1 1:5:2, NHA/2 3:25:10, NH/1 2:8 and NH/2 1:9. The wet-oxidized (steam, 1150°C, by 2 hours) thermal silicon dioxide masking layer stands up for less than 12 minutes to NHA/1 and NHA/2 and for less than 8 minutes to NH/1 and NH/2 etchants. Mask annealed in nitrogen in 1150°C for 30 minutes stands up for 60 minutes to NHA and 18 minutes to NH etchants. A sandwich of silicon dioxide layer, as above, covered with a 0.1 μm thick LPCVD silicon nitride layer, is completely resistant to NHA and NH etchants. A small addition of silicon, dissolved in NHA prior to the etch process, stabilizes properties and allows one to obtain a mirrored etched surface. The unintended effect of micro masking of patterns during NHA etching is observed (Fig. 3.36). More controllable etching is obtained in NH but both pattern shape and etch rate depend on agitation (Fig. 3.37). Etch rate depends on the method and intensity of agitation, and is below 5 μm/min for ultrasonic agitation and about 7–8 μm/min for a mechanically stirred or unmixed solution. Deeply etched in NH/1 etchant patterns are smooth; angles formed between back-side walls and the front surface of the wafer are about 90° (Fig. 3.38).

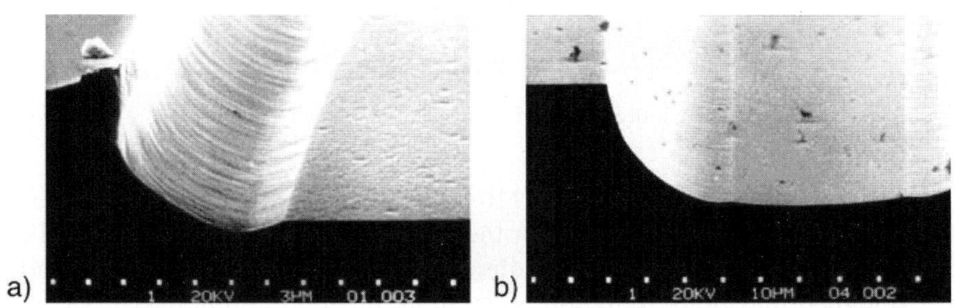

Fig. 3.36. Micromasking effects, NHA 1:5:2, a) channel after 20 minutes etch, b) after 60 minutes [116].

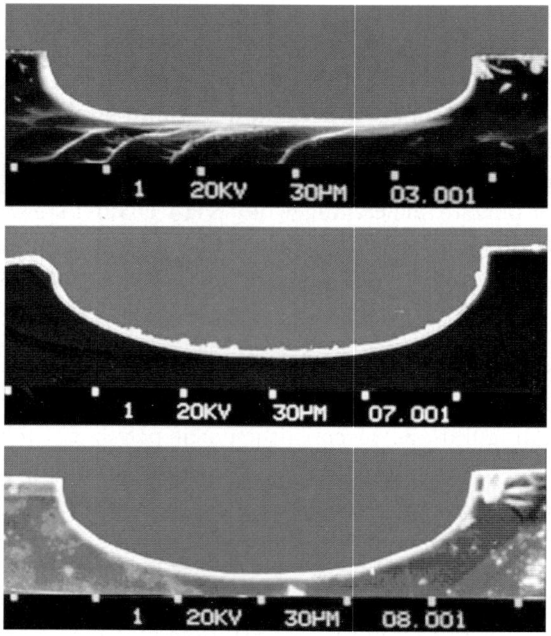

Fig. 3.37. Cross-sections of an isotropically etched pattern, NH/1 solution, 5 min, 20°C. From top: ultrasonically mixed, mechanically stirred, without agitation [116].

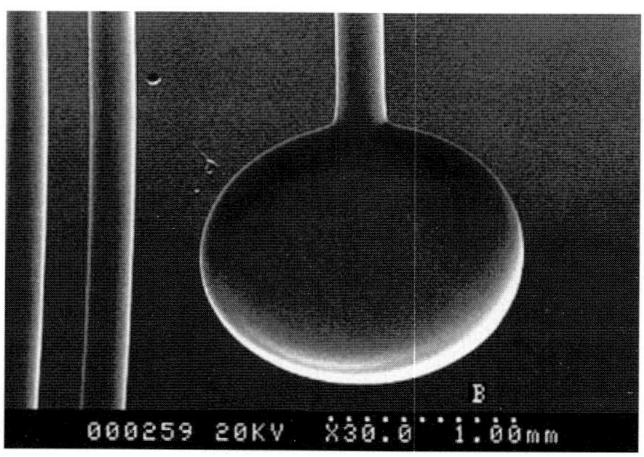

Fig.3.38. Example of a pattern deeply etched in silicon, NH/1 solution [116].

The above-cited results of paper [116] have clearly indicated that well-known NHA solutions cannot be applied for the deep isotropic etching of silicon. Much better results (see Fig. 3.39) may be obtained in NH (65% HF plus 40% HNO_3 1:9 volume proportion). Agitation of the solution, in order to improve chemical reaction products exchange and stabilization of the temperature, plays a most important role here.

Fig. 3.39. Perfectly etched U-shaped channels of a capillary column – example of deep silicon isotropic micromachining [116].

3.3. BASIC MICROMECHANICAL CONSTRUCTIONS

Basic three-dimensional micromechanical constructions anisotropically wet etched in silicon are as follows: square, rectangular, flat, bossed or surface-patterned membranes, grooves, beams and bridges, including beams with seismic mass, cavities, holes, tips. These micro constructions may be manufactured as single devices, as well as in the form of an array. They are applied in varying types of microsystems (Fig. 3.40, Table 3.11). The shape, planar dimensions and 3-D structure of the micro constructions depend on the properties of etch process – discussed earlier – and on the shape and dimensions of a mask (etch window) and, finally, on its alignment to the important crystallographic directions.

3.3.1. Membranes

Deeply etched, square or rectangular thin membranes are very often used as mechanically active parts of micromechanical sensors and actuators. Membranes are used as mechanical supports at the varying stages of more complex microsystems manufacturing, as prefabricated components or/and static and movable (acting) parts of microsystems. Membranes can be produced before or after fabrication of the microelectronic components of microsystems (Fig. 3.41a,b). The first method is very popular in laboratory-scale manufacturing of microsystems. Silicon wafer, after deep micro machining, can be further processed,

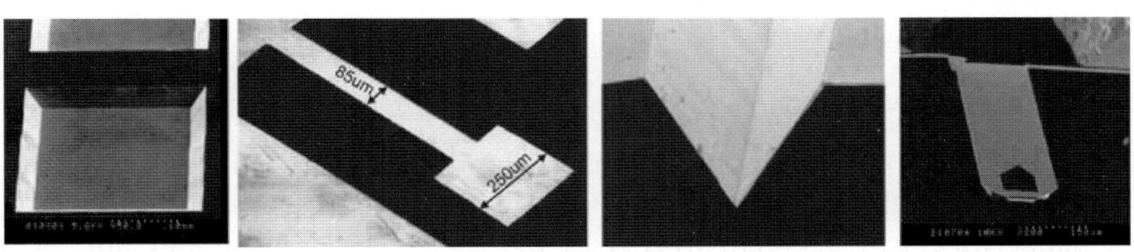

Fig. 3.40. Silicon three-dimensional, wet anisotropically etched microconstructions: membranes, seismic mass suspended on beams, V-groove, sharp tip at a beam.

Table 3.11. Anisotropically wet etched silicon micro constructions and their application

Micro-constructions	A few chosen examples of microsystems in which micro-constructions can be applied
Flat membrane	Piezoresistive pressure sensors, capacitive pressure sensors, radiation sensors, bolometers, photolithographic masks, microchemical sieves, valves and pumps, jet-engines, cantilevers for the support of small micromechanical details
Corrugated and bossed membrane	Pressure sensors with overload protection, force sensors, prefabricated products for the fabrication of accelerometers, accelerometers, pressure switches, valves and pumps
Grooves	Positioners (couplers) for optical fibers, microfluidic devices, drug dosers, chemical and biochemical separators
Beam, assembly of beams, beam with seismic mass and assemblies of beams with seismic mass (masses), membrane on beams	Heads of tunnel and atomic forces microscopes, force sensors, vibration meters, inclinometers, vibration analyzers, accelerometers, light modulators, vibrating mirrors, adaptive reflectors, chemical sensors, micro switches,, flow meters, mass meters, guides of optical fibers
Cavities, holes, hole arrays	Chemical micro reactors, DNA analyzers, heads of ink-jet printers, dosing systems for microchemistry and biochemistry, nebulizers, injection dishes, carburettors and atomizers, nerve regenerators
Tip(s), array of tips	Electron field emitters, micro syringes, measurement tips of probes of tunnel and atomic forces, microscopes, SNOM microscopes

and treated as a prefabricated wafer. The biggest technological problem here is an unintentional residual formation of via-holes in a membrane, caused by residual over-etching of a silicon wafer, involved by mask imperfections, crystal defects or mechanically induced shocks. Resists and resins, etching solution, etc. may penetrate through the failed membrane at the opposite side of the processed wafer and/or may block vacuum chucks of transportation/alignment equipment. The second method is widely used in mass-scale production, although microelectronic circuits fabricated onto the front side of the wafer must be carefully protected against destruction caused by the etching solution while the membrane is being micromachined.

3.3.1.1. Flat membranes

The repeatable, precise micromachining of hundreds of excellent, uniform membranes on a single wafer (Fig. 3.41c), as well as simultaneous micromachining of many wafers in a batch process, is the difficult and important aim of a fabrication procedure of many sensors (pressure, accelerometers, etc.). Planar dimensions of membranes are most often in the range of a part of millimeter

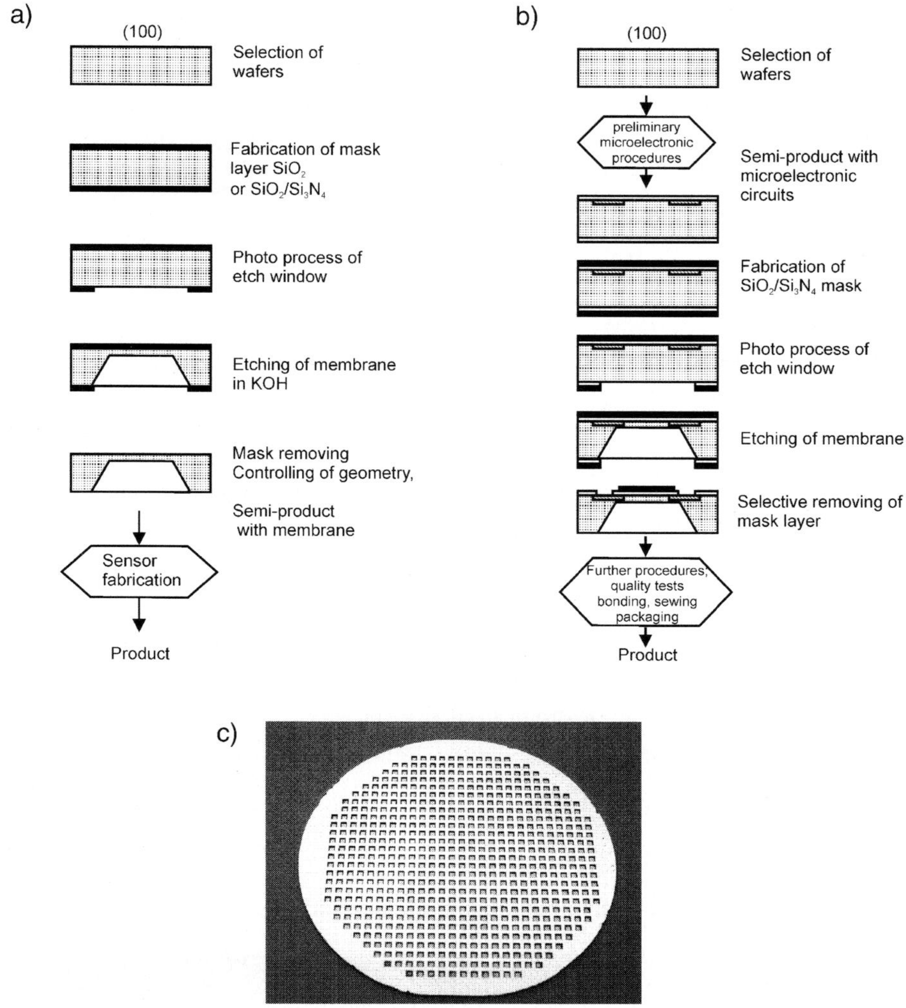

Fig. 3.41. Silicon membrane in a micromachining/microelectronics process flow-chart of typical pressure sensor: a) membrane micromachining before microelectronic procedures, b) process flow-chart of micromachining after microelectronic procedures, c) an example of a micromachined silicon wafer with an array of membranes of pressure sensors (EMSi process).

to a few millimeters, while their thickness lies in the range of a few dozen micrometers. Desired dimensions: the thickness m and the length of edges of membranes a_0, quality of surface of a membrane and back-side walls of an etched cavity, have to be carefully controlled.

The following material and process factors are very important: substrates of good quality, with suitable thickness and front/back surface parallelism; kind and composition of etching solution ensuring smooth, shining etched surface and good uniformity, and selectivity of an etch process, well controlled, stable high value of the etch rate V_{100} and sufficient stable basic anisotropy factor $V_{100}:V_{111}$.

Membranes have to be etched through a mask window with the edge length W, which results in the fabrication of the cavity a-wide. The pattern of the membrane mask (in the form of a square or rectangle) is aligned in such a way, that its edge is parallel to the flats produced in the direction $\langle 110 \rangle$ on the (100) wafer (Fig. 3.42,1).

For a substrate thickness d and a membrane thickness m the edge length W is given by:

$$W \approx a_0 + \sqrt{2}(d - m) - 2u, \tag{3.9}$$

where the lateral side under-etching of (111) plane u is given by:

$$u = \frac{V_{111} \cdot t}{\sin \Theta}, \tag{3.10}$$

$\Theta = 54.74°$ is the angle formed by (100) and (111) planes and t is the etch time. Depth of the etched pattern D is given by:

$$D = d - m = V_{100} \cdot t. \tag{3.11}$$

The membrane thickness m depends mainly – as mentioned earlier – on the etch rate V_{100}, and ratio of V_{100} and V_{111}, the thickness d of a silicon wafer, mask dimensions (for a square mask on W) and on etch time t. The etch time t of etching of membrane of the desired thickness m is given by equation 3.12:

$$t = \frac{d - m}{V_{100}}. \tag{3.12}$$

The parameter t determined from equation 3.12 is used as the main indicator

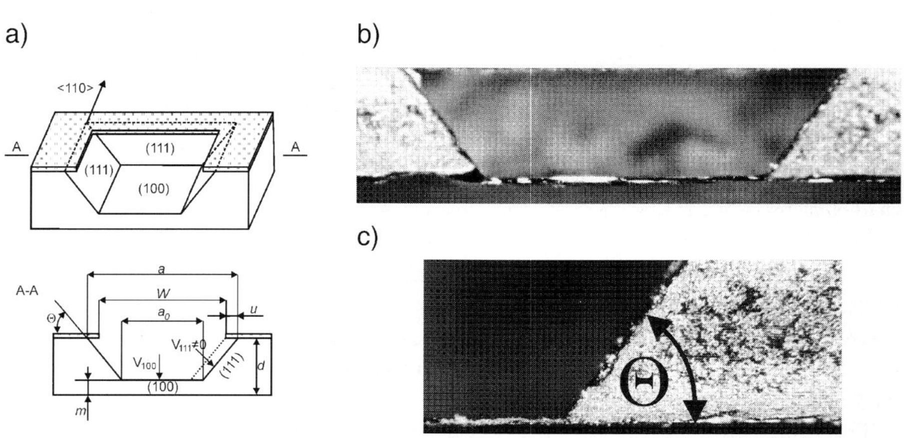

Fig. 3.42. Flat silicon membrane: a) three-dimensional projection, crystallographic planes and directions marked, b) cross-sectional view A-A of a membrane 20 μm thick, etched in the 380 μm thick substrate (microscopic photo picture), c) detail, $\Theta = 54.74°$.

of the etch stop. Incorrect etch time t influences the thickness of membrane m. For average value of etch rate $V_{100} = 1$ μm, typical for KOH etching, and 3″ silicon wafer, about 360 μm thick, the etch time of 20 μm-thick membrane formation equals 340 minutes. So, 1 minute prolongation of the process, which is approximately only 0.3% of the total etch time t changes at about 5% of the thickness m of a typical membrane.

The next important reason for the variation of thickness of membranes is the statistic variation of thickness of silicon wafers. The variation of thickness of the single or double-side polished wafer equals (usually) at least ± 2.5 μm for 3″ wafers, which results in the similar variation of the membrane thickness for the fixed etch time. The total thickness variation of serial, double side-polished wafers, used in a batch process of membrane fabrication, can reach ± 20 μm. This results in a serious variation of the membrane edge length, which may reach ± 28 μm. Such a big variation of planar dimensions of membranes is inadmissible in many micromechanical devices, e.g. in piezoresistive pressure sensors. That is why the wafers must be selected within the subgroups of similar thickness, matching with a suitable mask and the desired membrane geometry.

Membrane thickness and quality evaluation

Parameter m can be determined by different methods. One of them, simple and sufficient in laboratory practise, is the color method. This method utilizes the fact that the color of light transmitted through a thin layer of silicon depends strongly on its thickness. For white light illumination, a 3 μm-thick membrane is yellowish-brown, the color changes to orange-reddish for 8–10 μm, to red for 15 μm, to cherry-red and maroon for 20 and 30 μm. The thickness of the investigated membrane is evaluated by comparison of its color to the color of a set of membranes of different, but precisely known, thickness. The total error of the thickness determination of 10–25 μm-thick membranes, by the naked eye, is ± 2 μm. An evaluation of the wafer scale variation of thickness of membranes may be much more accurate, ± 1 μm accuracy may be obtained, because the eye is very sensitive at comparing colors at small distances.

Another laboratory method for evaluating the thickness of thin and thick silicon membranes is microscopic observation of the etched pattern. First, a microscope should be focused on the upper, front surface of a silicon wafer, next on to the bottom surface of the etched cavity. Knowing the wafer thickness, and movement of the translation stage of a microscope, it is easy to evaluate parameter m. This method is very useful, but its accuracy depends on the mechanical quality of the used equipment.

The thickness of thick membranes (more than 25 μm thick) may by measured by any of the standard mechanical methods – micrometric screw or profile meters.

The quality of membranes depends on process-induced fabrication mistakes and the quality of a substrate. The most common fabrication mistakes are: wrong alignment of mask edges to the chosen flat, which deforms the edge of etched pattern and causes forming of relief on walls (111) (Fig. 3.43a); rough

Fig. 3.43. Technological mistakes – examples of defective and perfectly (f) fabricated membranes.

surface of a membrane and hillock formation, being an effect of low concentration and/or temperature; or wrong selection of etching solutions (Fig. 3.43b); and mask over-etching (poor selectivity of the process) (Fig. 3.43c). Residual defects of etching may involve point perforations of a membrane.

The crystallographic quality of a whole silicon wafer should be perfect. If it is not, all the defects of wafer structure are "projected" on the bottom surface of the etched membrane during the long-lasting etch process, which worsens the morphological quality of the surface and causes etched pinholes to appear.

The polished surface of a wafer needs to be flat and parallel, otherwise the thickness of membranes will vary across a wafer.

Membrane quality is often evaluated visually. The evaluated features are: shape, homogeneity of color, smoothness of surface (plane (100)) and smoothness of side surfaces formed by walls (111). The membranes correctly etched should be perfectly smooth (Fig. 3.43f), without the "skin–orange effect" (Fig. 3.43e).

Thickness control by etch-stop

The simple processes of etching of micro constructions done in laboratory conditions are controlled by the so-called time etch-stop method (Fig. 3.44a). In this method, for known etch rate V_{100}, substrate thickness d and membrane thickness m, time of etching t is defined from equation 3.11. Usually, a typical process of micromachining of membranes lasts for hours. Many occasional factors [instability of an etchant temperature, variation in time exchange of products of chemical reaction of silicon dissolution (mixing), ageing of an etchant, operator mistakes, time control errors, etc.] have an influence upon

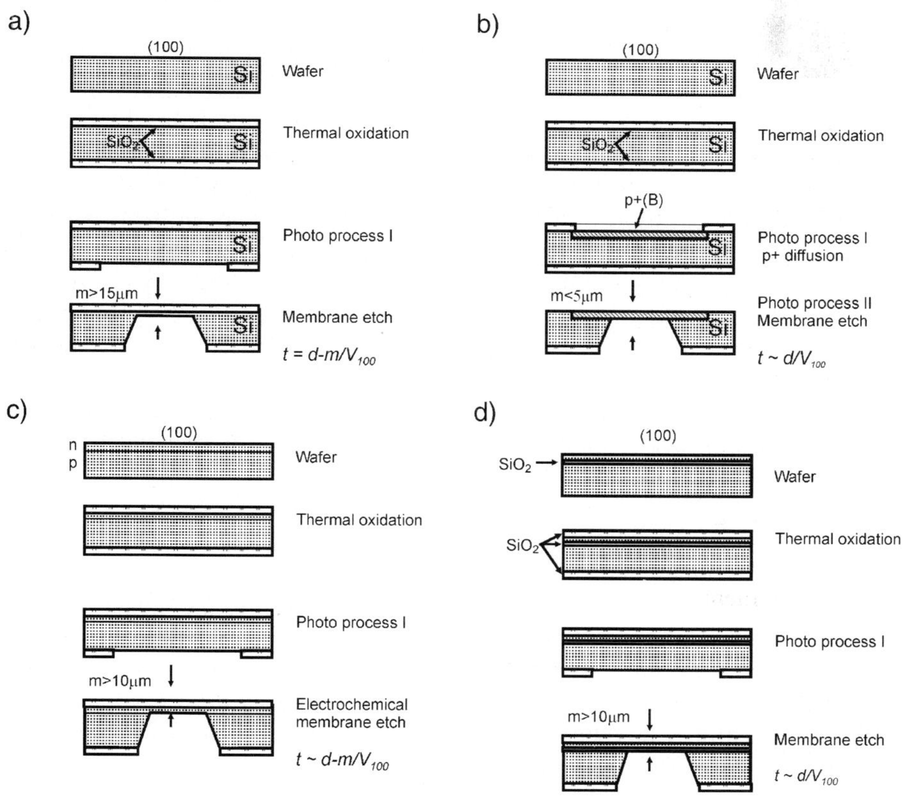

Fig. 3.44. Etch-stop methods: a) time control, b) stop-diffusion, c) electrochemical, d) buried oxide.

process repeatability and, as a final result, on the thickness of membranes. Different semi-automatic methods of controlling of the etch process, which ensure repeatable thickness of membranes, have been evaluated. They are:

- stop-diffusion etch-stop (Fig. 3.44b)
- *p-n* etch-stop (electrochemical etching) (Fig. 3.44c)
- oxide-stop layer (buried oxide) (Fig. 3.44d)

In the stop-diffusion etch-stop method a thin silicon layer is heavily doped with boron. The etch-front stops at the p^+ region, whose thickness corresponds to the desired thickness of a membrane m. Time of etching t has to be estimated from equation 3.11 for $m = 0$ plus a few minutes. The stop-diffusion cannot be successfully applied in a fabrication of large-area membranes, because strong boron doping introduces high mechanical stresses in silicon. Therefore, the large-area membranes tend to deform (corrugate, warp) [49]. Moreover, strong boron doping can involve texturing of the surface of membranes. This etch-stop method is very useful for manufacturing the very thin (1–5 μm-thick) but small-area flat membranes (Fig. 3.45).

The method of *p-n* etch-stop does not have these drawbacks. In this method the front of electrochemical etching of *p*-type silicon substrate stops at the *n*-type silicon layer. Thickness of the *n*-type epitaxial layer corresponds to the desired thickness m of a membrane. Etching proceeds through the *p*-type silicon, from the back-side of the wafer. Etch-stop signal is taken from current–time characteristics of the etch process. The repeatability of the thickness of membranes in this method reaches ± 1 μm, the influence of the variation of thickness of substrates on the membrane geometry may thus be negligible.

The oxide-stop method of membrane thickness control is applied for membranes formed at the SOI (Silicon on Insulator) substrates with a micron-thick silicon oxide layer buried under the top silicon layer. The thickness of the top silicon layer is adjusted according to the expected thickness m of a membrane. Etching proceeds through the substrate and stops at the oxide. The thickness of membranes is defined by the thickness of the top layer very precisely adjusted by the SOI substrate manufacturing producers. SOI substrates are expensive, so this method of membrane thickness control is used only sporadically.

3.3.1.2. Bossed and corrugated membranes

A silicon-bossed membrane (Fig. 3.46) contains convex corners. Convex corners formed by high-index crystallographic planes are etched faster than concave corners formed by (111) planes. As a result, the shape of the etched pattern differs from the designed one because convex corners are under-etched. This is illustrated by an example from Fig. 3.47. In this example a control L-shaped pattern of a mask with four windows was used to etch a deep cavity. After a few hours' etching in KOH convex corners recessed – under-etched. The shape of concave corners remained rectangular.

As is commonly stated in the literature, high-index crystallographic (133), (211), (212), (311), (321) and (411) dissolve in alkaline solutions much faster

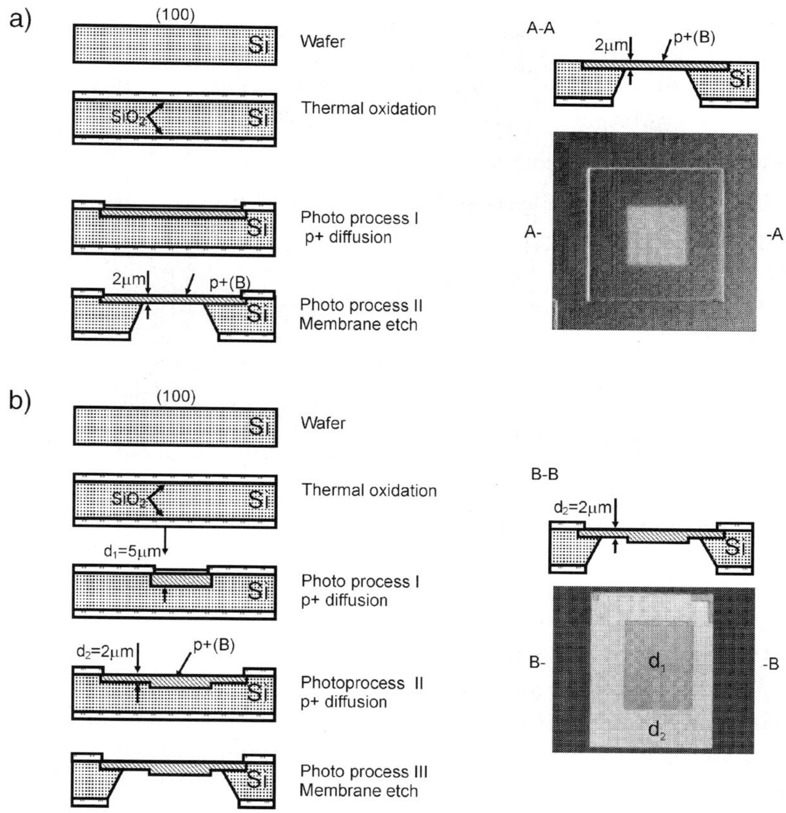

Fig. 3.45. Stop-diffusion – a method of fabrication of thin membranes (on the left) and their cross-section and appearance (on the right): a) membrane 200 μm × 200 μm × 2 μm, b) membrane with different thickness $d_1 = 5$ μm, $d_2 = 2$ μm.

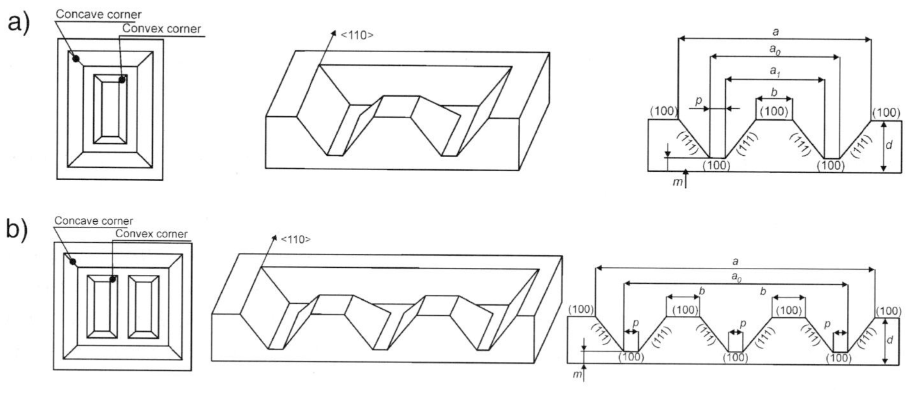

Fig. 3.46. Bossed membranes – geometry: a) single boss, b) double boss.

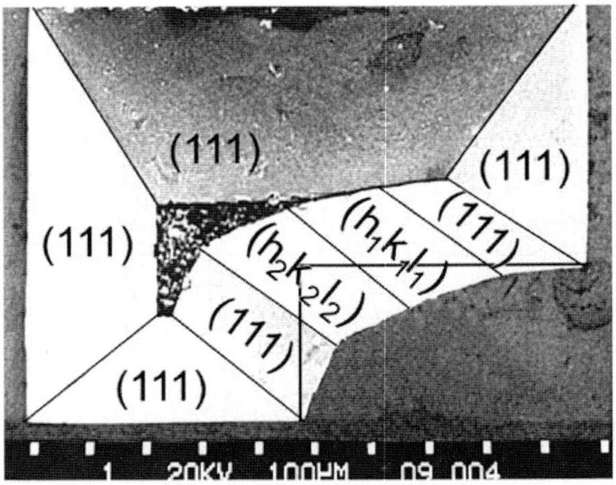

Fig. 3.47. L-shaped pattern anisotropically etched for 6 hours; concave corners formed by planes (111) are perfect, convex corners deformation (under-etching effect) can be seen; the shape of applied mask is marked.

than (100) or (110) depending on the etch rate of the particular plane, observed for the composition used and the temperature of solution. Etch rates of high-indexed planes, as well as the compensation of the under-etching, have been investigated in many papers [55, 78, 117–132]. There is no agreement as to which plane plays the most important role. In paper [132] planes (133) and (212) were considered to be responsible for under-etching effects. Experiments made for 10M KOH with the addition of IPA have shown that etch rates V_{133} and V_{212} compared to etch rates of planes (100) and (110) are in the relation:

$$\frac{V_{100}}{V_{110}} = 0.4, \qquad \frac{V_{100}}{V_{133}} = 0.58, \qquad \frac{V_{100}}{V_{212}} = 0.5. \tag{3.13}$$

It was also noticed in paper [132] that the final shapes of etched figures were the superposition of planes (100), (110), (111), (133), (313), (212), (122). The high-index planes etch rates in 30% and 40% pure KOH and in 27% KOH + 2% IPA were analyzed in paper [125]. Authors used the modified "wagon wheel", cross-shaped and rectangle-shaped patterns. A pictorial representation of the results is given in Fig. 3.48.

In publication [118] it was stated that the under-etching of convex corners in 15% to 50%, KOH for 60°C to 100°C, are defined first of all by planes (411). Simplified geometric relations between crystallographic planes (411) and plane (100), (111), significant for analyzing the under-etching phenomenon, are presented in Fig. 3.49a.

Let us assume that ANG is a profile of the convex corner of a mask. Two planes (411) undercut the etched corner under the mask, moving the etching

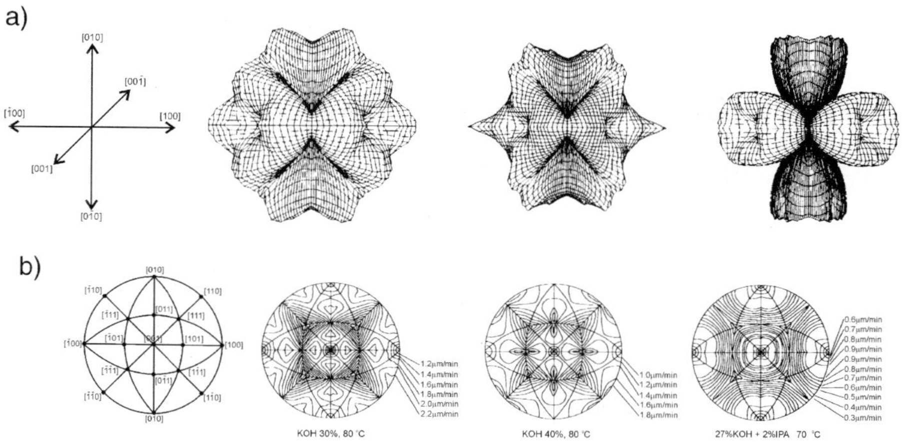

Fig. 3.48. Graphic representation of silicon etch rates for some chosen etching conditions: a) three-dimensional "view", b) stereographic projection at (100) plane [125].

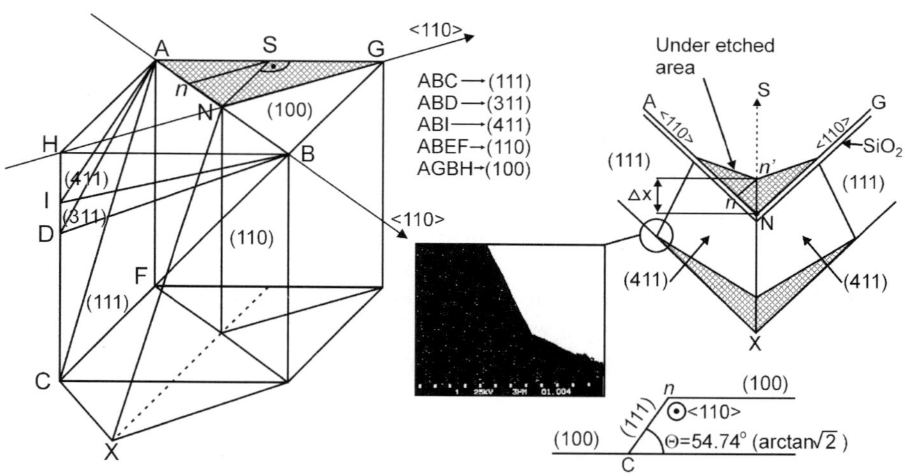

Fig. 3.49. Under-etching effect; planes (111), (110), (100), (311), (411) at the convex corner of the deeply etched in silicon (100) wafer 3-D micromechanical structure.

front in direction ⟨140⟩. The angle between planes (411) and (100) equals

$$\phi = \arccos\left(\frac{1}{\sqrt{18}}\right) = 76.37^\circ.$$

For the planned etch depth, expressed by a product of etch rate of plane (100) and time, that is $V_{100} \cdot t$, the convex corner will recede by

$$nn' = \frac{V_{411} \cdot V_{100} \cdot t}{V_{100}} \Big/ \sqrt{2} \sin\phi \cos\beta, \tag{3.14}$$

where: $\beta = \arctan(1/4)$.

Hence

$$nn' \approx 0.75 \left(\frac{V_{411}}{V_{100}} \right) V_{100} \cdot t. \tag{3.15}$$

For 40% KOH at 80°C the proportion V_{411}/V_{100} is about 1.34 (where V_{411} is etch rate of plane (411)), V_{321}/V_{311} equals about 1.74.

The recession of convex corners, caused by the fast etching of plane (411), is well presented in Fig. 3.49b. A star-shaped pattern on wafer (100) at the "cross-roads" was formed by two perpendicular V-grooves etched in 40% KOH at 80°C.

Under-etching compensation

The principle of convex corners under-etch compensation is based on the intended deformation of the mask at the corners – use of so-called compensation patterns. The shape and dimensions of the deformation are selected experimentally; they depend on the type of etchant and etch depth (Fig. 3.50). Usually the compensation pattern is designed so that high-index crystallographic planes are etched first. The most common are square, triangular-square and triangular; rectangular, fork, T or cross patterns are used sporadically.

The triangular-square and triangular compensation patterns, selected for 360 μm-deep etching in 250 g KOH, 800 g H_2O, 200 g IPA at 80°C are shown in Fig. 3.51.

Triangular-square compensation ensures the complete under etch compensation of convex corners at the "face" surface of a silicon wafer. The corners shape is not entirely compensated at the bottom of the etched cavity. An advantageous feature of the described compensation is a small field of the compensation pattern, which allows etching of bossed-type membranes with small planar dimensions ($a_0 \sim 1000$ μm).

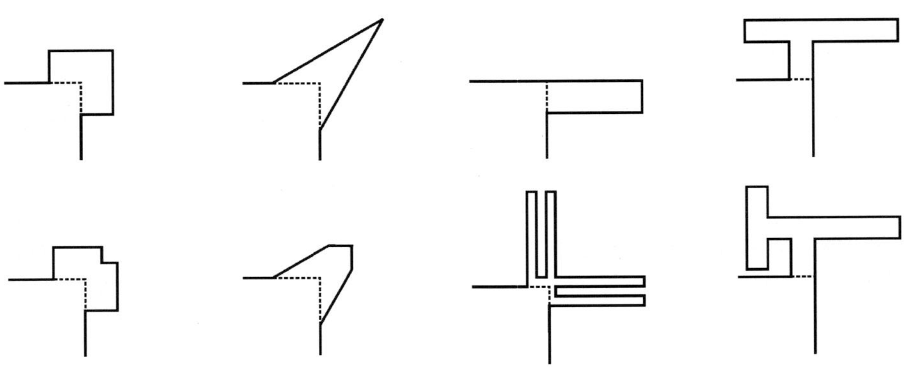

Fig. 3.50. Most-used compensation patterns [118, 120, 122, 124].

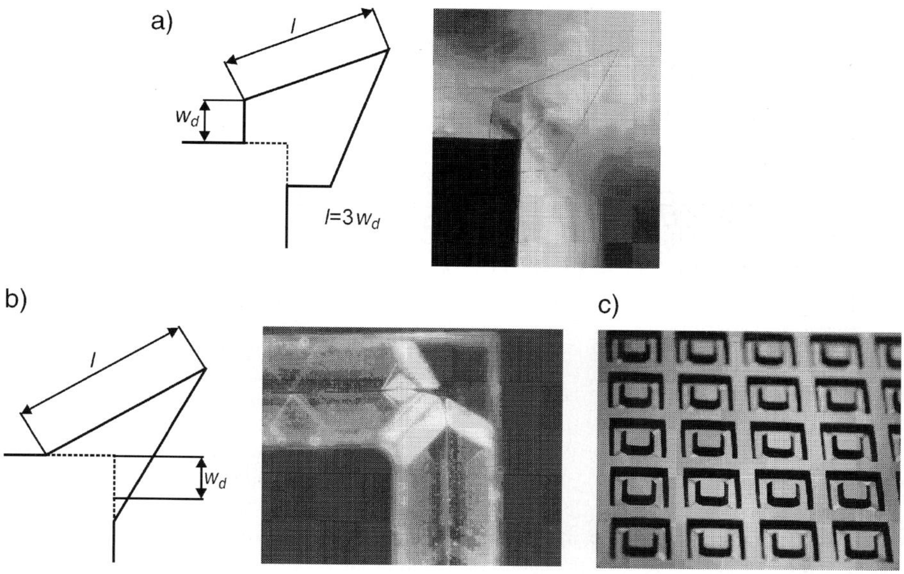

Fig. 3.51. Compensation of convex corners: a) triangular-square compensation and the result of etching, b) triangular compensation and the result of etching, c) array of bossed membranes (courtesy of M.Sc. Jerzy Jazwinski from the Institute of Electronic Technology of Warsaw).

Triangular compensation does not ensure complete compensation of the convex corners. This type of compensation is anyway technically better than the compensation of the triangular-square patterns, because for triangular compensation the shape of etched concave corners is more regular in the cross-section of the etch profile, although the compensation is only partial. The triangular pattern is often applied in the production of bigger bossed membranes ($a_0 \sim 2000$ μm), designed for micromechanical pressure sensors.

3.3.1.3. Other types of membranes

Corrugated membranes

Corrugated membranes (Fig. 3.52), more flexible than flat membranes, are made of thin silicon, silicon nitride, or a sandwich of silicon nitride on silicon oxide layers. They were invented by Jerman [133] and used as a stress release package, in micro actuators (valves) or in very sensitive sensors (accelerometers*).

Corrugated membranes are usually fabricated on the double-side polished (100) silicon substrate. First, the corrugation is etched through a dielectric mask (thermal oxide) at a front surface of the wafer. Most often the dry etch method

* Such sensors may be used, among other things, for the remote control and evaluation of the health state of animals by evaluation of their mobility [134].

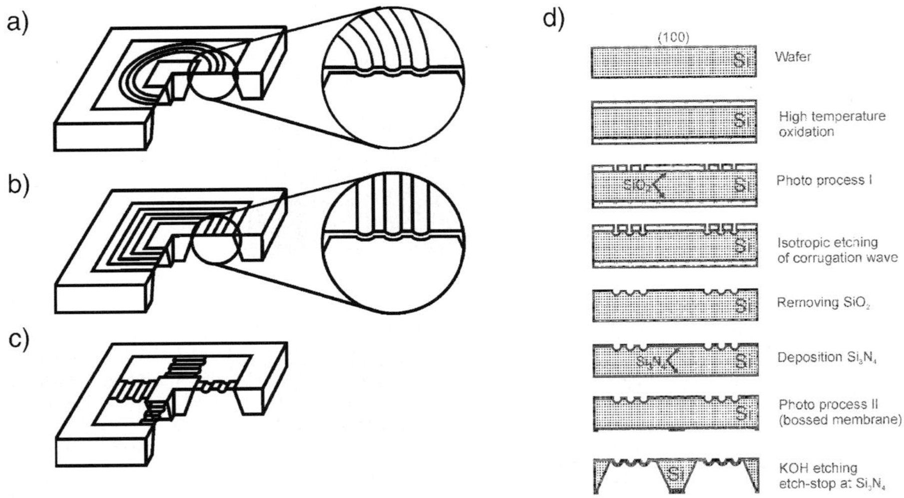

Fig. 3.52. Corrugated membranes: a) round corrugation, b) square corrugation, c) suspension of seismic mass of accelerometer, d) fabrication process flow-chart.

PERIE (Plasma Enhanced Reactive Ion Etching) in SF_6, or wet isotropic wet etch in a $HF\text{-}HNO_3{:}H_2O = 100{:}3{:}40$ solution at room temperature are used here. After the formation of the corrugation, the mask layer is removed and the sharp pattern edges are smoothed by a short etching in a low-concentration $HF\text{-}HNO_3{:}H_2O$ solution. Next, a Si_3N_4 layer is deposited onto both sides of the substrate, back-side and photo processed. Then a bossed membrane is anisotropically wet etched from the back-side of the substrate. The etch process stops at the nitride layer.

Patterned membranes

Patterned membranes are used to concentrate mechanical stresses in highly-sensitive piezoresistive pressure sensors. Membranes are produced on the double side-polished (100) substrate (Fig. 3.53). First, a typical flat silicon membrane is etched. Next, a mask pattern is formed on a front surface of the wafer and a selective thinning procedure is carried out in KOH or by means of a dry-etch procedure.

Round membranes

The shape and dimensions of a membrane wet micromachined in a (100)-oriented wafer through circular window are determined by two factors: anisotropy and time of wet etching. At the beginning of etching, the sidewalls of etched pattern are formed by four (111) planes and the four transitory areas, built of many high-index planes with indeterminate hkl indexes (Fig. 3.54). Walls (111) are flat and shining, the transitory areas are rough. As the etching

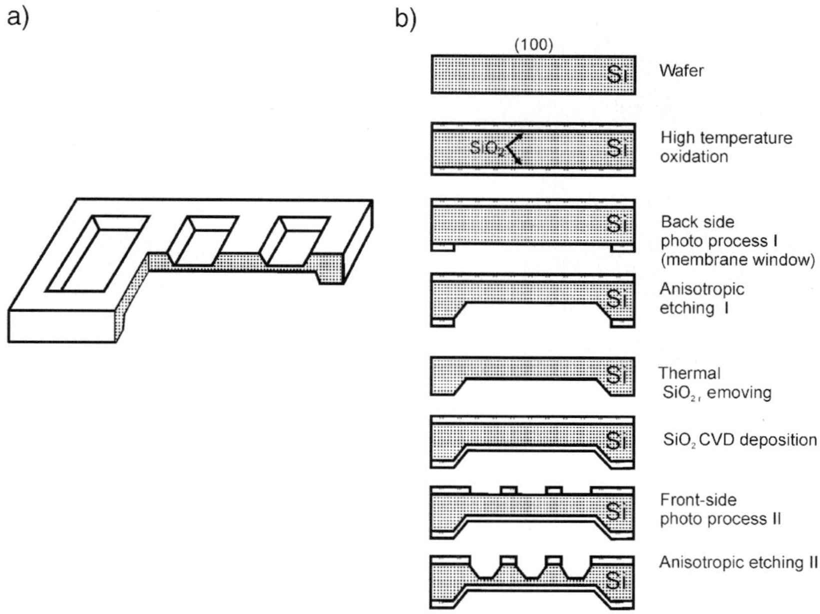

Fig. 3.53. Patterned membrane: a) half section view, b) process flow-chart.

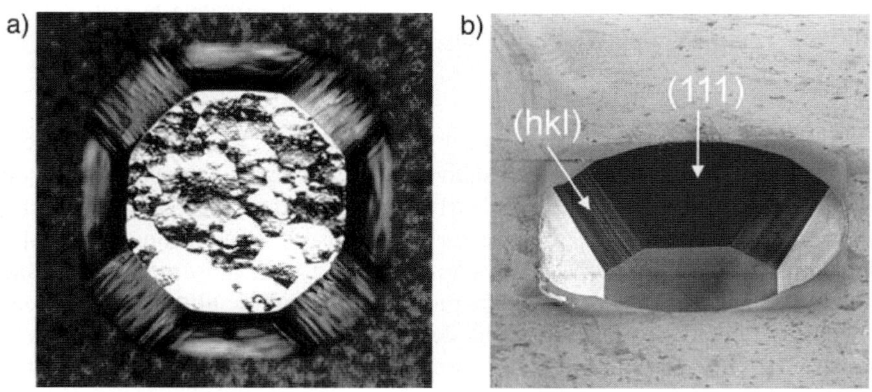

Fig. 3.54. The hollowed pattern etched by a circular mask after prolonged anisotropic wet etching: a) optical microscope view in reflected light, b) SEM picture.

proceeds, the transitory areas tend to disappear and the etched cavity becomes similar to a inverted truncated pyramid with a square base, whose sidewalls are (111) planes, while the bottom is formed by a smooth (100) plane. Then, dominant (111) planes tend to form the inverted regular pyramid. A similar effect happens for elliptical or irregular oval-shaped etch-windows. At the beginning of etching, the sidewalls of the pattern are formed by all of the planes

a) b)

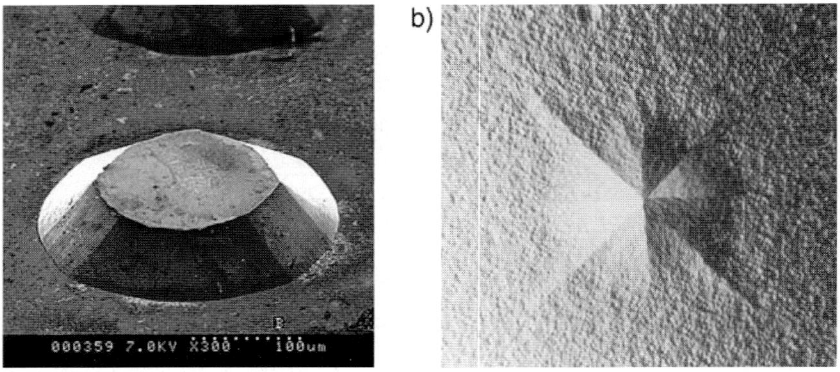

Fig. 3.55. The protruding pattern etched by an oval mask (10M KOH, 80°C): a) after about 2 h of etching, b) track left after 7 h of etching.

appearing in the chosen etching solution; the bottom is formed from a (100)-oriented plane. Next, (111) walls become dominant, transitory areas of the high-index planes disappear, and finally, the inversed pyramid cavity is formed by four (111).

For both types of circular/elliptical masks, the final etched pattern is reoriented according to ⟨110⟩ direction. This is the so-called self-alignment effect of silicon 3-D deep anisotropic wet etching. The patterns self-alignment appears for hollowed as well as for protruding patterns of any shape but, after a sufficiently long etching, the protruding figure will vanish (Fig. 3.55).

3.3.1.4. Application of membranes

Some important marketable micromechanical products, in which silicon membranes are applied, include piezoresistive pressure sensors, fabricated and sold in millions of pieces. It must be clearly said here, that there is no common specification of a typical membrane for such sensors. Dimensions and shapes of membranes vary; they may be square or rectangular according to the habits of pressure sensor producers. What is more, the type of membrane, their planar dimensions, thickness and structure depend on the technical parameters of pressure sensore: maximal measured pressure, sensitivity, over-pressure protection, etc. Membranes used for 50–200 kPa pressure ranges are flat, usually about 1 mm × 1 mm and dozens of micrometers thick. More sensitive, overload-protected sensors need thinner and/or bossed membranes, they are typically 2 mm × 2 mm, 10–20 μm thick.

In the most popular pressure sensors with a flat membrane (Fig. 3.56)* four identical monolithic piezoresistors, configured in the Wheatstone's bridge, are localized near the membrane edges. Under the influence of pressure membrane deflection, strong mechanical stress generated near the membrane edges – thanks

* For the reasons discussed in section 4.4.4.2.1, a die of the packaged silicon pressure must be bonded to a glass support, not shown in Fig. 3.56.

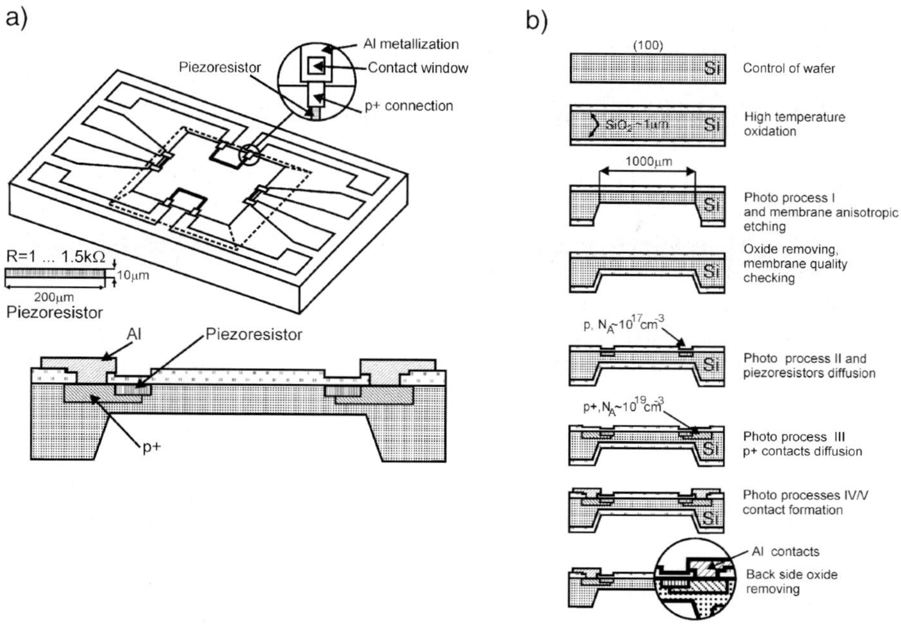

Fig. 3.56. Piezoresistive pressure sensor: a) scheme of construction, b) fabrication flow-chart.

to the extraordinary piezoresistive effect discovered by Smith in 1954 [135] – changes the resistance of piezoresistors. Two piezoresistors, parallel to a membrane edge (R_t), increase and two others, perpendicular to a membrane edge (R_l) decrease their resistance. As a result an output voltage signal is generated in the Wheatstone's bridge. The stress distribution and – being its result – the arrangement of piezoresistors depend on the shape and type of membranes (Fig. 3.57). These issues are the subject of many publications, e.g. papers [136–138], and are not discussed here.

Monolithic piezoresistors are fabricated either by selective diffusion or by implantation of impurities; the profile of the doping of the most used boron acceptor must be controlled very precisely. In some solutions polysilicon piezoresistors are used [139]. The resistors perpendicular to a membrane edge (R_l) are divided into shorter sections, usually two or more.

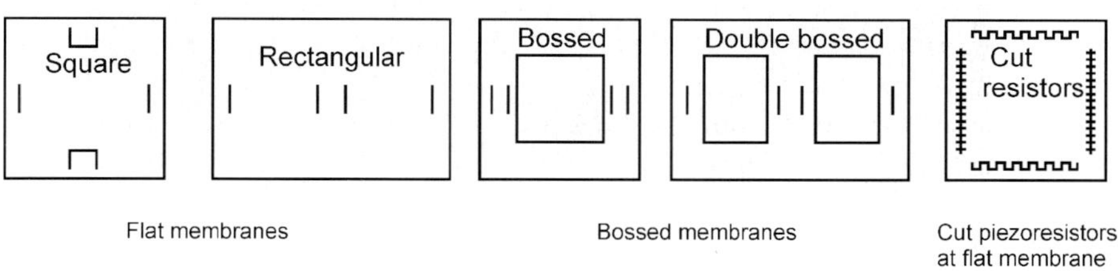

Fig. 3.57. Typical layouts of piezoresistors on membranes: from left to right; flat membranes, bossed membranes, the "cut" piezoresistors at flat membrane.

The control of dimensions of membranes, very accurate alignment of a back-side etched pattern of membranes to a front-side fabricated pattern of piezoresistors (and all co-working parts – e.g. connections, contacts, etc.) is the most important task in pressure sensor fabrication. A detailed discussion of this subject cannot be included in this book, but may be found elsewhere.

Examples of several types of structure of silicon piezoresistive sensors with flat and bossed membranes are shown in Fig. 3.57.

A sensor die of classic pressure sensor with a 1000 μm × 1000 μm, 20 μm thick membrane and classical configuration of piezoresistors (two half Wheatstone's bridges, consisting of a pair of piezoresistors parallel and perpendicular to the membrane edges) is shown in the Fig. 3.58 [140]. A sensor die equipped with newly developed "cut" piezoresistors is shown in Fig. 3.59. Piezoresistors are situated along the membrane edge. R_l or R_t piezoresistors

a)

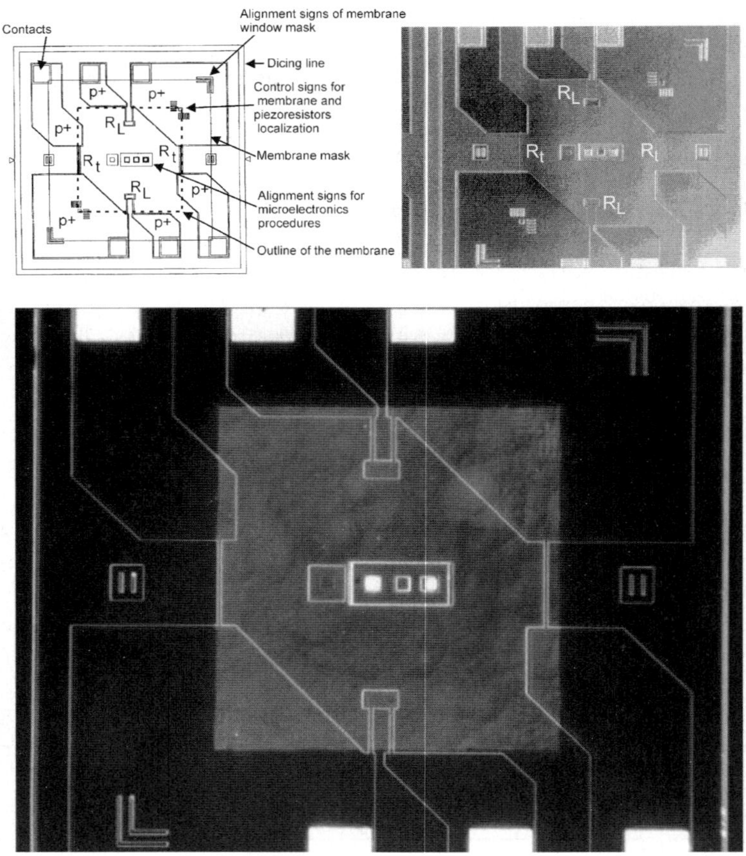

Fig. 3.58. A pressure sensor die, flat membrane, classical configuration of piezoresistors: a) layout and the structure top view, b) structure with membrane partly back-side illuminated; piezoresistors with p^+-type cramps and connections, and metal contacts can be seen; patterns for membrane and piezoresistor alignment are noticeable.

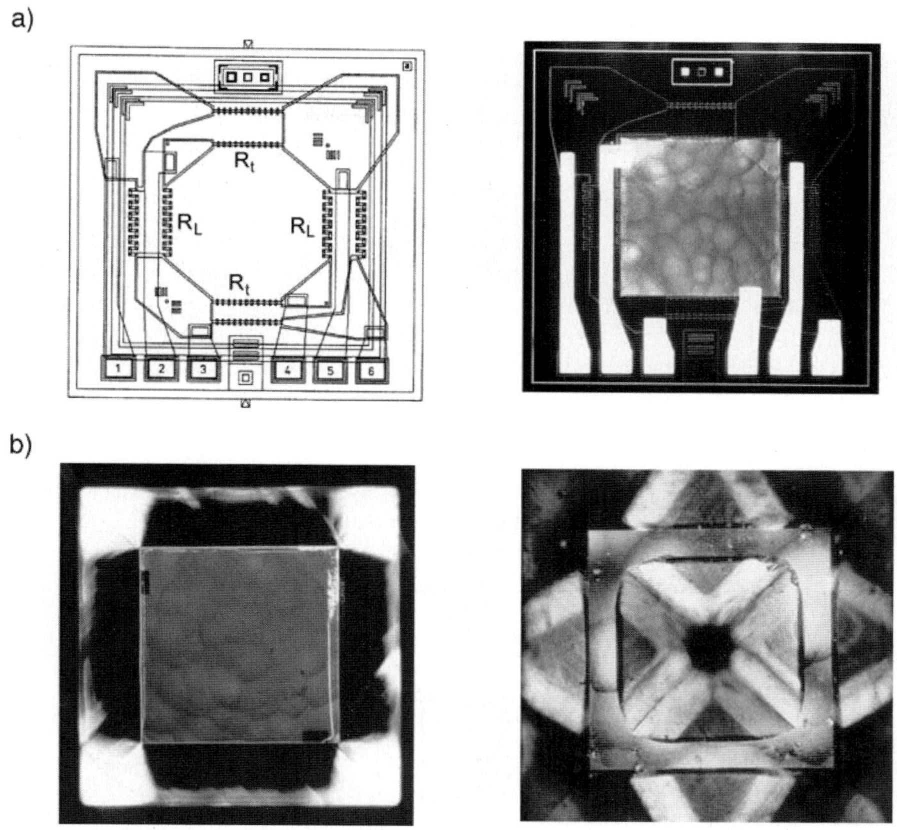

Fig. 3.59. A pressure sensor die, "cut" piezoresistors: a) layout and front view of the die, b) backside view of the die with flat (left) and bossed membrane (right).

are made by proper configuration of similar p and p^+ fields. The described construction of piezoresistors is extremely favorable, because they are symmetrically subjected to stresses. The pair of piezoresistors can be situated precisely on the edge of a membrane, which ensures very high tensometric sensitivity (Fig. 3.60). This arrangement of piezoresistors can be applied in sensors with a bossed-type membrane without loss of sensitivity [141, 142]. The "cut" piezoresistors make the technology of sensors easier, and at the same time ensure their high sensitivity. The same configuration of "cut" piezoresistors may be applied at flat and single-boss membranes. They have found their application among other things in highly-sensitive vibration sensors for the control of intelligent machining tools [143].

Silicon membranes can be used in pressure sensors in many ways. For instance, in widely applied pressure sensors of Yokogawa [144, 145], two ferromagnets are fabricated inside vacuum cavities made in a silicon flat membrane (Fig. 3.61). This extraordinary micromechanical structure is used in precise, overload-protected, differential pressure sensors. Ferromagnets, excited by

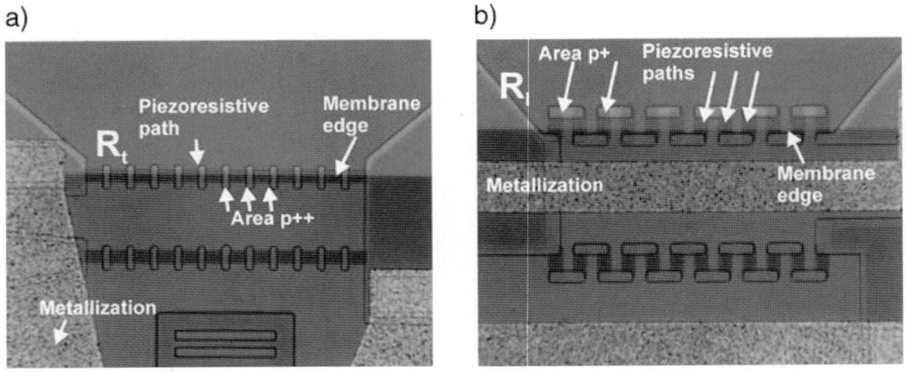

Fig. 3.60. Piezoresistors at membranes – details: a) classical solution; piezoresistor R_t parallel to the membrane edge (tensile stress across the conductive path) and piezoresistor R_l perpendicular to the membrane edge (tensile stress along the conductive path) together with the cramp, translucent membrane is visible, b) "cut" piezoresistors R_t and R_l. Note that the piezoresistor width is 10 μm, its distance from the membrane edge must not be larger than 15 μm.

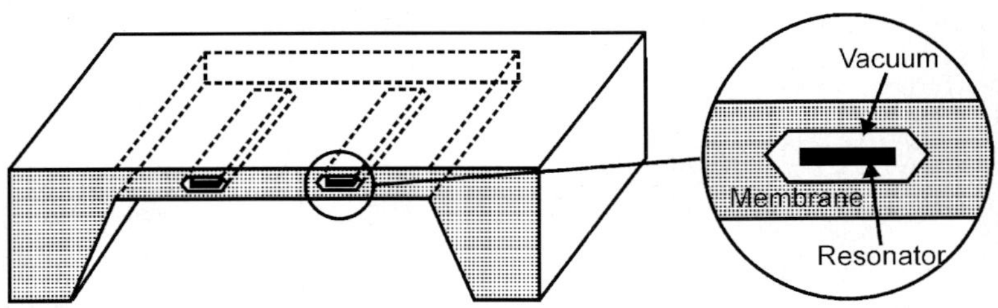

Fig. 3.61. Silicon flat membrane with vibrating ferromagnets [145].

an external magnetic field, vibrate on the first-order resonance frequency. Pressure, causing deflection of the membrane, stiffens its construction, which leads to a shift in the resonance frequency of vibrations. After suitable conversion in an electronic system, this shift is a measure of pressure value.

Silicon membranes are often applied in optical pressure sensors, which belong to the subgroup of optical-micro-electro-mechanical MEOMS devices. They can be designed in many different ways (Fig. 3.62). An example of such an optical pressure sensor is shown in Fig. 3.63 [146]. The sensor die consists of two bonded silicon chips. In the first chip a 20 μm-thick 6 mm × 6 mm membrane with a central boss, covering $\frac{1}{4}$ of its surface, is formed in 7.5M KOH with IPA at 80°C. Simultaneously, the V-groove in the frame surrounding the membrane and in the boss is formed. Dimensions of the V-groove are precisely adjusted to the typical optical fiber external diameter ED = 150 μm in such a way that the corn of the fiber – guiding a light from an external light source (halogen lamp, laser, etc.) – is aligned to a light-sensitive *p-n* diode fabricated near the edge of the frame. In the diode, a photoelectric voltaic signal is generated.

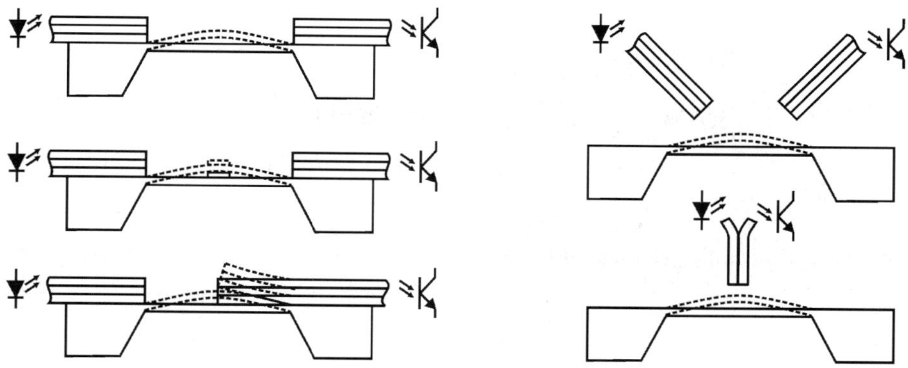

Fig. 3.62. Deflected silicon membrane modulating the light; possible constructions.

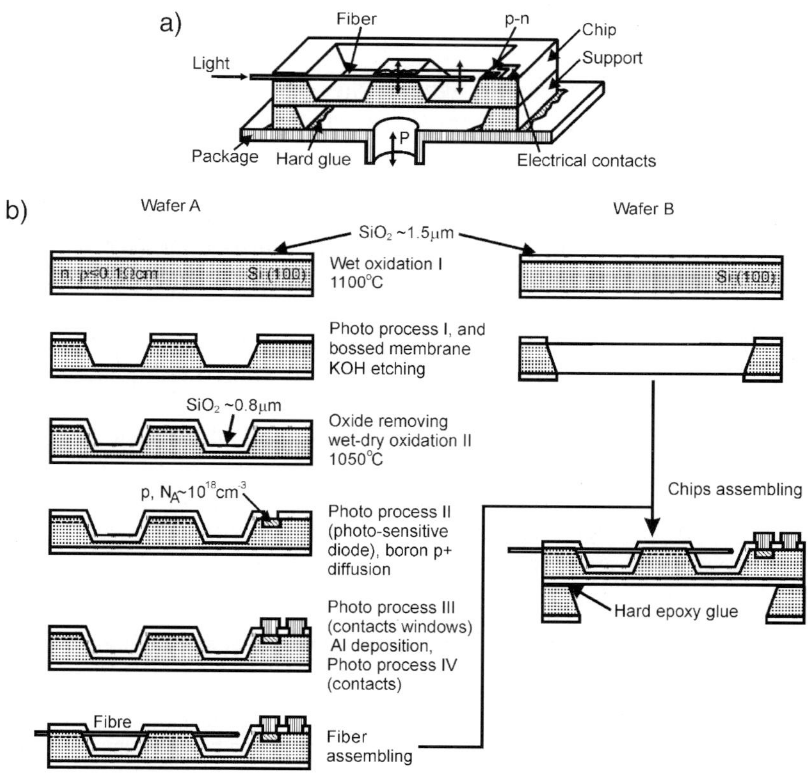

Fig. 3.63. Pressure sensor with optical fiber: a) scheme of construction, b) scheme of fabrication.

The membrane deflected by a gas pressure (Fig. 3.64a), moves up-and-down the tip of the fiber. The movement of optical fiber changes the photoelectric voltaic signal proportionally to the intensity of light, e.g. to pressure (Fig. 3.64b).

The main advantage of the described construction is complete spark safety

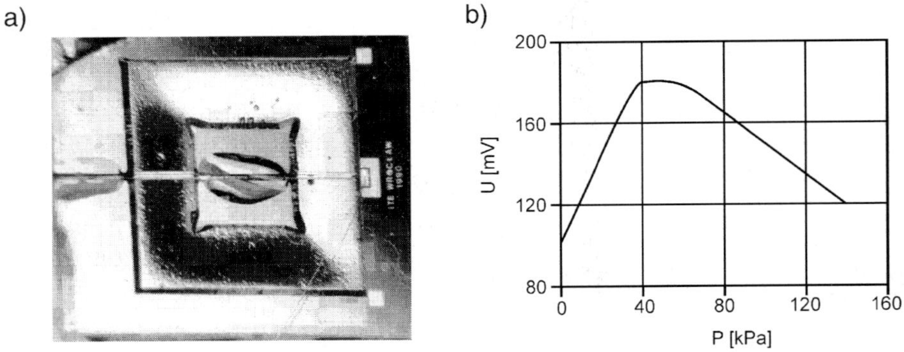

Fig. 3.64. Pressure sensor with an optical fiber: a) deflected membrane, $p \sim 60$ kPa, b) sensor characteristic.

of the sensor. Power generated in light-sensitive diode does not exceed 10 μW for the maximal output signal. A light source can be distanced from the pressure measuring point. Some parameters of the sensor are listed in Table 3.12, and its pressure characteristic is presented in Fig. 3.64.

Silicon membranes are used in highly sensitive bolometers (Fig. 3.65) [147–150], in miniature chemical filters [151] or in ink-jet printers [152, 153]. Three constructions of the ink-jet printers with silicon membranes are shown in Fig. 3.66. In an older solution a piezoceramic actuator moves a thin silicon membrane, which for a moment increases pressure in an ink container and causes a squirt of ink drops. In more developed constructions a monolithic heating resistor, supplied with current pulses, heats the ink and therefore leads to evaporation of liquid microvolume and formation of a vapor bubble with relatively high pressure near the micro hole made in a thin, flat silicon membrane. Finally, there is the squirt of an ink drop. Sets of many jets are employed in printing heads [152].

Table 3.12. Selected parameters of the pressure sensor with an optical fiber

Parameter	Value
Dimensions of structure [mm]	10 × 8.5 × 0.5
Maximal output signal [V]	0.7
Sensitivity [mV/100 kPa]	+240 for <40 kPa −80 for $p > 60$ kPa
Overload capacity	3 ×
Maximal work pressure [kPa], (of membrane 20 μm)	150 (200)
Electrical supply	none

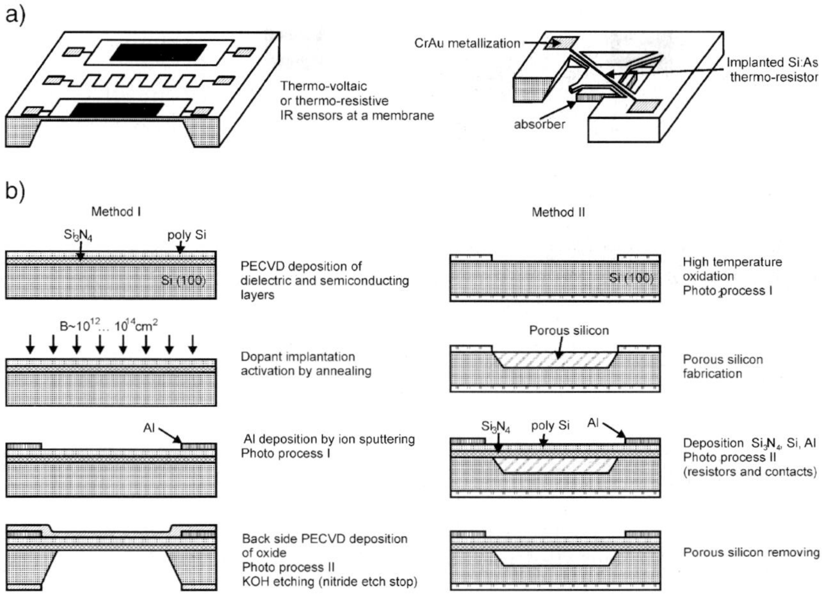

Fig. 3.65. Micromechanical bolometers: a) scheme of construction, b) two methods of fabrication [150].

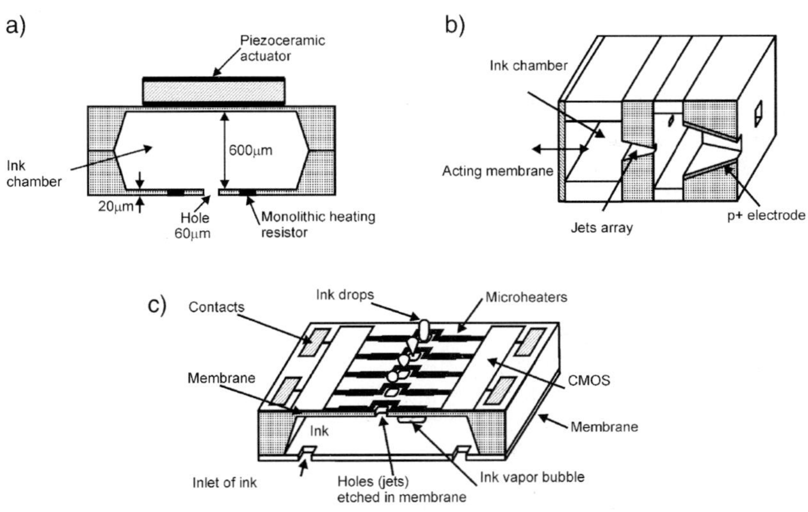

Fig. 3.66. Ink-jet heads: a) the simplest solution – top silicon membrane piezo-actuated, ink drop formed by a hole etched in bottom silicon membrane, b) head with a tandem of holes etched in two thicker membranes, thin membrane actuated by an external actuator, c) integrated printing head with ink vapor microbubble formation [152].

3.3.2. V-grooves, vials and holes

Grooves

The second important group of microconstructions etched anisotropically in silicon are grooves. The shape of grooves depends on the dimensions of the etch window made in a mask and its alignment to the crystallographic axis (directions).

The V-groove can be defined as the groove aligned to the ⟨110⟩ direction on a (100) wafer, which has a V-shaped cross-section, is narrow but long, its lateral and front walls are (111) planes, inclined to the front surface of the (100) wafer at angle $\Theta = 54.74°$ (Fig. 3.67).

The V-grooves are formed if the width of the mask window W satisfies equation 3.16:

$$W \leq W_{kr} \approx \sqrt{2}(d - m) - 2u, \tag{3.16}$$

where: W_{kr} = maximal window width ensuring the formation of V-groove, d = thickness of a wafer, m = distance from the bottom of V-groove to the back surface of a wafer, u = lateral under-etching.

In the configuration presented in Fig. 3.67, the fast etched (100) bottom fades, while slowly etched (111) walls develop. (111) planes after time t will come to contact and the etch process will be practically terminated:

$$t = \frac{d_n - \frac{W}{\sqrt{2}}}{2V_{111}} \tag{3.17}$$

where d_n is depth of a V-groove or

$$t = \frac{W}{\sqrt{2}(V_{100} - 2V_{111})}, \tag{3.18}$$

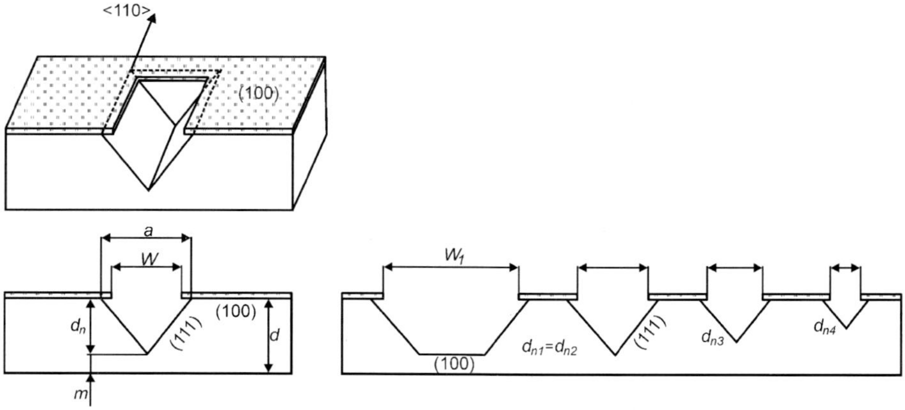

Fig. 3.67. V-grooves at the (100) wafer.

As stated earlier, for the majority of etching solutions and etching procedures, etch rate V_{100} is significantly higher than V_{111}. Assuming that:

$$\frac{V_{111}}{V_{100}} \approx \frac{1}{100},$$

the depth of V-groove is dependent only on the width of mask W:

$$d_n = \frac{\sqrt{2}}{2} \frac{W}{\left(1 - 2\frac{V_{111}}{V_{100}}\right)}, \tag{3.19}$$

the following formulas can be obtained:

$$d_n = \frac{\sqrt{2}}{1.96} \cdot W \tag{3.20}$$

hence:

$$W = \frac{1.96 \cdot d_n}{\sqrt{2}}. \tag{3.21}$$

Significant extension of etching time in comparison to a value resulting from equation (3.18) did not affect the V-groove dimensions. Consequently, it is possible to obtain grooves with different depths in a single mask etching process (Fig. 3.67b) or a multi-level structure of V- and U-grooves which can be etched utilizing a window with variable width W (Fig. 3.68).

Etching of narrow V-grooves requires intensive stirring of the solution. Weak stirring does not carry away the gaseous hydrogen from the bottom of the groove, where the micro masking of silicon by hydrogen bubbles occurs (Fig. 3.69).

More complex cross-sections of grooves are attained in a two-step fabrication process. In the first step silicon is locally melted by a NdYAG laser light beam scanned along the ⟨100⟩ direction. Then fused regions, with damaged crystallographic structure are etched in the process of mask anisotropic etching (Fig. 3.70) [154, 155]. On the (110) wafer grooves bordered by (111) walls inclined to the wafer surface at angle 35° evolve, and are inclined to each other at angles of 70° and 110°. On the (100) substrate the (111) sidewalls are inclined to the wafer surface at an angle of 54.74°, and to each other at angles of 110° and 70°. The depth of grooves corresponds to the average depth of melted area, and is not determined by alignment of the (111) walls.

Holes and vials

Fabrication of a hole in a (100) substrate is seemingly simple. According to the rules that have been presented above, the planar dimensions of mask need to be selected in order to etch the silicon wafer straight through. This method cannot be applied to produce precise holes with diameter equaling less than a

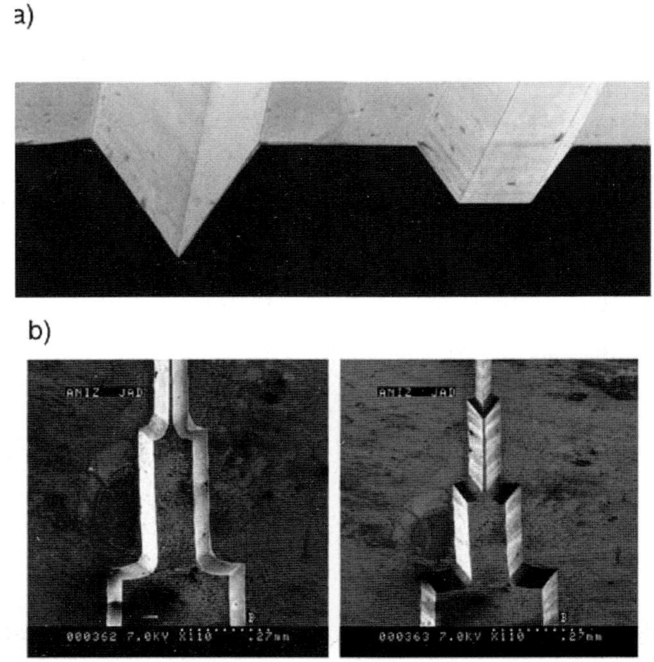

Fig. 3.68. Grooves: a) V-groove, _/ groove, b) cascade of grooves.

few micrometers. Such holes, and especially their arrays, are etched anisotropically in a silicon wafer by means of the boron etch-stop technique. A heavy doping of the p^+ region, determining the shape of holes, is performed on the front-side of substrate. Next, anisotropic silicon etching through a mask is carried out from the wafer back-side. The mask dimensions are selected to match the substrate thickness (taking into consideration its dispersion of thickness) in order to match the patterns precisely (Fig. 3.71). Using this method it is possible to fabricate the hole and vials arrays in substrates with diameters of many inches, most often exact to ± 1 μm, with the edge radius $r < 0.5$ μm. Packing density depends only on substrate thickness. The diameter of holes and vials is dominantly determined by the diameter of the window in the mask. The surfaces of holes formed in such a way are often covered with layers with a very high resistance to grinding with, e.g., diamond or silicon carbide.

So-called vials, regular inverted micro-pyramids, are etched similarly to V-grooves, but the mask windows have to be square. The square window width W needs to satisfy equation (3.21), then the dimensions and shape of 3-D micromachined vials are controlled by a self-stopping effect. Such cavities are applied on a relatively large scale in microchemistry, biochemistry and micropharmacology.

Fig. 3.69. Bottom of V-groove; the imperfectly etched (111) surface is a result of crystallographic defects of the silicon substrate, the micro masking effect can be distinguished, SEM picture.

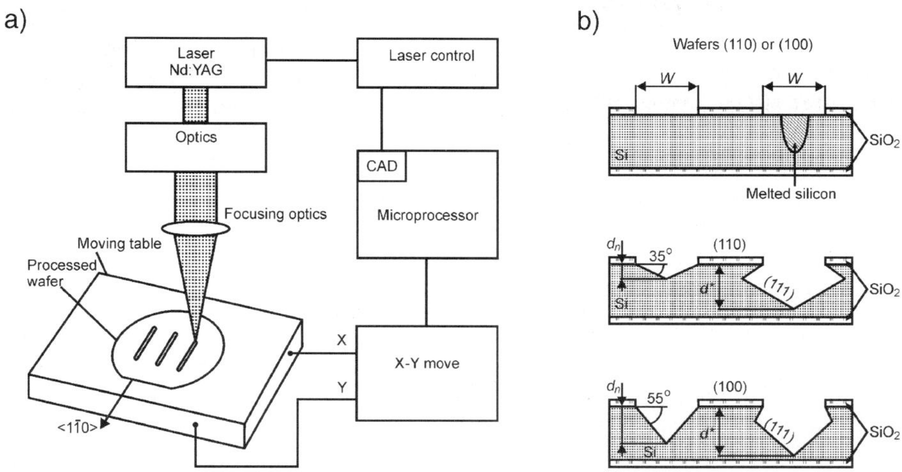

Fig. 3.70. Grooves in silicon – etched utilizing local laser fusing of silicon: a) scheme of fabrication, b) shapes on wafers (100) and (110). After paper [154].

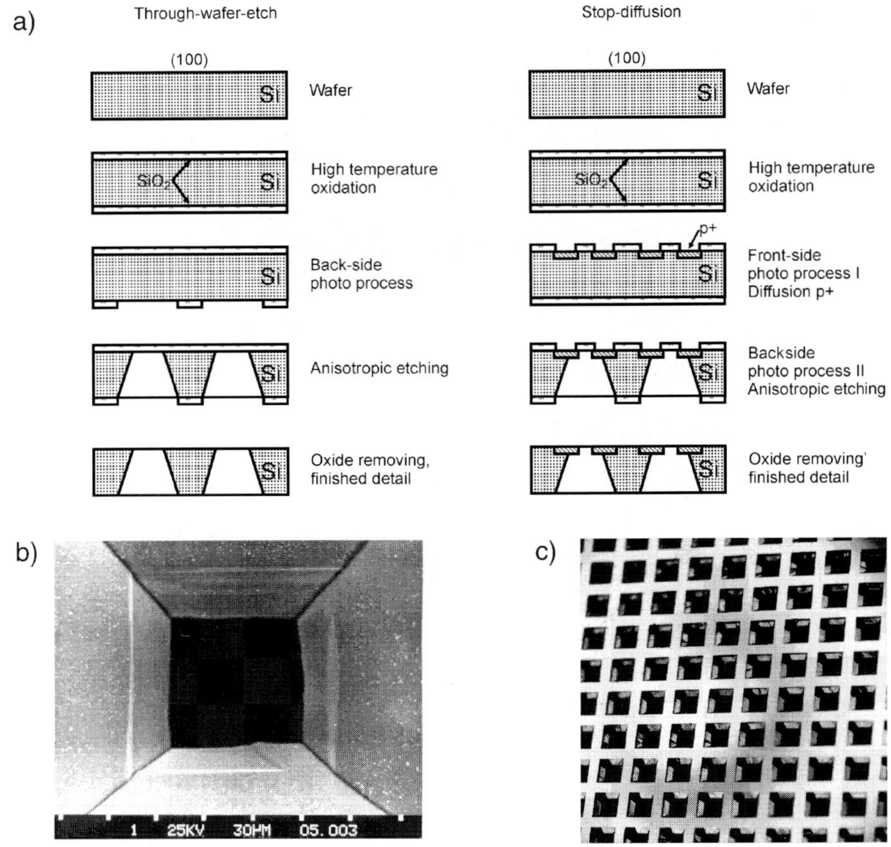

Fig. 3.71. Holes and vials: a) two fabrication methods: on the left – etching method, on the right – boron etch-stop technique, b) hole 60 μm × 60 μm, crystallographic imperfections created and unfavorable change in pattern shape, c) array of vials.

3.3.2.1. Application of V-grooves, holes and vials

V-grooves are commonly utilized in integrated optical microsystems MOEMS, chemical microsystems μTAS and micromechanical sensors for the positioning of optical fibers and components of MOEMS [156–163] (Fig. 3.72). They have been used for aligning of optical fibers in telecommunications, where sets of many prisms with V-grooves connecting the optical fibers are applied (Fig. 3.73a), as well as of a single optical fiber, which can be positioned in a V-groove by silicon beams, which are etched by means of electrochemical contact-less etching in 7M KOH [157], or by clamps made of silicon nitride [162] (Fig. 3.73b).

An example of an application of V-grooves taken from the author's studies is shown in Fig. 3.74. The micromachined silicon structure with three parallel V-grooves with different lengths is applied as a basic mechanical construction

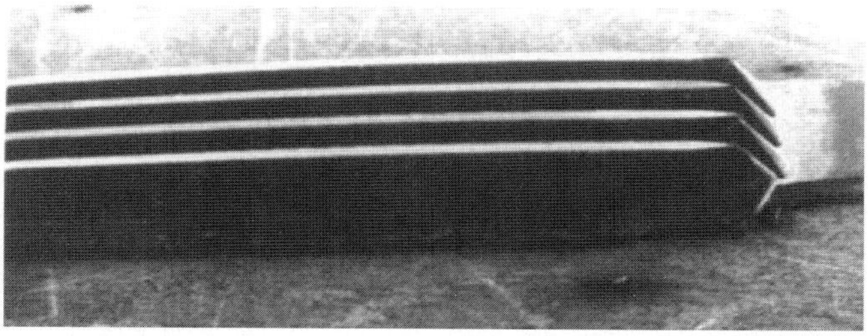

Fig. 3.72. Silicon structure with V-grooves for positioning of optical fibers.

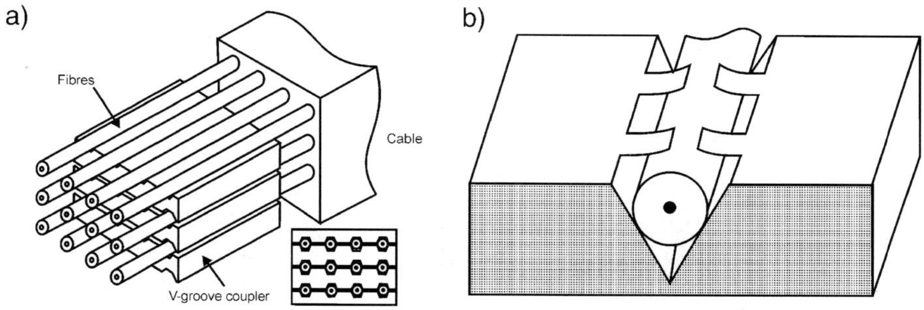

Fig. 3.73. Couplers of optical waveguides with V-grooves: a) coupler of many optical fibers (Bell Lab), b) coupler with clamps [157].

of the vibration sensor. V-grooves are etched in 5M KOH + IPA at temperature of 80°C. Mask shape is designed in such a way that the bottom of the V-grooves is situated 30 μm above the flat (100) bottom of the etched cavity. Optical fibers are assembled in V-grooves in such a way that the tip of every optical fiber is aligned to a light-sensitive diode located near the edge of the etched pattern (Fig. 3.75). Optical fibers are connected to a light source, for example a laser diode or a concentrated light beam from a halogen illuminator. Thanks to the movement of the free part of the optical fiber being influenced by vibrations, the degree of illumination of the photodiode changes proportionally to the degree of deflection of the fiber tip, which results in the formation of a photovoltaic signal variable in time.

The frequency characteristics of the vibration sensor exhibited a strong resonance near 46 Hz (for length of fiber free part equaling 6 mm) (Fig. 3.76).

The other construction of sensor with etched grooves, taken from the author's experience, is presented in Fig. 3.77 [164]. The sand hourglass-shaped cavity with narrow connecting groove is wet anisotropically etched in a (100) silicon substrate. A mercury micro-ball with an exactly known diameter is placed in the cavity. The sufficiently high impact force parallel to the groove shifts the ball along the device. Below the force limit the ball stays in the cavity. Gradation of groove widths ensures discrimination of indications and establishes the range of acceleration values, corresponding to the maximal force during the impact.

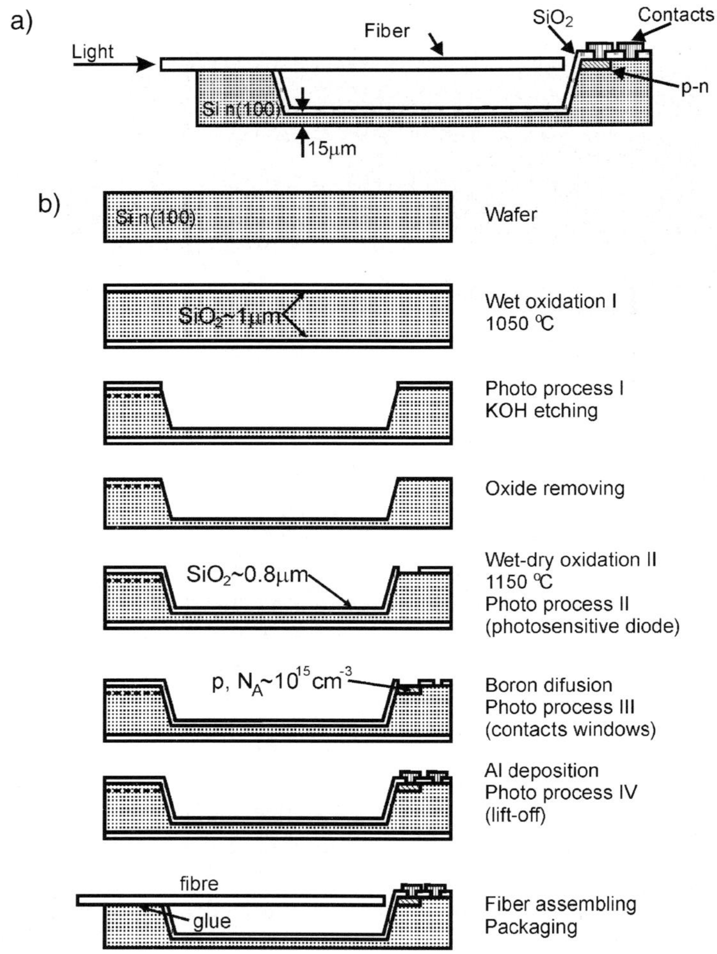

Fig. 3.74. Vibration sensor: a) cross-section of the structure, b) scheme of fabrication.

Mercury micro-balls are fabricated using the mold press method. First cavities in the shape of inverted pyramids (vials) are etched in a (100) wafer using the effect of etching self-stopping. On this prepared surface a big mercury drop is situated that is pressed onto the substrate utilizing a smooth and flat, clean glass plate. The high surface tension of mercury means that, after squeezing out the excess mercury, micro-balls with desired diameters are formed in the cavities (Fig. 3.78). By use of this method micro-balls with diameter ϕ in the range from 5 to 150 μm might be produced. The repeatability of this procedure is better than $\pm 1\mu m$.

The range of sensor can be widened or narrowed by the selection of a micro-ball diameter and a groove width. A single sensor die may consist of many sand hourglass-like cavities, differing in groove width as well as in globule diameter,

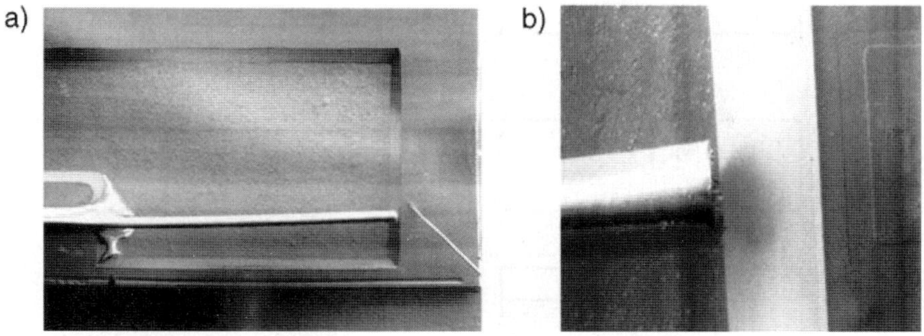

Fig. 3.75. Vibration sensor – details: a) side view, optical fiber mounted freely above the bottom of the etched pattern can be seen, b) close-up: distance from optical fiber tip to pattern edge equals 35 μm.

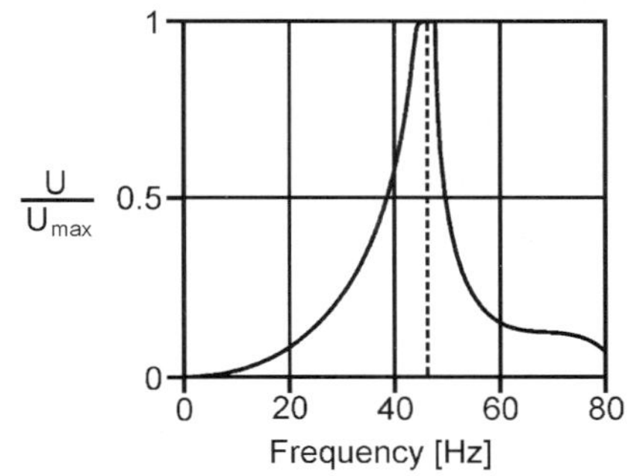

Fig. 3.76. Frequency characteristics of vibration sensor.

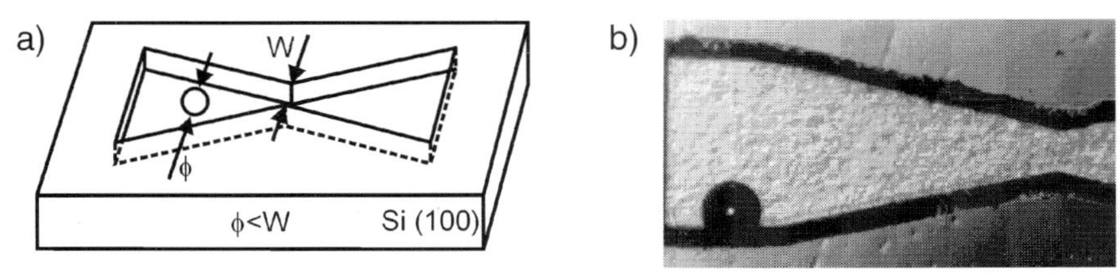

Fig. 3.77. High acceleration sensor: a) construction, b) structure – a Hg micro-ball is visible.

arranged in a star-shaped configuration. This configuration can allow a rapid simultaneous evaluation of both impact force and its direction. Because of the fact that "erasing" of a sensor after the impact requires force at least equal to the force which shifts Hg globules, such devices can find their application in automobile black-boxes – crash recorders*.

Grooves etched anisotropically in silicon in KOH, more rarely in EDP, through the windows in oxide or nitride layers, are commonly applied in many

* This solution was recognized as very promising in the periodical Technische Rundschau (Switzerland) [165].

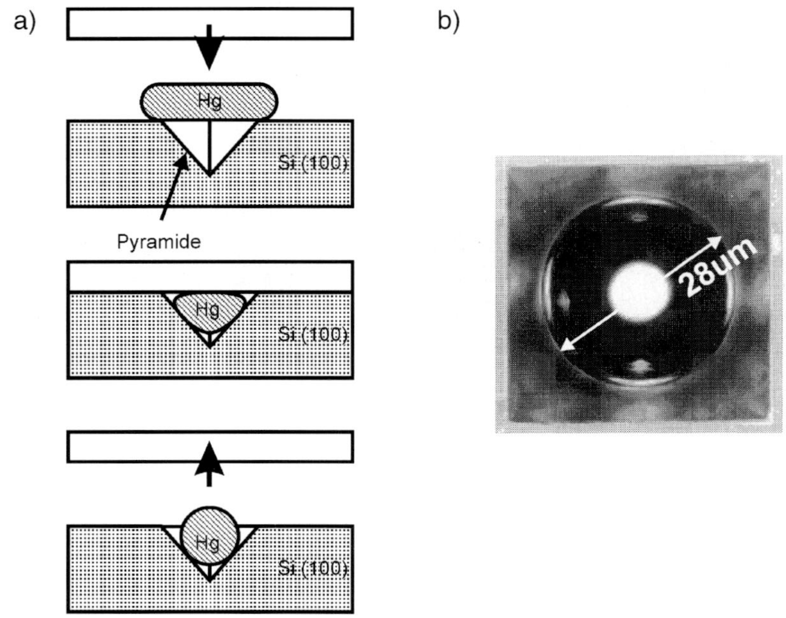

Fig. 3.78. Hg micro-balls: a) three steps of the fabrication procedure, b) Hg micro-ball in silicon mold.

gas flow/mass meters [165–170] and in chemical microsystems [171–179], including such unique applications as genetic studies [177, 178], monitoring of glucose [172], micro dialyzers [174] and in devices for storing and manipulating embryos [175] or sensing of nanovolumes of bio-species [179].

An array of vials formed by (111) walls etched at a (100) substrate is used – as mentioned earlier – as the subminiature reaction vessels (Fig. 3.79). The vessels (connected by micro canals, which due to the capillary micro pumping enable the transport of liquid analytes [180]) ensure carrying out thousands of chemical processes that proceed rapidly in precisely measured volumes in the range of pico- and nanoliters. On a single wafer 3″ in diameter there can be up

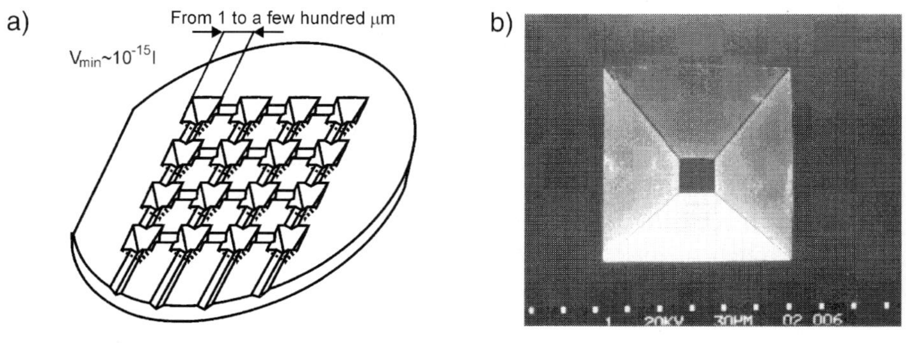

Fig. 3.79. Micro- and nanovials: a) silicon bio-fluidic-platform – idea, b) a nano-vial.

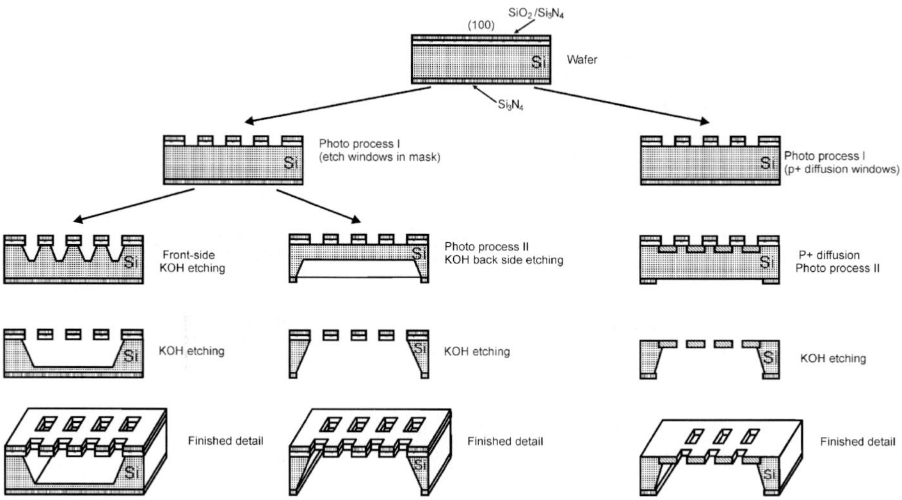

Fig. 3.80. Three ways of fabrication of micromechanical filters.

to 500 000 vessels produced – micro- and nanovials. Silicon platforms with micro- and nanovials are employed in combinatorial chemistry and in genetic research.

Thin silicon membranes with micro via-holes are used as filters in miniature chemical reactors [151, 178]. The technology of filters of this type utilizes wet anisotropic etching, connected with the technique of etch-stop diffusion (Fig. 3.80) or wet double-side etching.

Holes and their arrays etched anisotropically in (100) silicon wafers are used in ink-jet printers [152, 153], injection devices and fog generators, micro- and pico-volumes dosing systems, and microsystems for total chemical analysis*. Arrays of holes etched in silicon are also applied in the research of active control of turbulences (vortex control) in fighter aircraft [181] (Fig. 3.81).

Silicon micro hole arrays have found a particularly spectacular application in nerve regenerators [182–184]. Thin silicon membrane with an etched array of small holes a few micrometers in diameter is used in these regenerators (Fig. 3.82). Holes are surrounded by contact pads connected to each other by metal paths with on-chip integrated amplifiers, situated near the holes. Amplifiers are joined to contacts close to the edge, allowing the attachment of external connections (similar to the classic systems integrated inside the casing). Many dendrites of the broken nerve grow through the holes. Then dendrites join the contacts by physical contact, which ensures sufficient electrical connection. The nerve signals are amplified in *on-chip* amplifiers and transmitted to

* Very detailed information on the application of micromechanical silicon constructions in biomedical and chemical microsystems can be found in materials of the series of *Micro-Total-Analysis Conference μTAS* beginning from 1994, published among others by former Kluwer Academic Publishers.

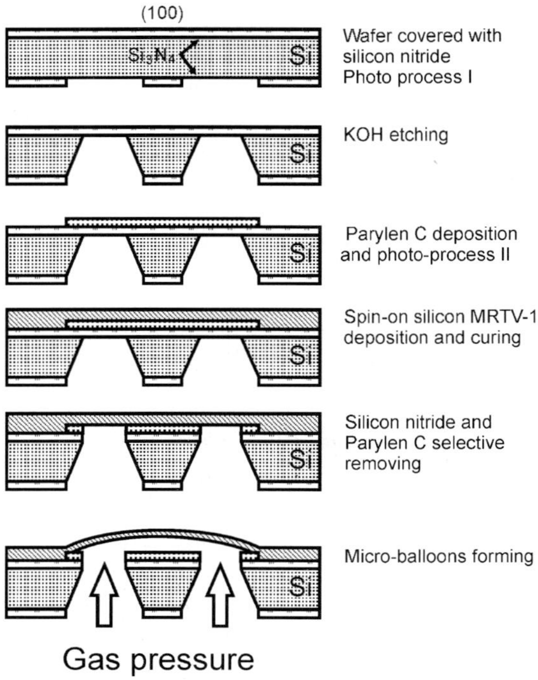

Fig. 3.81. Silicon vortex controller for fighter aircraft [181].

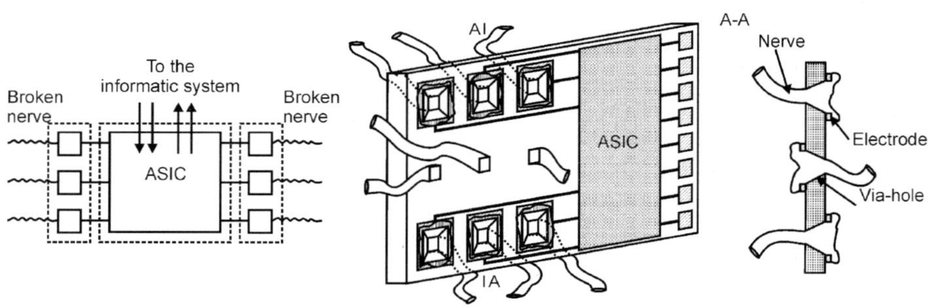

Fig. 3.82. Nerve regenerator – after papers [183] and [184].

external computer systems and devices. In these devices, at the beginning by trial and error, the natural trajectory of the nerve signal is reproduced. The signal, after proper adjustment of level, is let into the regenerator and passes to the next parts of nerve connection.

The 8–10 μm thick membrane of a regenerator is etched in KOH using an Si_3N_4 mask. Holes are etched using the wet–dry method. The diameter of holes equals 10 or 40 μm, which allows their integration into a system, ensuring 4000 or 121 connections [184].

Arrays of precise anisotropically wet etched in (100) wafer grooves and vials have been successfully used for the formation of microlenses and arrays [185, 186]. First, a pattern of reverse-pyramid cavities limited by (111) walls is etched in KOH. Next, a mask layer is removed and mask-less etching in KOH is continued. Under-etching effects (mainly fast etching of (411) crystallographic planes) led to the formation of the very regular circular cavity, used as a mold for microlens formation by the hot embossing method (Fig. 3.83) or – as will be presented in later sections of this book – by re-flow of anodically bonded glass and polishing.

3.3.3. Movable constructions: wheels, gears, beams and seismic mass

Silicon movable constructions are usually fabricated at double-side polished substrates. Micromechanical details can be wet etched anisotropically either by use of a single-step or a multi-step etch, which can be provided from any side of the wafer. Three-dimensional patterns can be formed on both sides of a substrate by means of one-side or double-side photolithography followed by etching procedures. Simultaneous etching from both wafer sides is very often used.

Sides of the substrate are chosen to be a front-side, a face, and the back-side, a back. First the front-side pattern is produced. Then a pattern made on the back-side of the wafer has to align with the pattern fabricated earlier on the front-side. There are two methods to obtain this: alignment of two masks and

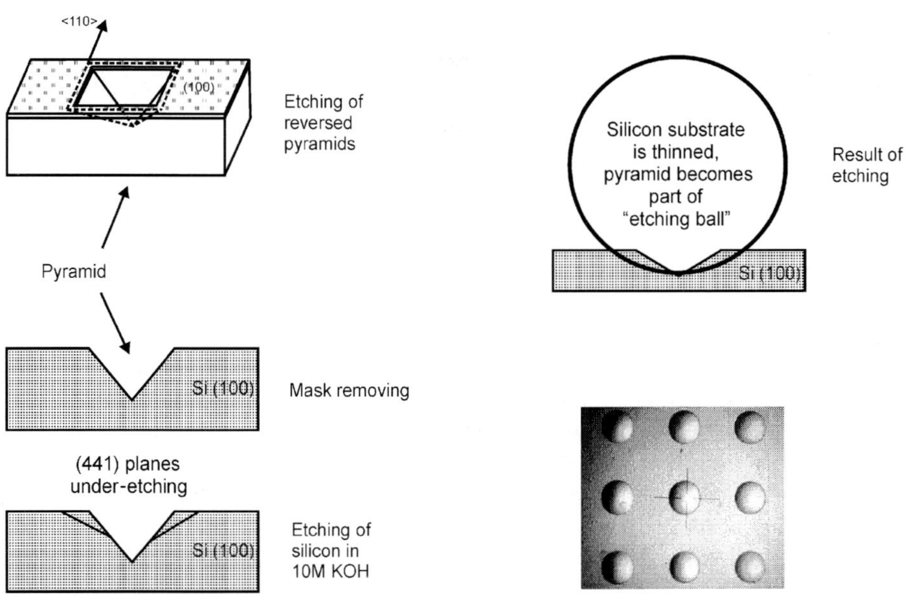

Fig. 3.83. Mask-less double steps silicon mold etching for microlens fabrication; process flow-chart and array of molds etched in 10M KOH at 80°C.

formed patterns from either side of the wafer, or aligning the back-side mask to the previously formed front-side pattern.

Both methods need double-side photolithography, which is a unique process, used very rarely in classical planar silicon microelectronics technology.

Methods of masks alignment are various:

- projection method, in which two masks are aligned by observation of optical or electronic images, a chosen mask is aligned to a pattern or/and to the primary flat on a wafer,
- infrared light method, utilizing infrared illumination cross-penetrating through the silicon wafer, used for alignment of a mask and micromachined patterns,
- method of mechanical holders and positioners, two masks are aligned and fixed in holders; next, a chosen mask is aligned to a pattern or to the primary flat on a wafer.

The alignment of masks to the existing three-dimensional deeply etched patterns is very difficult. The edge of the pattern is barely visible. In addition, it is not possible to observe simultaneously the front-side of wafer and the bottom of the pattern, because none of the aligners provide a sufficient depth of focus. In exceptional circumstances, when a deeply etched pattern is visible in a light transmitted through the silicon, alignment of a mask to the already-fabricated pattern can be performed, in standard equipment, merely by direct observation.

The most precise double-side alignment of masks ensures the projection method. The accuracy of alignment usually equals ± 3 to ± 5 μm. The infrared method is not so accurate; the alignment error varies from ± 5 to ± 10 μm, depending on substrate thickness. This method cannot be applied for the alignment of wafers, which are too thick ($d > 400$ μm) or are covered with metallic layers, because of the limitation of transmission of infrared radiation. The method of mechanical holders is very accurate but small imperfections at wafer edge can destroy its precision, and that is why it is used reluctantly. From the author's point of view the most practical, convenient method of double-side photolithography is fabrication of the specially designed alignment signs on either side of the wafer, localized beyond the micromachined area, prior to 3-D micromachining. Projection or mechanical holder methods may be applied for unprocessed substrates (the infrared method cannot be used here, as a wafer is usually too thick). Masks have to be aligned to proper signs side-by-side.

Movable constructions may be micromachined – following the photolithography stage – by use of one-side or double-side wet processes. The one-side process usually needs one simple deep anisotropic wet etching and a chosen etch-stop technique – most often the stop-diffusion technique. Double-side etched movable microconstructions are produced by applying a more complex mixture of procedures: wet anisotropic etching, etch-stop methods, isotropic wet etching, or dry reactive ion etching (RIE or PERIE). Examples of simple and complex procedures, which can be utilized for fabrication of a large choice of silicon structures, will follow. The applicability of the process is shown here mainly in the example of the fabrication of small silicon/metal mechanical parts and moving micromachines, which are very "photogenic", although their application in the technique is negligible. The method described can also be applied for the fabrication of more important microconstructions.

Single-side etching, stop-diffusion technique

In this method a pattern is front-side heavily doped with boron. The silicon oxide layer is removed from a wafer front (the back-side oxide layer is protected), and a wafer is etched in hot KOH. Thanks to the stop-diffusion effect, the pattern heavily doped with boron is not etched; non-doped superfluous silicon is etched away. Usually, important edges of the pattern formed at the (100) wafer (p^{++}doped beams or bridges) are disoriented 45° from the ⟨110⟩ direction. In this situation the main back-side walls of the 3-D pattern are formed by fast-etched (100) walls, and release of the free-standing details is faster.

In practice this method can be applied as follows: First, a (100), one-side polished silicon wafer is oxidized in steam at 1100°C for 3 hours. Next, diffusion windows are fabricated in the oxide at the front, polished side. Then, a boron diffusion is made. In the technological procedure described here, the spin-on solid source of impurity doping is applied. First a sol emulsion containing boron oxide is front-side spin-on deposited, baked in 80°C, and annealed at high temperature (~1100°C) in oxygen for a period of time sufficient to allow formation of the etch-stop boron concentration ($N_A > 7 \times 10^{20}\ cm^{-3}$) at least 3 μm below a face of the wafer. After doping, the thermal oxide is removed in buffered 10% HF:water and superfluous silicon is etched in 7M KOH at 80°C for a few hours. Etching finishes when all of the fabricated beams became movable. Finally, the wafer is rinsed in DI water and dried.

This method is very useful for beams and sets of beams, surface springs (Fig. 3.84) as well as small mechanical detail fabrication (gears, wheels)

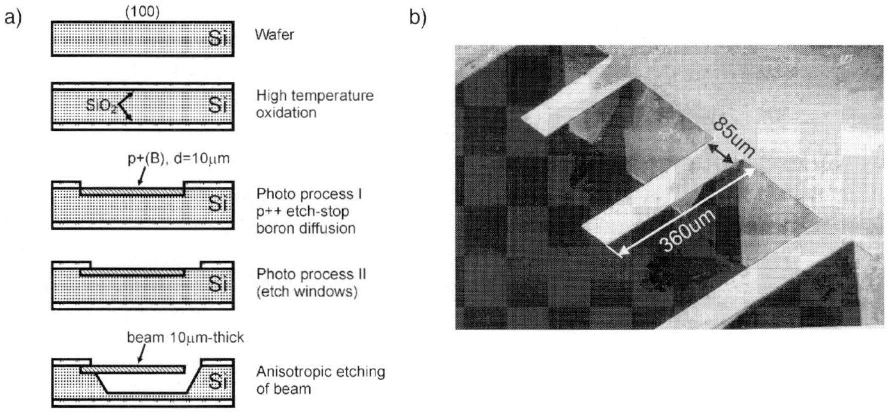

Fig. 3.84. Silicon beams: a) fabrication procedure, b) released silicon beams, SEM picture.

(Fig. 3.85). At the final step of manufacturing mechanical details they have to be separated from the silicon wafer, "fished" from the KOH and rinsed in DI water. Finally, DI water with the "swimming" wheels was spilled onto a piece of laboratory paper and details are dried under an infrared lamp.

Double-side wet anisotropic etching

In this method the flat, thin silicon membrane, a few micrometers thick, is micromachined in a (100), double-side polished wafer. Next a heavily doped p^+ pattern is formed at the front-side of the wafer. The shape and dimensions of the p^+ pattern correspond to the expected micromechanical details. Next the membrane is etched away; precisely formed details are collected from the solution.

In the example described here (Fig. 3.86) double-side polished silicon substrate is wet-oxidized in steam at 1150°C and back-side photo-patterned followed by etching of the 20 µm-thick membrane in 7M KOH at 80°C. After a second

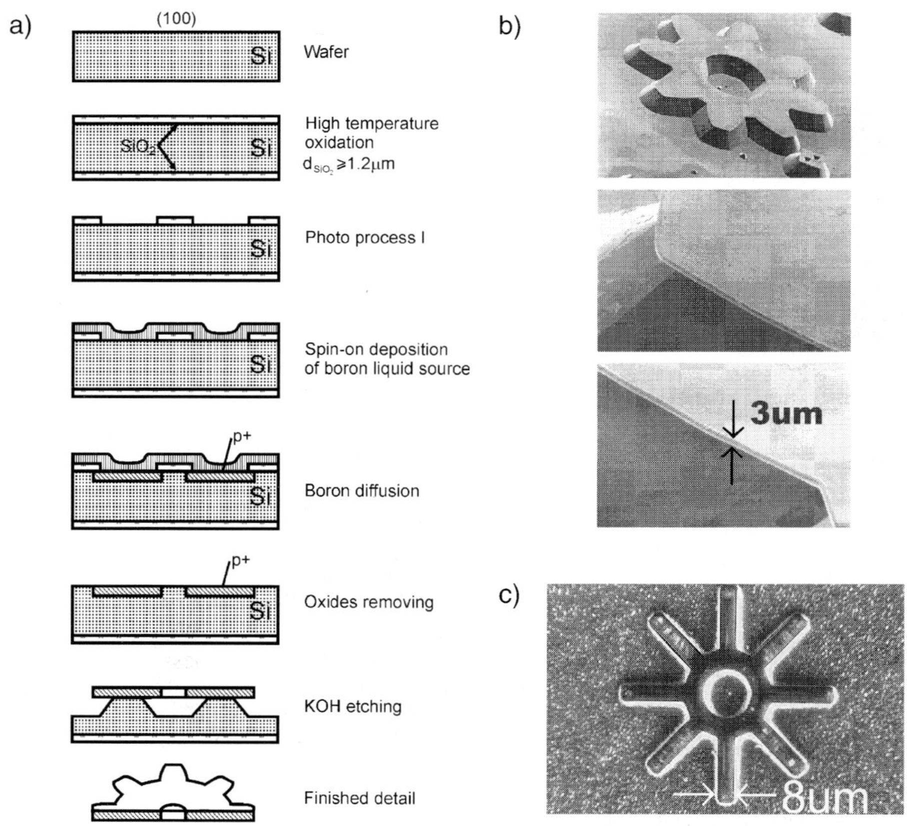

Fig. 3.85. Silicon toothed wheels and turbines: a) fabrication process, b) toothed wheel $\phi = 980$ µm after 4 hours of etching, c) released turbine with eight blades.

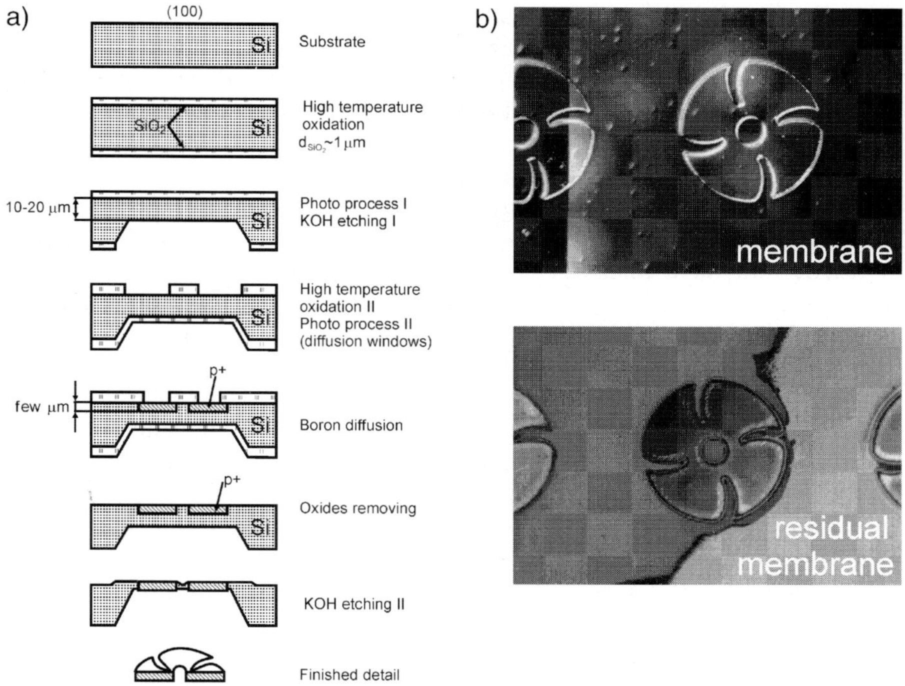

Fig. 3.86. Mechanical details: silicon micro turbines $\phi = 98$ μm in diameter: a) course of the process, b) p^+ pattern visible on thin silicon membrane (~ 5 μm thick) and the detail immediately before releasing.

oxidation the pattern of diffusion windows corresponding to the expected details is formed, and boron diffusion from the BN solid dopant source at 1200°C is done. The etch-stop boron concentration has to be obtained a few micrometers below the face surface of the wafer. Following this, the superfluous membrane material is etched away in 7M KOH at 80°C and the details are released. The "fishing" of very small details released after many hours of etching and floating freely in the solution causes the biggest difficulties in this process. Two methods are applied: in the first one the etching is carried out in a platinum sieve with very small meshes; after the release, details are washed in deionized water and dried under an infrared radiator. In the second method the silicon membrane is thinned in an etching bath to reach a few micrometers thickness. Next, etching proceeds on blotting paper in KOH drops. Finally, released details are washed in deionized water and dried.

Side-by-side anisotropic wet–dry etching

The term "wet–dry technique" indicates the etching processes applied successively: wet isotropic/wet anisotropic followed by dry etching. There are many methods and equipment systems for dry etching. The process can be performed in the gas phase physically, chemically or as a combined physical-chemical etch. A detailed discussion of dry etching of deep patterns can be found elsewhere*.

The wet–dry process combines the advantages of wet and dry etching. In this method a thin membrane is back-side wet etched in a double-side polished (100) silicon wafer. Next, the front-side of the wafer is covered again with the masking layer and then patterned; after this the superfluous region of the membrane is dry etched away. Due to the fact that the membrane is not thick (a few dozen micrometers), silicon etching from the front-side of the wafer can be carried out in a relatively short time, with no special DRIE equipment needed. Parameters of the process: type of atmosphere, discharge power, plasma polarization potential, pressure, gas flows, anisotropy degree, etc., have to be adjusted according to the process specification. In our own studies isotropic dry etching in the two-electrode configuration of the GIR 350 etch system (ALCATEL, Annecy, France) has been used under the following conditions: pressure 20 Pa, flow $SF_6 + O_2 = 12$ sccm + 2 sccm, discharge power $P_{RF} =$ 150 W, plasma polarization voltage $U_{DC} = 100$ V, which ensure a gentle lateral profile of etched pattern (Fig. 3.87a) and fast silicon etching in the direction perpendicular to the surface of the wafer. The average silicon etch rate under these conditions equals $V_{TR} = 1.37$ µm/min. Deep dry anisotropic etching has been obtained through a positive resist mask AZ1350 or magnetron-sputtered Al thin-film layer at a pressure of 4 Pa, discharge power $P_{RF} = 150$ W and voltage $U_{DC} = 260$ V, for gas flows given above. Attained silicon etch-rate is low ($V_{TR} = 0.3 \div 0.46$ µm/min); but the side edges of the pattern are very sharp (Fig. 3.87b). The process anisotropy equals from 5:1 to 10:1. Such conditions of etching are selected in order to avoid the unfavorable hardening of the AZ1350 mask that makes its removal difficult.

Two types of the wet–dry etching technique can be distinguished (Fig. 3.88). In the first type a pattern is selectively front-side dry etched before back-side

a) b)

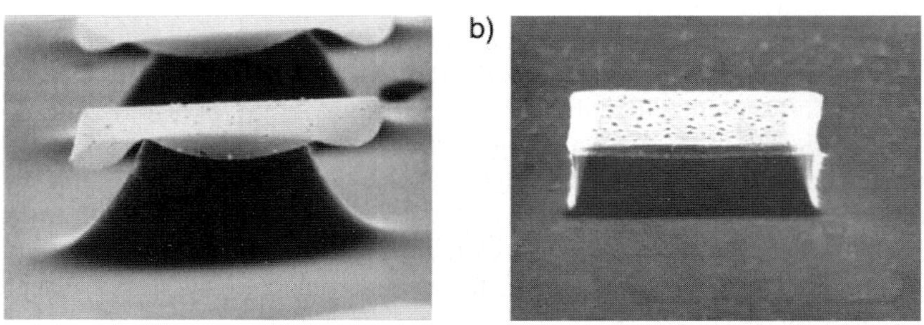

Fig. 3.87. Profiles of dry silicon etching: a) isotropic etching, $V_{TR} \sim 1.37$ µm/min, isotropic under-etching of mask can be distinguished, b) anisotropic etching, $V_{TR} \sim 0.3 \div 0.46$ µm/min. The process was elaborated by dr A. Górecka-Drzazga from the Faculty of Microsystem Electronics and Photonics of Wroclaw University of Technology.

* See Marc Madou, "Fundamentals of microfabrication", chapter 2. CRC Press, Boca Raton, (Fl)–New York, USA, 1997.

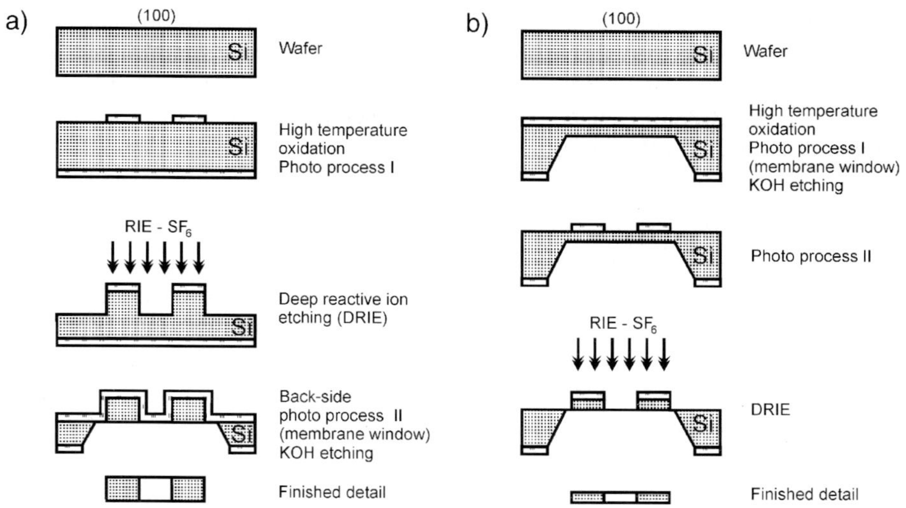

Fig. 3.88. Two types of wet–dry techniques; examples of small silicon details fabrication procedures: a) first technique, b) second technique.

etching of a membrane and releasing of details; in the second type of the process, a dry-etched pattern is formed after wet etching of a membrane.

Technological utilization of both procedures has been shown in the fabrication of small silicon micromechanical toothed gear wheels (Fig. 3.89).

Side-by-side wet anisotropic–wet isotropic etching

In this method, first a thin membrane is etched in KOH in the double-side polished (100) silicon substrate and thickly oxidized in steam. Next, a thick layer of the material resistant to isotropic etching is deposited and a pattern of a detail is formed from the front-side of the wafer. Then the useless silicon is etched away in HNA 50:3:8 (HF:HNO_3:CH_3COOH). The process of isotropic etching is performed on the filter paper, a drop of HNA is positioned at the front-side of the wafer from a burette. After silicon is completely removed, details are "hung" on the remaining residual silicon oxide. The oxide is removed in a buffered solution of HF, fed in drops as before. After etching of the oxide,

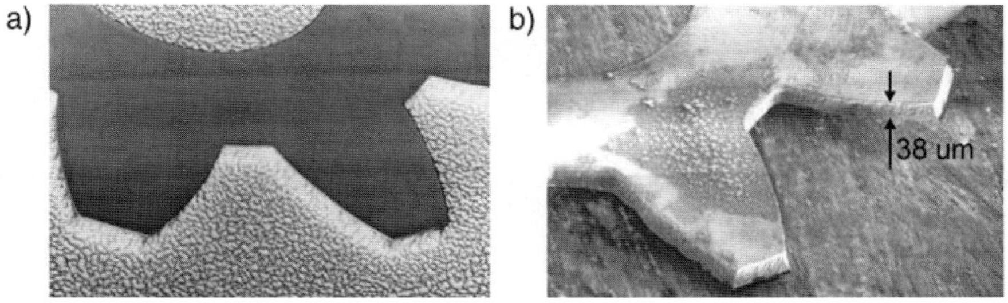

Fig. 3.89. Silicon toothed 980 μm wheel: a) after plasma etching, b) released.

details must be washed in deionized water and dried under an infrared radiator. The procedures described above have been used to fabricate aluminum microturbines with thickness $d = 3$ μm and external diameter $ED = 98$ μm (Fig. 3.90), which have been employed in a micro flow meter and compressed air micro engine (described in other sections of this book).

Wet anisotropic–back-side dry isotropic etching, a projection technique

In this technique, first a flat or bossed membrane is wet etched from the back-side of a wafer. Next, an aluminum layer is deposited on the back-side of the substrate so formed. The layer – after projection photolithography (hence the name of the described technique) – functions as a mask for dry etching (RIE) (Fig. 3.91a). Then the superfluous silicon is removed by use of rapid isotropic dry etching. This method is especially suitable for the releasing of the movable components of micromechanical sensors, because the front-side of the etched wafer, containing fragile electronic elements, which can be susceptible to etching, is protected by direct contact with the metal table of the etching equipment.

In our own studies etching in any oxygen/SF_6 mixture in a two-electrode configuration of the GIR 350 equipment of ALCATEL (Annecy, France) has been applied for plasma excited by frequency $f = 13.55$ GHz, pressure 4 Pa, discharge power 100 W and applied flows: $SF_6 = 125$ sccm and $O_2 = 45$ sccm.

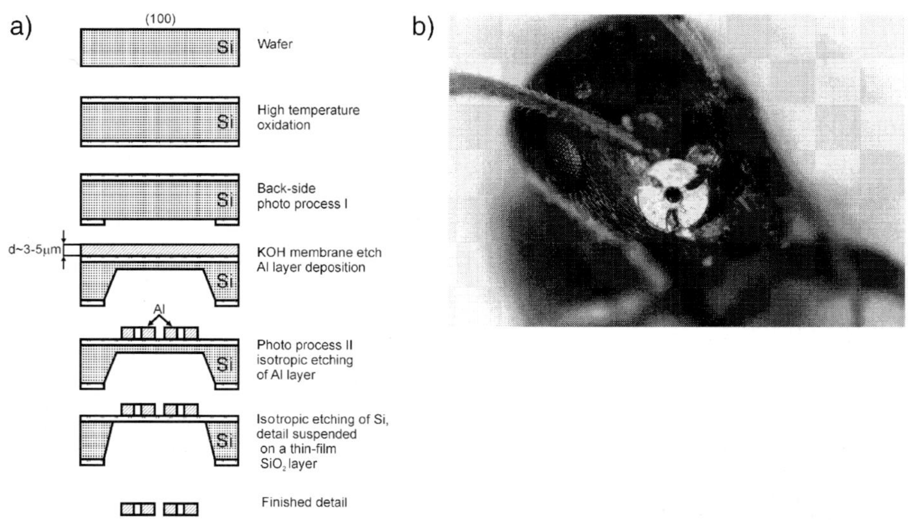

Fig. 3.90. Aluminum micro turbines $\phi = 98$ μm: a) fabrication method, b) turbine situated on the "nose" of an ant*.

* This useful insect is often utilized by researchers in order to illustrate the micromechanical scale, e.g. by UCLA: www.mems.uc.edu., Frauenhofer Institute from Karlsruhe: www.pmt.fzk.de.

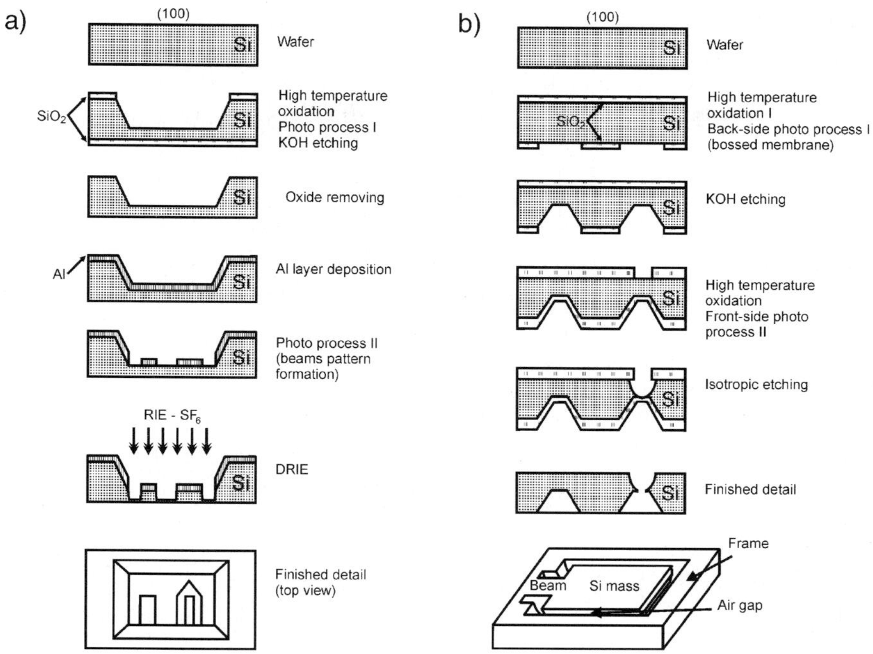

Fig. 3.91. Two fabrication methods of beams and vibratory masses on silicon beams: a) method of projection, b) wet method.

Wet anisotropic–wet isotropic etching, front-side masking technique

First the semi-finished product is back-side etched. After that, the dielectric thin-film masking layer is deposited onto the wafer. A pattern is formed and etched from the front-side of a substrate (Fig. 3.91b). A silicon oxide layer is sufficient for KOH etching, an AZ1350 masking layer can be applied for a short-time etching in $HF:HNO_3:CH_3COOH$ 1:3:8, an Si_3N_4 mask can be used in the both cases.

3.3.3.1. Applications of movable constructions

Micro gear with toothed wheels

Mechanism of the micro gear consisting of two toothed wheels $\phi = 980$ μm and housing 2 mm × 3 mm with two axles and pneumatic canals was produced by selective dry and wet etching [187]; 1 μm thick silicon oxide served as a mask. Toothed wheels with thickness 38 μm (Fig. 3.92) were fabricated by means of double-side wet–dry etching method. The axles and canals in housing were dry isotropically etched in SF_6. After fabrication of the component parts, the micro gear was assembled manually under the microscope (Fig. 3.93).

Beam/mass accelerometers, light modulators

Movable constructions with beams and masses made of silicon are used in microsystem technology for different purposes (Table 3.13). Most often they

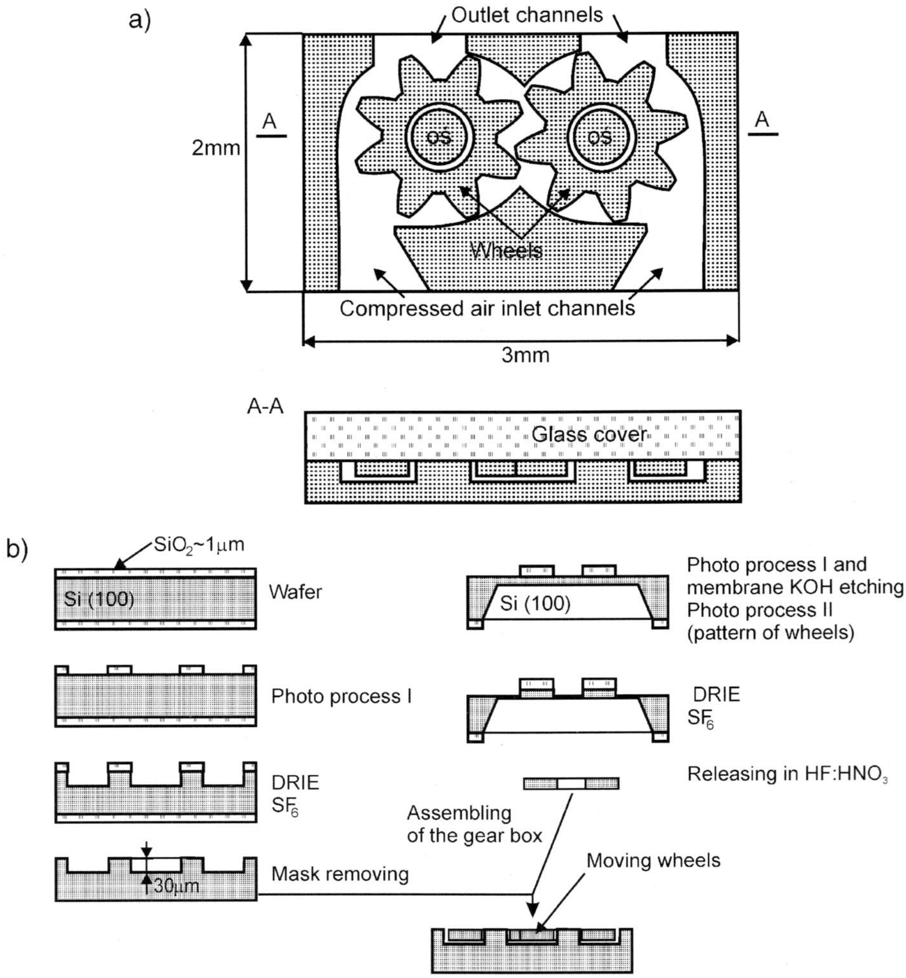

Fig. 3.92. Silicon micro gear: a) scheme of construction, b) scheme of fabrication.

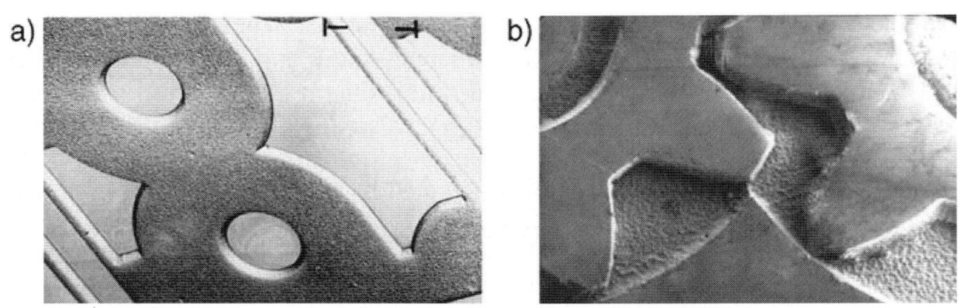

Fig. 3.93. Silicon micro gear: a) body after dry etching, b) completed mechanism.

Table 3.13. Application of silicon three-dimensional structures in micromechanical devices (on the basis of paper [189])

Type of device	Type of structure	Fabrication method
pressure sensor	M	Deep anisotropic etching
Accelerometer	M	Etch-stop diffusion and deep anisotropic etching
Pressure sensor	MB	Ditto
Pressure sensor	MB	double-side deep anisotropic etching
Pressure sensor	BDP	Electrochemical etching or, under-etching under the polysilicon bridge, or etch-stop diffusion with deep anisotropic etching
Accelerometer	BDP	Under-etching (4 crossed beams)
Accelerometer	D/BDP, T/BDP	Double-side, wet–dry deep anisotropic etching
Accelerometer	MBW/D/BDP, MBW/4/BDP	Double-side, wet–dry deep anisotropic etching
Vibration sensor	BJP	Etch-stop diffusion, deep anisotropic etching
Modulator	BJP N	Etch-stop diffusion, deep anisotropic etching or wet–dry method

M, membrane; MB, membrane hung on beams; BDP, double-side supported beam; D/BJP or BDP, binary beams ditto; T/BJP or BDP, triple beams ditto; MBW, membrane (mass) on beams, BJP, one-side supported beam.

find their application in accelerometers for motorization [20, 188], in airbags and traction control systems.

Examples of structures are shown in Fig. 3.94. The projection technique, described earlier, is particularly useful in the technology of accelerometers, because it allows the release of movable elements of these sensors in the final step of their technology without applying an additional front-side masking procedure.

A seismic mass suspended on four beams is used in the piezoresistive silicon accelerometer (Fig. 3.95). In this device shifting of the mass, caused by an external force, results in the formation of strong compressive-tensile stresses in the beams. The monolithic piezoresistors arranged in the Wheatstone bridge are fabricated in beams in the field of these stresses. An output signal of the Wheatstone bridge is proportional to stress and is a measure of the acceleration value.

Sets of beams have been utilized in the production of the first micromechanical picture projector [1, 20] (Fig. 3.96), as well as in the tunnel microscopes

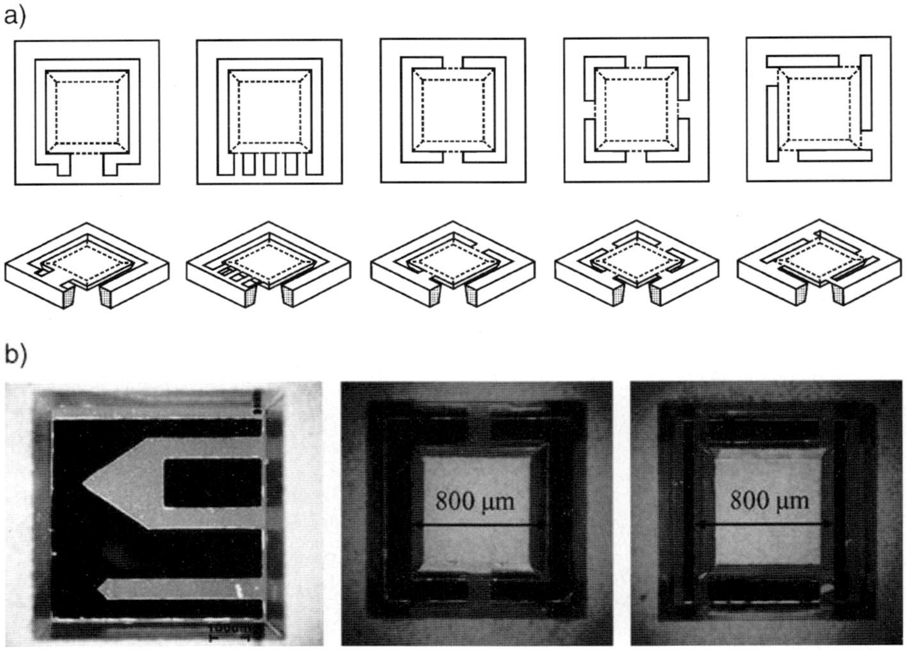

Fig. 3.94. Beams and arrays of beams with seismic masses: a) typical geometric array, b) constructions obtained by means of the projection method (a beam deflected under the pressure of the tip is visible), c) constructions fabricated using the wet–wet method. Fig. 3.98c obtained by courtesy of M.Sc. Jerzy Jaźwiński from the ITE in Warsaw. Picture "c" was supplied by dr hab. Iwo Rangełow from the Technical University in Kassel (Germany).

and microscopes of atomic forces [190], vacuum meters [191], switches [1, 5], flow meters [192] and artificial noses [193].

The picture projector consists of a set of electrically actuated beams of equal lengths. Beams are produced in a (100) wafer using the method of wet anisotropic electrochemical etching. They are covered with metal and form condensers with the p^+ layer. The electrostatic force, which occurs in the condensers (for potentials of a few volts), is sufficient to deflect the beam. Movement of beams that are illuminated with laser light corresponds to the selection of lines, while movement of a galvanometric mirror corresponds to the selection of frame. In the original Petersen's experiment [1, 194] such a projector served for the formation of the inscription "Micromechanics in Silicon" on the windowpane of his laboratory.

A thin silicon membrane, supported by a prism and hanging on two silicon torsion beams, that make its movements possible, has been applied in the new constructions of light modulators and picture generators [195, 196]. The membrane, whose upper surface is covered with a reflective layer, is deflected from the balance point by an electrostatic attraction force (Fig. 3.97). Vibrations of the structure deflect the light beam, which – if there is a suitable attraction of

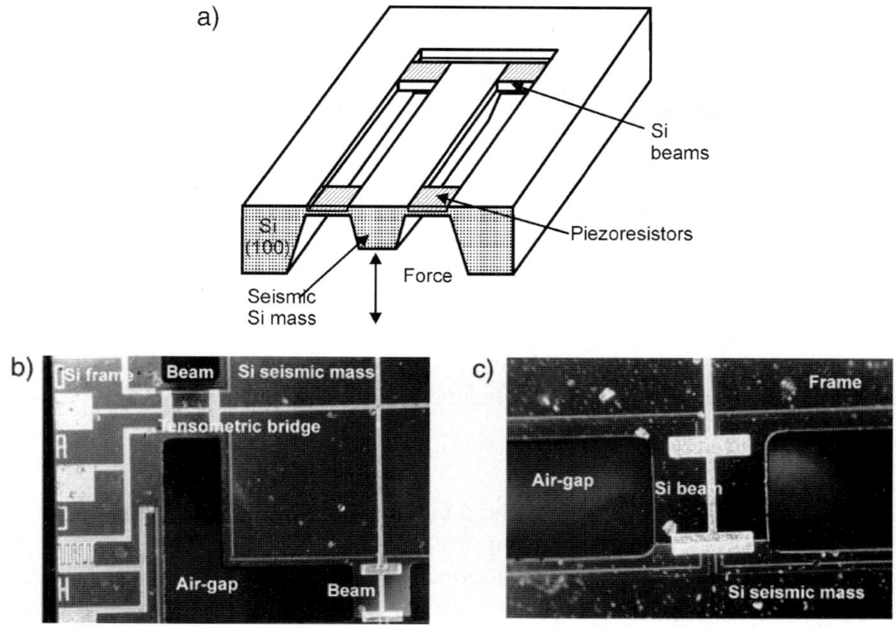

Fig. 3.95. Accelerometer: a) scheme of construction, b) front-side view, c) beam. The structure was fabricated in the ITE in Warsaw, and obtained by courtesy of dr Jan Łysko.

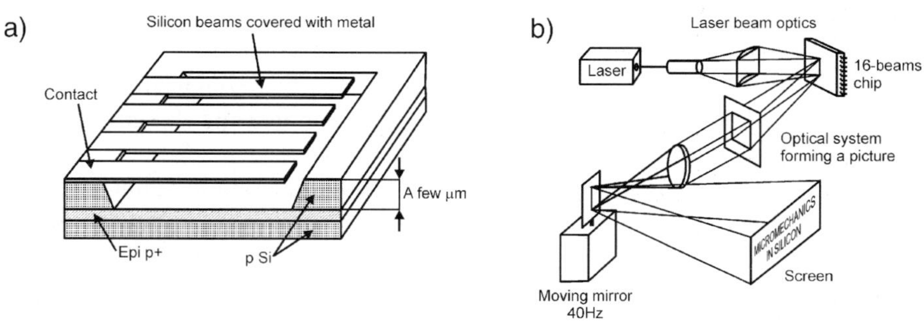

Fig. 3.96. Picture projector with silicon beams: a) idea of projector construction, b) projection assembly [1].

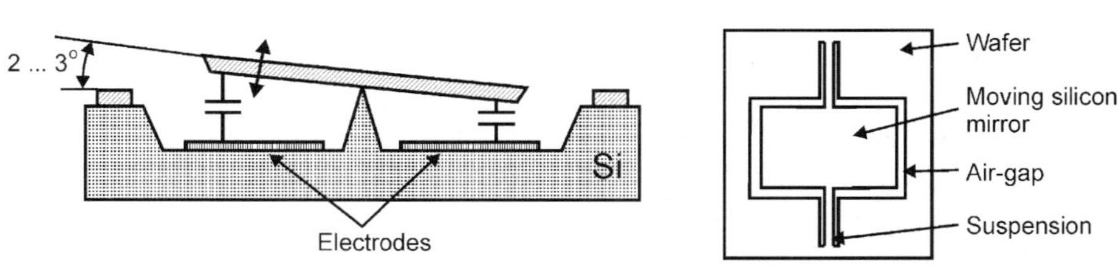

Fig. 3.97. Vibrating structure [1, 194].

vibrations – can lead to the projection of pictures. A similar technological solution has been used in vacuum meters, where the membrane vibration is suppressed by the viscosity of ambient air, which is directly proportional to the value of negative pressure [198].

Silicon movable constructions fabricated by means of the wet–wet and wet–dry method, using etch-stop diffusion, have been applied in pressure sensors designed for working under extreme conditions in gas and petroleum wells [197, 198]. Deep-etched flexible silicon electrodes have been used in spectacular work on the artificial eye [199, 200], as well as in the innovative studies on the production of the "smart skin", which allows active control of von Karman's vortex (*vortex active control*)* [201, 202]. Studies continue on the application of silicon beams in new-generation neurological probes [203], as well as in the construction of surgical tools of high precision [204, 205].

3.3.4. Tips, array of tips

Silicon tips and arrays of tips can be fabricated on (100) substrates by means of wet anisotropic etching, and on wafers with any orientation using wet isotropic etching in a NHA solution (aqueous solutions of ***N****itric,* ***H****ydrofluoric,* ***A****cetic acids*), or dry anisotropic or isotropic etching. Wet and dry etching processes are often applied together [206].

In the first technological step, a silicon substrate is covered with a masking layer and then a suitable window pattern is formed in the mask. A thermal oxide mask can be used for wet etching, while a sandwich of two or three layers: oxide–positive resist, oxide–Al or CrAu, oxide–positive resist–metallic layer should be applied for dry etching. Next, the so-called precursor is etched. Since the summit of the precursor is not sharp, in the next technological step it has to be sharpened (the Grove-Deal effect†) by high-temperature oxidation. This method sharpens both convex and concave tips (Fig. 3.98).

In the example discussed below, an array of tips is etched in hot concentrated KOH with IPA, by a 0.3–0.5 μm thick, silicon dioxide mask. Square areas protected by the mask evolved into convexities formed by (111) walls and (or) rapidly etched walls with high-index walls. The shape of tip precursor depends on two factors: shape and dimensions of the mask and the anisotropy of the etch. The dimensions of edge a_p of the square mask are chosen in this way, that for a given height of precursor H, walls forming the precursor have to converge in the geometric midpoint of the square (Fig. 3.99).

Then:

$$a_p = 2u \tag{3.22}$$

that is

$$a_p = 2\,V_{hkl} \cdot \cos \Theta_{hkl} \cdot t, \tag{3.23}$$

---

* It has been estimated that flying micro objects covered with such "skin" can reach speed of 20M (M – Mach number) in earth's atmosphere.

† Grove-Deal effect: a strongly curved silicon surface oxidizes slower than a flat one.

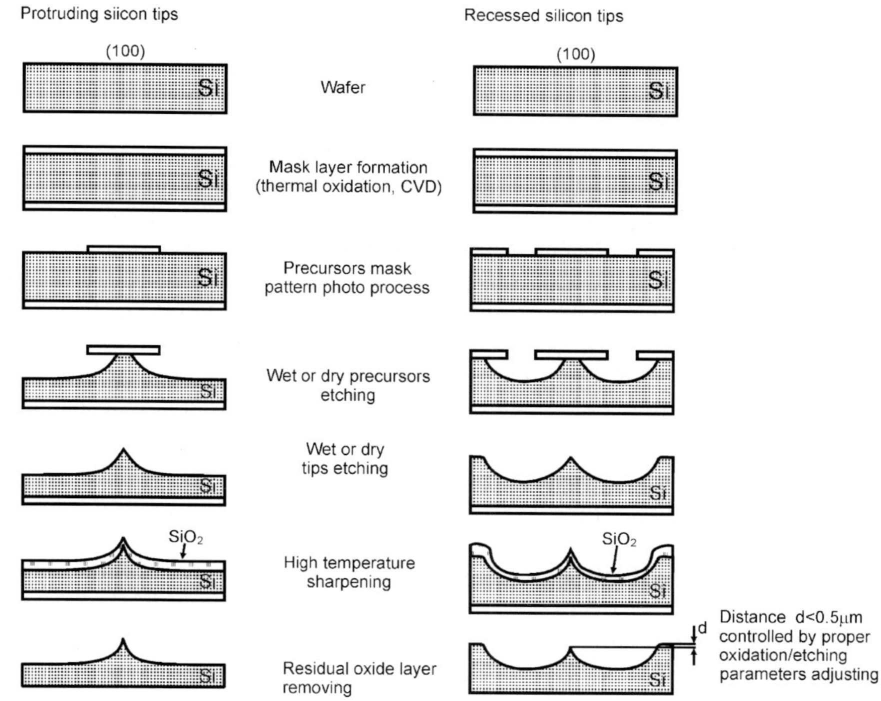

Fig. 3.98. Fabrication of tips (scheme); on the left a convex tip, on the right a concave tip.

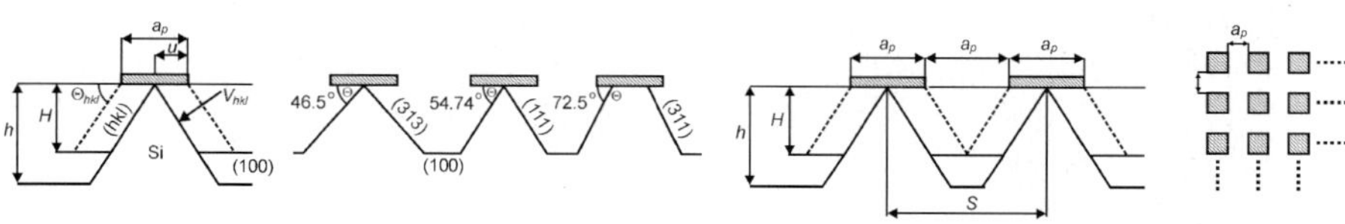

Fig. 3.99. Geometry of convex tips on a (100) substrate.

where: $V_{\text{hkl}} =$ etch rate of (hkl) plane, $\Theta_{\text{hkl}} =$ angle between the (hkl) plane and surface of the wafer, $t =$ etching time, and because

$$\cos \Theta_{\text{hkl}} = \frac{H}{\sqrt{(h^2 + k^2 + l^2)}}, \tag{3.24}$$

thus

$$a_{\text{p}} = 2V_{\text{hkl}} = \frac{H}{\sqrt{(h^2 + k^2 + l^2)}} \cdot t. \tag{3.25}$$

The height of the tip can be calculated from equation (3.26):

$$h = V_{100} \cdot t, \tag{3.26}$$

or from formula (3.27):

$$h = \frac{a_p}{2} \cdot \frac{V_{100}}{V_{hkl}} \cdot \frac{\sqrt{(h^2 + k^2 + l^2)}}{H}. \tag{3.27}$$

Since the etch rate of crystallographic planes, that appear during the etching of precursor islands in KOH or in KOH and IPA, is dependent on the composition of the solution, it is possible to "tune" the tip geometry with nanometric precision by adjusting the etching conditions only.

The minimal array mesh, that forms a mask for etching the tip array, results from the arrangement of (111) planes. The step between masking squares equals twice the length of its edge. Hence, for the tips with expected height equaling 2.8 μm the array of 250^2 tips per 1 mm^2 can be obtained.

The other method of array of silicon protruding tip manufacturing has been discussed in paper [207]. In this method the first technological step consisted in forming tip precursors (Fig. 3.100) by means of silicon plasma etching for 5 minutes in $SF_6 + Cl_2$. Applied flows equaled 24 sccm and 65 sccm, RF power $P_{RF} = 150$ W, pressure $p = 100$ Pa, plasma potential $U_{DC} = 260$ V. At this technological stage an adequately high etching selectivity of the mask against silicon was maintained, because a too-low selectivity of etching would lead to defects in the mask (the toreador's effect in Fig. 3.104b). Following this, precursors were

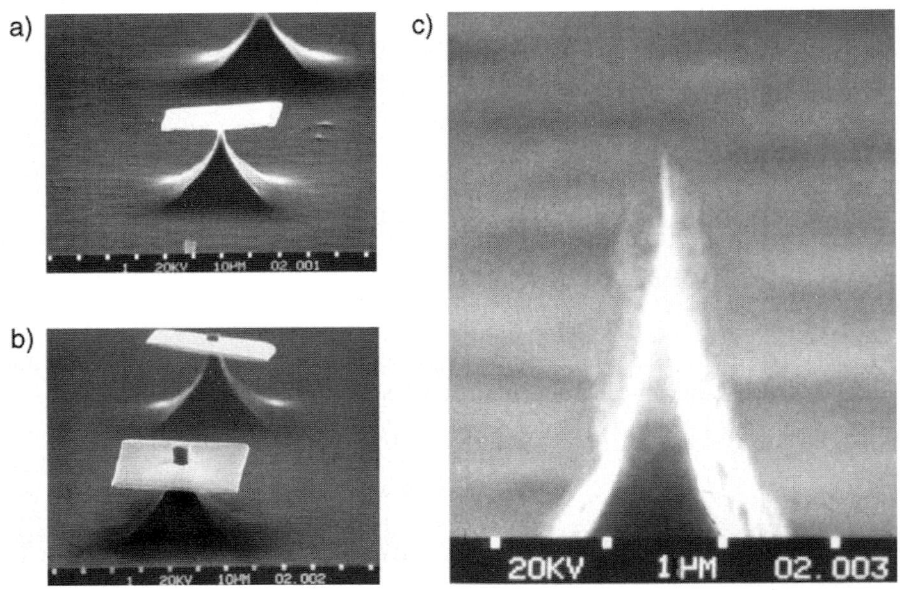

Fig. 3.100. Precursors and protruding tips wet–dry etched on a (100) wafer: a) precursor etched through a square mask, b) the toreador's effect on the micro scale, c) super sharp ($r < 10$ nm) silicon tip – the thin layer of silicon oxide sharpening the tip can be seen. Courtesy of dr Anna Górecka-Drzazga from the Faculty of Microsystem Electronics and Photonics of Wrocław University of Technology.

etched in NHA, at 20°C, for 30 seconds. Then the residual oxide mask was removed in a BHF solution and the tips were sharpened for 20 minutes by wet thermal oxidation in steam at 1050°C.

Combined procedures of wet and dry etching can be applied to produce a silicon cathode with an extracting gate of the FEA-G-type (***Field Emitter Array-Gated***) (Fig. 3.101). The first technological step is a sandwich of SiO_2/Al thin-film layers forming; next plasma etching of etch windows in $SF_6 + Cl_2$ (conditions of the process and utilized device were as described above). Following this, the precursors are etched in a NHA solution for 5 minutes at 20°C, cleaned in BHF and washed in deionized water. Finally, tips are sharpened in steam for 30 minutes at 1050°C. Then about 1 μm thick silicon dioxide is PECVD deposited on the finished tips. After that a sandwich of chromium and gold layers (0.12/0.8 μm thick respectively) is deposited and an access hole is selectively etched. As a result, each silicon tip is surrounded by a round-shaped metal gate insulated from silicon by a SiO_2 spacer (Fig. 3.101b,c).

The complex process of fabrication of a silicon beam with a tip localized at the free movable end (Fig 3.102), uses at least three anisotropic etching processes, followed by wet and dry isotropic etching [208].

Manufacturing of this device starts from thick wet thermal oxidizing of the (100) silicon wafer, photo patterning and 10M KOH, 55°C back-side etching of the 30 μm-thick membrane. After the next oxide formation and photo-process,

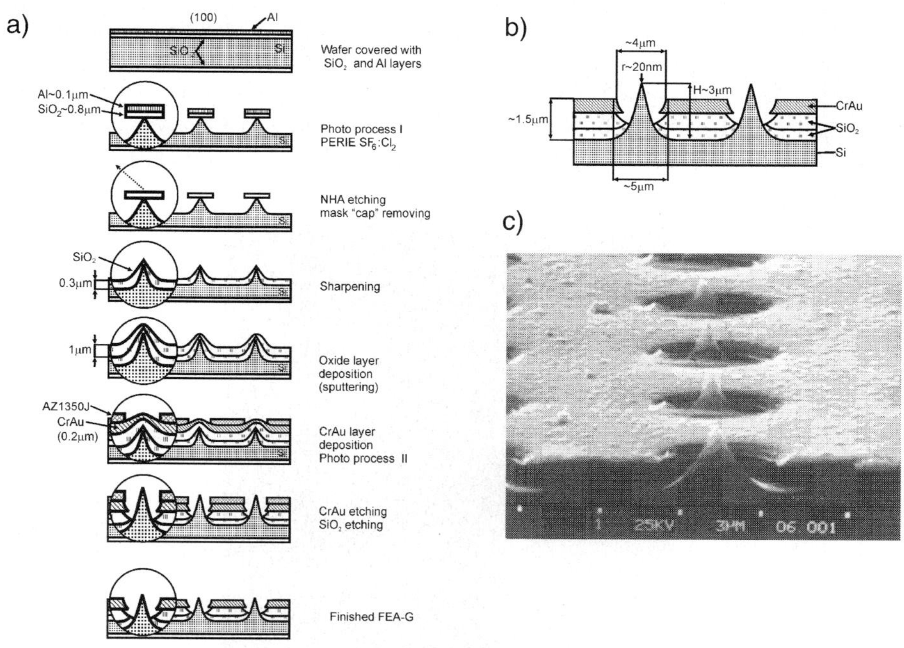

Fig. 3.101. Silicon tip array with an extracting gate – FEA-G: a) fabrication method, b) cross-section picture, c) FEA-G $Si/SiO_2/CrAu$, 110^2 tips per 1 mm^2 [207].

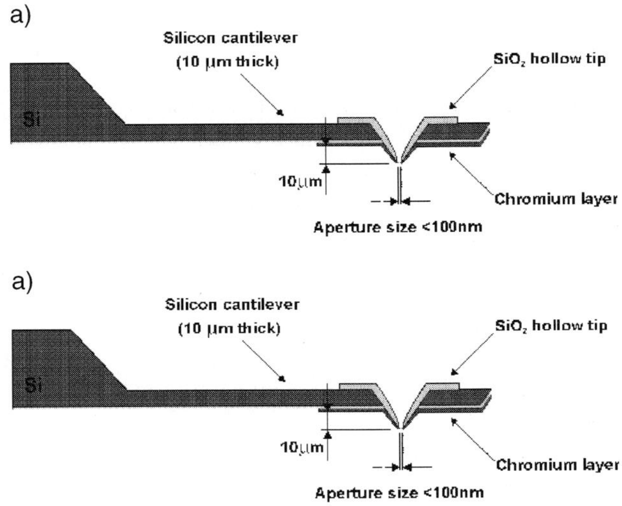

Fig. 3.102. Silicon beam with a tip: a) cross-section, b) whole structure.

a 20 µm deep reversed pyramid is back-side etched in KOH (5–7M, 80°C). The wafer is thickly oxidized (sharpening procedure), photo-processed again and the pattern of beam is pre-formed in 10M KOH at 55°C (etch depth 10 µm). Following this, a Cr 1 µm-thick layer is back-side magnetron-sputtered and patterned. Next, the thickness of the wafer is decreased by dry SF_6 etching. This process is stopped when the summit of the protruding tip, formed from SiO_2, is about 10 µm above a beam surface. Finally, a small hole is formed at the summit by shallow etch of SiO_2 in buffered HF, the old Cr layer is removed and a new one is back-side deposited and the device is finished (Fig. 3.103).

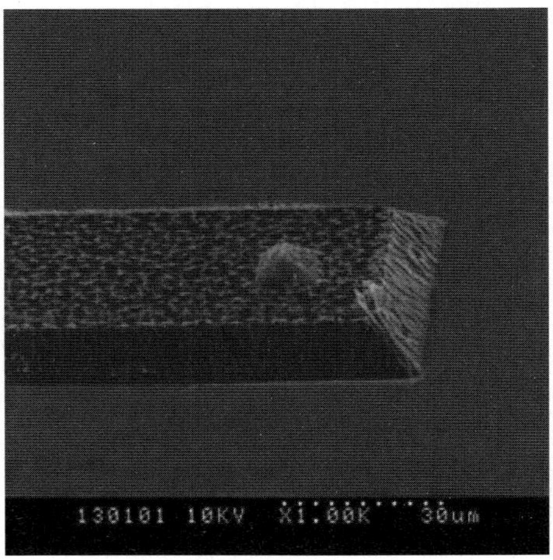

Fig. 3.103. Silicon beam with pyramid-like SiO_2 tip.

Arrays of protruding tips with extracting gates are being intensively researched as future electron sources in flat FED-type displays, power switches, microwave generators, etc. Beams with tips are commonly applied in tunneling (STM) and atomic force scanning microscopes (AFM), and scanning near-field optical microscopes (SNOM). These issues are widely reported in the literature of the subject. Although these problems are fascinating from both informative and utilitarian points of view, they will not be discussed further here, because they are outside the scope of this book.

3.4. LITERATURE

[1] K. Petersen, Silicon as a mechanical material. *Proceed. IEEE, El. Dev.*, 70(5), 1982, 420–457.

[2] J.B. Agnell, S.C. Terry, P.W. Barth, Silicon micromechanical devices. *Sc. Am.*, 44, 1983, 44–54.

[3] J. Bryzek, Impact of MEMS technology on society. *Sensors and Actuators A*, 56, 1996, 1–9.

[4] J.C. Greenwood, Silicon in mechanical sensors. *J. Phys.*, 21, 1988, 1114–1128.

[5] K. Petersen, Dynamic micromechanics on silicon, techniques and devices. *IEEE Tran. El. Dev.*, ED-25, 1978, 1241–1250.

[6] G.L. Pearson, Deformation and fraction of small silicon crystals. *Act. Metalurgica*, 5, 1957, 187–191.

[7] J. Wolf, *Silicon Semiconductors Data.* Pergamon Press, New York, 1969.

[8] J. Wortman, R.A. Evans, Young's modulus, shear modulus and Poisson's ratio in silicon and germanium. *J. Appl. Phys.*, 36, 1965, 153–156.

[9] J. Frühauf, E. Gärtner, E. Jänsch, Silicon as a plastic material. *J. Micromech. Microeng.*, 9, 1999, 305–312.

[10] H.J. McSkimin, P. Andreatch, Elastic modulus of silicon vs. hydrostatic pressure at 25°C and 195.8°C. *J. Appl. Phys.*, 35, 7, 1964, 2161–2165.

[11] S.P. Nikanarov, Yu.A. Burenkov, A.V. Stepanov, Uprugije swojstwa kriemni. *Fiz. Twierd. Tieła*, 13, 1971, 3001–3004.

[12] X. Ding, W.H. Ko, J.M. Mansour, Residual stress and mechanical properties of boron doped p^+ silicon films. *Sensors and Actuators A*, 21–23, 1990, 866–871.

[13] H.J. McSkimin, Measurement of elastic constants at low temperatures by means of ultrasonic waves – data for silicon and germanium single crystals and for fused silica. *J. Appl. Phys.*, 24, 8, 1953, 988–997.

[14] J.J. Hall, Electronic effects in elastic constants of n-type silicon. *Phys. Rev.*, 161(3), 1967, 756–761.

[15] F. Maseeh, S.D. Senturia, Plastic deformation of highly doped silicon. *Sensors and Actuators, A*, 21–23, 1990, 861–865.

[16] W. Chu, Effect of oxidation on residual stress distribution through the thickness of p^+ silicon films. *Techn. Dig. IEEE Solid-State Sensors and Actuators Workshop*, 1982, 90–93.

[17] L. Csepregi, Micromechanics: a silicon microfabrication technology. *Microelectronics Eng.*, 3, 1985, 221–234.

[18] M. Mehregany, Silicon microactuators, in: *Advances in actuators*, A. Dorey, J. Moore (eds.). IOP Publishing, 1995, 135–178.

[19] X. Ding, W.H. Ko, Y. Niu, W. He, A study on silicon-diaphragm buckling. *Proceed. IEEE Transducers*, 1990, 128–131.

[20] J. Bryzek, K. Petersen, J. Mallon, L. Christell, F. Pouhmaradi, *Silicon sensors and microstructures.* Nova Sensor, Fremont, CA, 1991, USA.

[21] Special issue on three dimensional device structures *IEEE Tran. El. Dev.*, ED-25, 10, 1975.

[22] K.D. Wise, S.K. Clark, Diaphragm formation and pressure sensitivity in batch-fabricated silicon pressure sensor. *Int. Electron. Dev. Meeting (IEDM), Techn. Dig., IEEE*, 1978, 96–99.

[23] A.I. Stoller, The etching of deep vertical patterns in silicon. *RCA Review*, 6, 1970, 271–275.
[24] Special issue on solid-state sensors, actuators and interface electronics. *IEEE Trans. El. Dev. ED-26*, 12, 1979.
[25] B. Puers, E. Peeters, W. Sansen, Wet and dry etching experience for sensor micromachining. *J. Micromech. Microeng.*, 13, 1991, 443–446.
[26] G. Delapierre, Micro-machining: A survey of the most commonly used processes. *Sensors and Actuators A*, 17, 1989, 123–138
[27] J.M. Crishal, A.L. Harrington, A selective etch for elemental silicon. *J. Electrochem. Soc.*, 109, 1962, 71C.
[28] R.M. Finne, D.L. Klein, A water amine complexing agent system for etching silicon. *J. Electrochem. Soc.*, 114, 1967, 965–970.
[29] H.J. Declerg, L. Gerzberg, J.D. Meindl, Optimization of the hydrazine water solution for anisotropic etching of silicon in integrated circuit technology. *J. Electrochem. Soc.*, 122, 1979, 545–552.
[30] M. Mehregany, S.D. Senturia, Anisotropic etching of silicon in hydrazine. *Sensors and Actuators A*, 13, 4, 1988, 375–390.
[31] Yong Ping Xu, Ruey Shing Huang, Anodic dissolution and passivation of silicon in hydrazine. *J. Electrochem. Soc.*, 137, 1990, 948–953.
[32] J. C. Greenwood, Ethylene diamine-catechol-water mixture shows preferential etching of p-n junctions. *J. Electrochem. Soc. Electrochem. Techn. Lett.*, 116, 9, 1969, 1325–1326.
[33] A. Reisman, M. Berkenblit, S.A. Chan, F.B. Kaufman, D.C. Green, The controlled etching of silicon in catalyzed ethylenodiamine-pyrocatechol-water solutions. *J. Electrochem. Soc.*, 126, 1979, 1406–1415.
[34] A. Bough, Ethylenodiamine-pyrocatehol mixtures-water mixtures shows etching anomaly in boron-doped silicon. *J. Electrochem. Soc.*, 118, 1971, 401.
[35] N.F. Raley, Y. Sugiyama, T. van Duzer, (100) silicon etch-rate dependence on boron concentration in ethylene-diamine-pyrocatechol-water solutions. *J. Electrochem. Soc.*, 131, 1984, 161–171.
[36] G. Kaminsky, Micromachining of silicon mechanical structures. *J. Vac. Sci. Technol.*, B-3, 1985, 1015–1024.
[37] H. Linde, L. Austin, Wet silicon etching with aqueous-amine gellates. *J. Electrochem. Soc.*, 139, 1992, 1170–1174.
[38] U. Schnakenberg, W. Benecke, P. Lange, TMAH etchants for silicon micromachining. Tech. Digest, 6th Int. Conf. Solid St. Sensors and Actuators (Transducer 91) San Francisco CA, USA, 1991, 815–818.
[39] A. Merlos, M.C. Acero, M.H. Bao, J. Bausells, J. Estere, TMAH/IPA anisotropic etching characteristic. *Sensors and Actuators A*, 37–38, 1993, 737–743.
[40] H. Linde, L. Austin, Catalytic control of anisotropic silicon etching. *Sensors and Actuators A*, 49, 1995, 181–185.
[41] A. Merlos, M.C. Acero, M.H. Bao, J. Bausells, J. Estere, A study of undercutting characteristics in the TMAH-IPA system. *J. Micromech. Microeng.*, 2, 1992, 181–183.
[42] I. Zubel, M. Kramkowska, The effect of isopropyl alcohol on etching rates roughness of (100) Si surface etched in KOH and TMAH. *Sensors and Actuators A*, 93, 2001, 138–147.
[43] O. Tabata, R. Asahi, H. Funabashi, S. Sugiyama, Anisotropic etching of silicon in $(CH_3)_4OH$ solutions. Tech. Digest, 6th Int. Conf. Solid St. Sensors and Actuators (Transducer 91) San Francisco CA, USA, 1991, 811–814.
[44] P.J. Holmes, *The Electrochemistry of Semiconductors*, P.J. Holmes (ed.). Academic Press Ltd., London, 1962.
[45] D.B. Lee, Anisotropic etching of silicon. *J. Appl. Phys.*, 40, 1969, 4569–4574.
[46] J.B. Price, Anisotropic etching of silicon with KOH-H_2O-isopropyl alcohol. In *Semiconductor silicon*. The Electrochemical Soc. Softbond Symposium Proceed, Series, Princetown, New York, USA, H.R. Huft, R.R. Burgess (eds.), 1973, 339–353.
[47] K.E. Bean, Anisotropic etching of silicon. *IEEE Tran. El. Dev.*, ED-25, 1978, 1185–1193.

[48] E. Bassous, Fabrication of novel three dimensional microstructures by the anisotropic etching of (100) and (111) silicon. *IEEE Tran. El. Dev.*, ED25, 1978, 1178–1185.
[49] J.K. Kendall, Vertical etching of silicon at very high aspect ratios. *Ann. Rev. Materials Scientific*, 9, 1979, 373–378.
[50] W.H. Ko, D.G. Fleming, T.L. Poteat, Submicron accuracies in anisotropic etched silicon piece parts – a case study. In: *Micromachining and Micropackaging of Transducers*, C.D. Fung (ed.). Elsevier, 1985, 151–158.
[51] H. Seidel, L. Csepregi, A. Hauberger, H. Baumgartel, Anisotropic etching of crystalline silicon in alkaline solutions (Part I). *J. Electrochem. Soc.*, 137, 1990, 3612–3626.
[52] E. Herr, H. Baltes, KOH etch rates of high index planes in silicon. *Sensors and Actuators A*, 31, 1992, 283–287.
[53] H. Camon, Z. Moktadir, Atomic scale simulation of silicon etched in aqueous KOH solution. *Sensor and Actuators A*, 46–47, 1995, 25–28.
[54] H.L. Offereins, K. Kuhl, H. Sandmaier, Methods of fabrication of convex corners in anisotropic etching of (100) silicon in aqueous KOH. *Sensors and Actuators A*, 25–27, 1991, 9–13.
[55] I. Barycka, I. Zubel, Silicon anisotropic etching in KOH-isopropanol etchant. *Sensors and Actuators A*, 48, 1995, 229–238.
[56] E.D. Palik, H.F. Gray, P.B. Klein, A Raman study of etching silicon in aqueous KOH. *J. Electrochem. Soc.*, 130, 1983, 956–959.
[57] U. Schnekenberg, W. Benecke, B. Löckel, S. Ullerich, P, Lange, NH_4OH based etchants for silicon micromachining: influence of additives and stability of passivation layers. *Sensors and Actuators A*, 25–27., 1991, 1–7.
[58] U. Schnekenberg, W. Benecke, B. Löckel, NH_4OH-based etchants for silicon micromachining. *Sensors and Actuators A*, 21–23, 1990, 1031–1035.
[59] Ch. Ju, P.J. Hesketh, Measurements of anisotropic etching of a single crystal silicon sphere in aqueous cesium hydroxide. *Sensors and Actuators A*, 33, 1992, 191–196.
[60] T. Wang, S. Surve, P.J. Hesketh, Anisotropic etching of silicon in rubidium hydroxide. *J. Electrochem. Soc.*, 141(9), 1994, 2493–2497.
[61] M. Elwespoek, The form of the etch minima in wet chemical anisotropic etching of silicon. *J. Micromech. Microeng.*, 6, 1996, 405–409.
[62] P. Allongue, V. Costa-Kieling, H. Gerischer, Etching of silicon in NaOH solutions. *J. Electrochem. Soc.*, 140, 1993, 1009–1026.
[63] J. Dziuban, Microwave enhanced wet anisotropic etching of monocrystalline silicon. Proceed. Eurosensors XIII, 13rd Europ. Conf. on Solid-St. Transducers, September 12–15, den Haag, Holland, 1999, 337–338.
[64] H. Linde, L. Austin, Catalytical control of anisotropic silicon etching. *Sensors and Actuators A*, 49, 1995, 181–187.
[65] C. Moldovan, R. Iosub, D. Dascaln, G. Nechifor, Anisotropic etching of silicon in a complexant redox alkaline system. *Sensors and Actuators B*, 58, 1999, 438–449.
[66] K. Sato, M. Shikida, T. Yamashiro, M. Tsukanawa, S. Ito, Characterization of anisotropic etching of single crystal silicon: surface roughening as function of crystallographic orientation. IEEE Workshop MEMS 98, January 25–29, Heidelberg, Germany, 201–206.
[67] Ch. Merveille, Surface quality of (111) side-walls in KOH etched cavities. *Sensors and Actuators A*, 60, 1997, 2444–248.
[68] E.D. Palik, O.J. Glembocki, L. Heard jr, P.S. Burno, L. Tenerz, Etching roughness for (100) silicon surfaces in aqueous KOH. *J. Appl. Phys.*, 70(6), 1991, 3291–3300.
[69] Y.K. Bhatnager, A. Nathan: On pyramidal pretrusions in anisotropic etching of (100) silicon. *Sensors and Actuators A*, 36, 1993, 233–240.
[70] S.A. Campbell, K. Cooper, S.N. Port, D.J. Schiffrin, Inhibition of pyramid formation in the etching of Si p (100) in aqueous potassium hydroxide-isopropanol. *J. Micromech. Microeng.*, 5, 1995, 209–218.
[71] T.A. Kwa, R.F. Wolfenbuttel, Effect of solution contamination on etched silicon surfaces. *J. Micromech. Microeng.*, 5, 1995, 95–97.
[72] C. Merveille, J. Weber, Surface quality of (111) side-walls in KOH solutions. Proceed.

Eurosensors X, 10th Europ. Conf. on Solid-St. Transducers, Sept., 8–11, 1996, Leuven Belgium, 485–488.

[73] E.D. Palik, V.M. Bermudez, O.J. Glembocki, Ellipsometric study of orientation-dependent etching of silicon in aqueous KOH. *J. Electrochem. Soc.*, 132, 1985, 871–994.

[74] P.J. Hesketh, Ch. Ju, S. Gowda, E. Zanoria, S. Danyluk, Surface free energy model of silicon anisotropic etching. *J. Electrochem. Soc.*, 140, 1993, 1080–1085.

[75] M. Elwenspoek, On the mechanism of anisotropic etching of silicon. *J. Electrochem. Soc.*, 140, 1993, 2075–2080.

[76] C. Mihalcea, A. Hőlz, M. Kuhawara, J. Tominaga, E. Oesterschultze, N. Atada, Improved anisotropic deep etching in KOH solutions to fabricate highly speculative surfaces. *Microel. Engineer.*, 57–58, 2001, 781–786.

[77] I. Zubel, *Three dimensional silicon structures anisotropic (wet) etching for microelectronic applications*. Oficyna Wydawnicza Politechniki Wroclawskiej, Wroclaw, Poland, 2004,.

[78] I. Barycka, H. Teterycz, Z. Znamirowski, Sodium hydroxide solution shows selective etching of boron doped silicon. *J. Electrochem. Soc.*, 126(2), 1979, 345–346.

[79] V. Lehmann, K. Miatui, D. Feijo, U. Gösele, Implanted carbon, an effective etch-stop in silicon. *J. Electrochem. Soc.*, 138, 1991, L-3–L-4.

[80] M.C. Acero, J. Esteve, J. Montserrat, J. Bausells, A. Perez-Rodriguez, A. Romano-Rodriguez, J.R. Monante, Anisotropic etch-stop properties of nitrogen implanted silicon. *Sensors and Actuators A*, 45, 1994, 219–225.

[81] H. Seidel, L. Csepregi, A. Henberger, H. Baumgartel, Anisotropic etching of silicon in alkaline solutions (Part II). *J. Electrochem. Soc.*, 137, 1990, 3626–3632.

[82] E.D Palik, J.W. Faust jr, H.F. Gray, R.F. Greene, Study of the etch-stop mechanism in silicon. *J. Electrochem. Soc.*, 129, 1982, 2051–2059.

[83] E.D. Palik, V.M. Bermudez, O.J. Glembocki, Ellipsometric study of the etch stop mechanism in heavily doped silicon. *J. Electrochem, Soc.*, 132, 1985, 135–141.

[84] E.D. Palik, V.M. Bermudez, O.J. Glembocki, Ellipsometric study of bias dependent etching and the etch-stop mechanism for silicon in aqueous KOH. In: *Micromachining and Micropackaging of Transducers*, W.H. Ko, D.G. Fleming (eds.). Elsevier, 1985.

[85] H.A. Waggener, Electrochemically controlled thinning of silicon. *Bell. System Techn. J.*, 50, 1970, 473–475.

[86] M.J.J. Theusninssen, J.A. Appels, W.H.C.G. Yerhuylen, Application of preferential electrochemical etching of silicon to semiconductor device technology. *J. Electrochem. Soc.*, 117, 1970, 959–965.

[87] H.J.A. von Dijk, J. de Jonge, Preparation of thin silicon crystals by electrochemical thinning of epitaxially grown structures. *J. Electrochem. Soc.*, 117, 1970, 553–554.

[88] R.L. Meck, Electrochemically thinned N/N^+ epitaxial silicon method and application. *J. Electrochem. Soc.*, 118, 1971, 1240–1246.

[89] A.I. Stoller, R.F. Speers, S. Opresko, A new technique for etch thinning of silicon wafers. *RCA Review*, 1970, 265–270.

[90] T.N. Jackson, M.A. Tischler, K.D. Wise, An electrochemical p-n junction etch-stop for the formation of silicon microstructures. *IEEE Trans. El. Dev. EDL-2*, 2, 1981, 44–45.

[91] J.W. Faust Jr, E.D. Palik, Study of the orientation dependent etching and initial anodization of Si in aqueous KOH. *J. Electrochem. Soc.*, 130, 1983, 1413–1420.

[92] J. Glembocki, R.E. Stahibush, M. Tomkiewicz, A bias-dependent etching of silicon in aqueous KOH. *J. Electrochem. Soc.*, 132, 1985, 145–151.

[93] P.M. Sarro, A.W. van Herwaarden, Silicon cantilever beams fabricated by electrochemically controlled etching for sensor fabrication. *J. Electrochem. Soc.*, 133, 1986, 1724–1728.

[94] R.L. Smith, B. Kloeck, N.F. de Roij, S.D. Collins, The potential dependence of silicon anisotropic etching in KOH in 60°C. *J. Electroanal. Chem. and Interfacial Chem.*, 238, 1987, 103–113.

[95] M. Hirata, S. Suwanoro, H. Tanigawa, Diaphragm thickness control in silicon pressure sensors using an anodic oxidation etch stop. *J. Electrochem. Soc.*, 134, 1987, 2037–2041.

[96] M. Hirata, K. Suzuki, H. Tanigawa, Silicon diaphragm pressure sensors fabricated by anodic oxidation etch-stop. *Sensor and Actuators A*, 13, 1988, 63–69.
[97] Y. Lindenen, L. Tenerz, B. Hök, Fabrication of three dimensional structures by means of doping selective etching (DSE). *Sensors and Actuators A*, 16, 1989, 67–81.
[98] B. Kloek, S.D. Collins, N.F. de Roij, R.L. Smith, Study of electrochemical etch-stop for high precision thickness control of silicon membranes. *IEEE Trans. El. Dev.*, ED-36, 1989, 663.
[99] E.D. Palik, O.J. Glembocki, R.E. Stahibush, Fabrication and characterization of Si membranes. *J. Electrochem. Soc.*, 135, 1988, 3126–3134.
[100] H. Seidel, The mechanism of anisotropic electrochemical silicon etching in alkaline solutions. Techn. Dig. 5th Int. Conf. Solid St. Sensors and Actuators (Transducers 90), Yokohama, Japan, 1990, 86–87.
[101] H. Seidel, The mechanism of electrochemical anisotropic etching of silicon and its application. In: *Integrated Micromotion Systems*, F. Harashima (ed.). Elsevier, 1990, 51–68.
[102] D. Lapadatu, R. Puers, On the anodic passivation of silicon in aqueous KOH solutions. *Sensor and Actuators A*, 60, 1997, 191–196.
[103] G.K. Celler, L.E. Trimble, J. Frackoviak, C.W. Jergensen, R.R. Kola, A.E. Nivembre, G.R. Weber, Formation of monolithic masks for 0.25 μm X-Ray lithography. *Appl. Phys. Lett.*, 59(24), 1991, 3105–3107.
[104] E.L. Demar, C.M.A. Ashruf, P.J. French, P.M. Sarro, Thickness of membranes fabricated with galvanic etch-stop: uniformity and reproducibility. Proceed. Eurosensors XII, 12th Europ. Conf. on Solid-St. Transducers, 13–16 Sept., Southampton, United Kingdom, M.N. White (ed.), IOP Series 1998, 3–6.
[105] H. Teterycz, J. Dziuban, R. Walczak, *Method of silicon etching*. Patent application 353673, 29 April 2002.
[106] J. Dziuban, Microwave enhanced wet anisotropic etching of monocrystalline silicon. *Sensors and Actuators A*, 85, 2000, 133–138.
[107] J. Dziuban, R. Parosa, E. Reszke, A means for wet anisotropic etching of mono-crystalline semiconductor material and a device implementing this means. PCT No WO 00/34993, 15 June 2000.
[108] J. Dziuban, R. Walczak, Fast wet anisotropic etching process for deep micromachining of single crystal silicon. World Microtech. Cong. Proceed. MICRO., tec 2000, Expo 2000, Hannover, Germany, 609–613.
[109] J. Dziuban, R. Walczak, Etching microwave silicon [EMSi]-microwave enhanced fast deep anisotropic etching of silicon for micro-electromechanical systems. *Sensors and Materials*, 15(1), 2001, 41–55.
[110] R. Walczak, J. Dziuban, Microwave enhanced wet anisotropic etching of silicon utilizing a memory effect of OH activation – a remote E2MSi process. *Sensors and Actuators A*, 116, 2004, 161–170.
[111] H.R. Robbins, B.Schwartz, Chemical etching of silicon – I. The system, HF, HNO_3 and H_2O, *J. Electrochem. Soc.*, 106, 1959, 505–508.
[112] H.R. Robbins, B. Schwartz, Chemical etching of silicon – II. The system, HF, HNO_3, H_2O and $HC_2C_3O_2$, *J. Electrochem. Soc.*, 107, 1960, 108–111.
[113] B.Schwartz, H.R. Robbins, Chemical etching of silicon – III. A temperature study in the acid system, *J. Electrochem. Soc.*, 108, 1961, 365–372.
[114] B. Schwartz, H.R. Robbins, Chemical etching of silicon – IV. Etching technology, *J. Electrochem. Soc.*, 123, 1976, 1903–1909.
[115] A. Bochenschuetz, W. Krusemark, K. Loeher, W. Mussinger, Activation energies in the chemical etching of semiconductors in HNO_3–HF–CH_3COOH. *J. Electrochem. Soc.*, 114(9), 1976, 970–973.
[116] L. Nieradko, *Microelectronic methods of modification of separation properties of micromechanical, capillary chromatographic columns*. Thesis, The Wroclaw University of Technology, Wroclaw, Poland, 2001, 43–51
[117] B. Puers, W. Sansen, Compensation structures for convex corner micromachining in silicon. *Sensors and Actuators A*, 21–23, 1990, 1036–1039.

[118] H.L. Offereins, H. Sandmaier, K. Maruscyk, K. Kuhl, A. Plettner, Compensating corner undercutting of (100) silicon in KOH. *Sensors and Materials*, 3, 1992, 127–144.
[119] M.M. Abu-Zeid, Corner undercutting in anisotropically etched isolation contours. *J. Electrochem. Soc.*, 131, 1984, 2138–2142.
[120] E.G. van Hal, *Advanced packaging of ISFETS: design, encapsulation and bonding*, chapter 5: *Anisotropic etching of silicon in KOH and TMAH solutions*. Thesis, 1994, 99–113.
[121] M. Bao, Chr. Burrer, J. Estere, J. Baussels, S. Marco, Etching front control of ⟨110⟩ strips for corner compensation. *Sensors and Actuators A*, 37–38, 1993, 727–732.
[122] J. Dziuban, A. Górecka-Drzazga, I. Barycka, I. Zubel, Compensation of corners in bossed structures of pressure sensors. Proceedings of IVth Conf. COE 96, 13–16 May, Szczyrk, Poland, 1996, 174–177.
[123] D. Zielke, J. Fruhauf, F. Rößler, Simulation of the orientation dependent etching of complex mask structure. Tech. Dig. 6th Micromechanics Europe Workshop MME 95, Copenhagen, Danmark, 3–5 Sept., 1995, 164–167.
[124] I. Barycka, J. Dziuban, M. Kramkowska, I. Zubel, Compensation of convex corners in sensors with bossed structure etched in TMAH and TMAH/IPA solutions. *Mat. SPIE*, 4516, 2001, 56–65.
[125] J. Frühauf, K. Trautman, J. Wittig, D. Zieike, A simulation tool for orientation dependent etching. *J. Micromech. Microeng.*, 3, 1993, 113–115.
[126] D. Zielke, J. Frühauf, Determination of rates for orientation-dependent etching. *Sensors and Actuators A*, 48, 1995, 151–156.
[127] I. Barycka, I. Zubel, Silicon anisotropic etching in alkaline solutions I. The geometric description of figures developed under etching Si (100) in various solutions. *Sensors and Actuators A*, 70, 1998, 250–259.
[128] G.K. Mayer, H. Loffereins, A. Sandmeier, K. Kuhl, Fabrication of non-underetched convex corners in anisotropic etching of (100) silicon in aqueous KOH with respect to novel micromechanical devices. *J. Electrochem. Soc.*, 137, 1990, 3947–3951.
[129] E. Herr, H. Baltes, KOH etch rates of high-index planes from mechanically prepared silicon crystals. Tech. Dig. 6th Int. Conf. Solid St. Sensors and Actuators (Transducers 91), San Francisco, CA, USA, 24–28 June, 1995, 807–810.
[130] Xin Xin Li, M. Bao, S. Shen, Maskless etching of three-dimensional silicon structures in KOH. *Sensors and Actuators A*, 57, 1996, 47–52.
[131] I. Zubel, Silicon anisotropic etching in alkaline solution II. On the possibility of spatial structures forming on the course of Si (100) anisotropic etching in KOH and KOH + IPA solutions. *Sensors and Actuators A*, 84, 2000, 116–125.
[132] O. Than, S. Büttgenbach, Simulation of anisotropic chemical etching of crystalline silicon using a cellular automata model. *Sensors and Actuators A*, 1995, 85–89.
[133] J.H. Jerman, The fabrication and use of micromachined corrugated membrane. *Sensors and Actuators A*, 23, 1990, 998–992.
[134] D. Lapadatu, A. Pyka, J. Dziuban, R. Puers, Corrugated silicon nitride membranes on suspensions in micromachined silicon accelerometers. *J. Micromech. Microeng.*, 6, 1996, 73–76.
[135] C.S. Smith, Piezoresistance effect in germanium and silicon. *Phys. Rev.*, 94, 1954, 42–49.
[136] T. Pancewicz, R. Jachowicz, Z. Gniazdowski, Z. Ażgin, P. Kowalski, The empirical verification of the FEM model of semiconductor pressure sensor. *Sensors and Actuators A*, 76, 1999, 260–265.
[137] M. Bao, *Micro Mechanical Transducers, Pressure Sensors, Accelerometers and Gyroscopes*. Elsevier, 2000.
[138] M. Bao, W. Qi, Y. Wang, Geometric design rules of four terminal gauge for pressure sensor. *Sensors and Actuators A*, 18, 1989, 149–156.
[139] N. Lu, L. Gerzberg, C. Lu, J. Meindl, Modeling and optimalization of monolithic polycrystalline silicon resistors. *IEEE Trans. on Electron Dev.*, ED-28, 1981, 818–830.
[140] J. Dziuban, A. Górecka-Drzazga, U. Lipowicz, Silicon pressure sensor for biomedicine

applications. Proceed. Microelectronic 92, Warsaw, Poland, 1992, Proc. SPIE, Vol. 1783, 32–328.

[141] J. Dziuban, A. Górecka-Drzazga, U. Lipowicz, W. Indyka, J. Wąsowski, Self-compensating piezoresistive pressure sensor. *Sensors and Actuators A*, 41–42, 1994, 368–374.

[142] J. Dziuban, A. Górecka-Drzazga, J. Wąsowski, U. Lipowicz, Simple method of compensation of off-set voltage of piezoresistive pressure sensors and its temperature drift (in Polish). Proceed. of COE 94, Zegrze, Poland, 1994, 235–238.

[143] J. Thomas, R. Kühnlod, R. Schunpp, H. Ryssel, A silicon vibration sensor for tool state monitoring working in the high acceleration range. *Sensors and Actuators, A*, 85, 2000, 194–201.

[144] K. Ikeda, Silicon pressure sensor integrates resonant strain gage on diaphragm. Proceed. Transducers '89 and Eurosensors III, the 5th Int. Conf. On Solid-St. Sensors and Actuators, Montreux, Switzerland, June 25–30, 1989, 100–101.

[145] Catalogue of Yokogawa Co., The first and single digital pressure transducer. Yokogawa Austria GmbH, Franzensbruckestrasse 26, P.O.B 159, A 1021 Wien.

[146] J. Dziuban, A. Górecka-Drzazga, U. Lipowicz, Silicon optical pressure sensor. *Sensors and Actuators A*, 32, 1992, 628–631.

[147] W. Lang, K. Kuehl, A micro-thin bolometer for radiation thermometry of ambient temperature. Proceed. Transducers '89 and Eurosensors III, the 5th Int. Conf. On Solid-St. Sensors and Actuat. Montreux, Switzerland, June 25–30, 1989, 173.

[148] G.A. Racine, Low temperature operating silicon bolometers for nuclear radiation detector. Proceed. Transducers '89 and Eurosensors III, the 5th Int. Conf. On Solid-St. Sensors and Actuat. Montreux, Switzerland, June 25–30, 1989, 174–175.

[149] A.W. Herwaarden, Floating-membrane thermal vacuum sensor. *Sensors and Actuators A*, 17, 1989, 259–267.

[150] J.S. Skie, P.K. Weng, Fabrication of micro-bolometer on silicon substrate by anisotropy etching technique. Proc. Transducers 91, CH 2817, 627–630.

[151] P.B. Zou, J.T. Pang, Z.F. Wang, X. Qian, H.Q. Gong, M.K. Lim, Z.J. Li, Single chip fabrication of integrated fluid systems (IFS). Proceed. IEEE Workshop MEMS 98, Heidelberg, Germany, January 25–29, 1998, 448–453.

[152] P. Krause, E. Obermeier, W. Wehl, Backshooter – a new smart micromachined single-chip inkjet print head. Proceed. Transducers 95, Eurosensors IX, Stockholm, Sweden June 25–29, 1995, 325–328.

[153] D.J. Coe, M.G. Allem, B.L. Smith, A. Glecer, Addreasable micromachined jet arrays. Proceed. Transducers 95, Eurosensors IX, Stockholm, Sweden June 25–29, 1995, 329–331.

[154] M. Alavi, S. Buttgenbach, A. Schumacher, H.J. Wagner, Fabrication of microchannels by laser machining and anisotropic etching of silicon. *Sensors and Actuators A*, 32, 1992, 299–302.

[155] M. Alavi, Th. Fabula, A. Schumacher, H.-J. Wagner, Monolithic microbridges in silicon using laser machining and anisotropic etching. *Sensors and Actuators A*, 37–38, 1993, 661–665.

[156] Y. Bäcklund, Micromechanics in optical systems – with focus on telecom systems. *J. Micromech. Microeng.*, 7, 1997, 93–98.

[157] C. Strandman, Y. Bäcklund, Passive and fixed alignment of devices using flexible silicon elements formed by selective etching. *J. Micromech. Microeng.*, 8, 1998, 39–44M.

[158] Hoffmann, P. Kopka, T. Gross, E. Voges, Optical fiber switches based on full wafer silicon micromachining. *J. Micromech. Microeng.*, 9, 1999, 151–155.

[159] Ch. Gorecki, Optical waveguides and silicon-based micromachined architecture. In: *MEMS and MOEMS technology and applications*, P. Rai-Choudhury (ed.). SPIE Press, 2000, 209–300.

[160] M. Hoffmann, E.Voges, Bulk silicon micromachining for MEMS in optical communication systems. *J. Micromech. Microeng.*, 12, 2002, 349–360.

[161] Ch. Gorecki, M. de Labachelerie, L. Thierry, The role of silicon micromachining in optical fiber sensing technologies. *IEEE Sensors Journal*, 3(1), 2003, 121–130.

[162] R.M. Bostockë, J.D. Collier, R. Jones, D.F. Moore, J.E. Townsend, Silicon nitride microchips for the kinematic location of optic fibres in silicon V-shaped grooves. *J. Micromech. Microeng.*, 8, 1998, 343–360.

[163] M. de Labachelerie, N. Kaou, V. Armbruster, J.-C. Jeannot, P. Mollier, H. Porte, N. Devoldere, A micromachined connector for the coupling of optical waveguides and ribbon optical fibers. *Sensors and Actuators A*, 89, 2001, 36–42.

[164] J. Dziuban, A micromachined silicon accelerometer with a movable mercury ball of micrometer size. *MST News Poland*, 1996, nr 4, 9–11.

[165] F. Krull, Nutzen der dritten Dimension. *Technische Rundschau*, No. 11, 1995.

[166] T.S.J. Lammerink, Micro-liquid flow sensor. *Sensors and Actuators A*, 37–38, 1993, 45–50.

[167] S. Bouwstra, P. Kemna, R. Legtenberg, Thermally excited mass flow sensor. *Sensors and Actuators A*, 20, 1989, 213–223.

[168] M.A. Gajda, H. Ahmed, Applications of thermal sensors on membranes. *Sensors and Actuators A*, 49, 1995, 1–9.

[169] D. Maser, R. Lenggenhager, H. Baltes, Silicon gas flow sensor using industrial CMOS and bipolar IC technology. *Sensors and Actuators A*, 25–27, 1991, 577–581.

[170] J. Robadey, O. Paul, H. Baltes, Two-dimensional integrated gas flow sensors by CMOS IC technology. *J. Micromech. Microeng.*, 1995, 243–250.

[171] T. Laurell, L. Rosengren, A micromachined enzyme reactor in (110) oriented silicon. *Sensors and Actuators B*, 18–19, 1994, 614–617.

[172] T. Laurell, L. Rosengren, J. Drott, A micromachined glucose oxidase enzyme reactor. Proceed. μTAS 94 Workshop, Twente, 21–22 November, 1994, 227–231.

[173] A. Manz, E. Verpoorte, B.E. Raymond, C.S. Eftenhauser, N. Burggraf, H.M. Widmer, μTAS: Miniaturized chemical analysis systems. Proceed. μTAS 94 Workshop, Twente, 21–22 November, 1994, 5–27.

[174] S. Böhm, W. Olthuis, P. Bergveld, A μTAS based on microdialysis for on-line monitoring of critically relevant substances. Proceed. μTAS 94 Workshop, Twente, 21–22 November, 1994, 31–34.

[175] I.K. Glasgow, H.Ch. Zeringue, D.J. Beebe, S.-J. Choi, J.T. Lyman, M.B. Wheeler, Individual embryo transport and retention on a chip. Proceed. μTAS 94 Workshop, Twente, 21–22 November, 1994, 199–206.

[176] P. Dario, M.C. Carrozza, A. Benvenuto, A. Menciassi, Micro-systems in biomedical applications. *J. Micromech. Microeng.*, 10, 2000, 235–244.

[177] T.B. Taylor, P.M. St. John, M. Albin, Micro-genetic analysis systems. Proceed. μTAS 98 Workshop, Banff, 13–16 October, 1998, 261–266.

[178] H. Andersson, A. Ahmadian, W. van der Wijngaart, P. Nilsson, P. Enoksson, M. Uhler, G. Stemme, Micromachined flow-through filter-chamber for solid state phase DNA analysis. Proceed. μTAS 2000 Conference Enschede, 14–15 May, 2000, 473–476.

[179] S. Bargiel, A. Gorecka-Drzazga, J. Dziuban, P. Prokaryn, M. Chudy, A. Dybko, Z. Brzozka, Nanoliter spectrofluorometric detector for flow systems. Proceed. 17 European Conference on Solid-State Transducers Eurosensors XVII, Guimaraes, Portugal, 21–24 Sept., 2003, 394–395.

[180] J. Roeraade, M. Stjemström, A. Emmer, E. Litborn, U. Lindberg, Nanochemistry and nanoseparations of biomolecules, H.M. Widmer, E. Verpoorte, S. Bernold (eds.). Anal. Meth. Instrum., Spec. Issue, μTAS 96, Basel 19–22 November, 1996, 34–38.

[181] C. Grosjean, G. Lee. W. Hong, Y.C. Tai, C.M. Ho, Micro Balloon Actuators for Aerodynamic Control. Proceed, 11th Annual. Int. Workshop MEMS 98, January 25–29, 1998, Heidelberg, Germany, 166–171.

[182] E. Valderrama, P. Garrido, P. Heiduschka, A. Harsch, W. Göpel, Microfabrication and characterization of microelectrode arrays for in-vivo nerve signal recording. Proceed. Transducers 95, Eurosensors IX, Stockholm, June 25–29, 1995, 63–66.

[183] P. Dario, M. Cocco, G. Soldani, E. Valderrama, J.U. Meyer, T. Giesler, H.-J. Beutel, H. Scheithauer, M. Alavi, V. Bulker, Technology and fabrication of hybrid neural interfaces

for the peripheral nervous system. *Microsystem Technologies Berlin, Oct.*, 19–21, 1994, 417–426.
[184] K. Najafi, K. Wise, A high-yield IC compatible multichannel neuraling array. *IEEE Tran. El. Dev.*, ED-32, 7, 1985, 1206–1211.
[185] D.W. de Lima Monteiro, O. Akhzar-Mehr, P.M. Sarro, G. Vdovin, Single mask fabrication of aspherical optics using KOH anisotropic etching of silicon. *Optic Express*, 11(18), 2003, 2244–2252.
[186] G. Vdovin, O. Akhzar-Mehr, P.M. Sarro, D.W. de Lima Monteiro, M.Y. Loktev, Arrays of spherical micromirrors and molded lenses fabricated with bulk micromachining. In: *MEMS/MOEMS Advances in photonic communications, sensing, metrology, packaging and assembly*, U. Behringer, B. Courtois, A.M. Khounsary, D.G. Uttamchandani (eds.). Proc. SPIE 4945, 2003, 107–111.
[187] A. Górecka-Drzazga, J. Dziuban, U. Lipowicz, Plasma etch for micromechanics integrated (in Polish). Proceed. of IV Conf. ELTE 90, 11–14 Sept., 1990, Książ, Poland, 72–74.
[188] H.V. Allen, S.C. Terry, J.W. Knutti, Understanding silicon accelerometers. *Sensors*, September 1989.
[189] G. Stemme, Resonant silicon sensors. *J. Micromech. Microeng.*, 1, 1991, 113–125.
[190] M. Tortonese, H. Yamada, R.C. Barret, C.F. Quate, Atomic force microscopy using a piezoresistive cantilever. Proceed. IEEE Transducers '91, CH2817, 448–451.
[191] C.J. Welham, J. Greenwood, M. Bertioli, A lateral resonant pressure sensor fabricated via fussion bonding, wafer thinning and reactive-ion-etching. Proceed. Eurosensors XII, Southhampton, 13–16 September, 1998, UK, IOP Sensor-Series, 353–356S.
[192] Bouwstra, R. Laktenberg, H.A.C. Tilmans, Resonating micro-bridge mass flow sensor. *Sensors and Actuators A*, 21–23, 1990, 332–335.
[193] H.P. Lang, F.M. Battiston, M.K. Baller, R. Berger, J.-P. Ramseyer, P. Fornaro, E. Meyer, H.-J. Günterodt, C. Andreali, J. Brugger, M. Despont, P. Vettiger, J.-H. Fabian, T. Mezzacasa, L. Scandella, Ch. Gerber, J.K. Gimzewski, An electronic nose based on a micromechanical cantilever array. Proceed. of μTAS 98 Workshop 1998, Banff, Canada, October, 13–16, 57–60.
[194] K.E. Petersen, Silicon torsional scanning mirror. *IBM J. Res. Dev.*, No 24, 1980, 631–637.
[195] H. Kück, W. Doleschal, A. Gehner, W. Grundke, R. Mehler, J. Paufler, R. Zeltmann, G. Zimmer, Deformable mirror devices as faze modulating high resolution light valves. *Sensors and Actuators A*, 52–54, 1996, 536–541.
[196] J.B. Sampsell, The digital micromirror device and its application to projection displays. Proc. Transducers 93, Yokohama, Japan, 7–10 June 1993, 24–27.
[197] B. Mischke, Mikromechanik. *Radio Fernsehen Elektronik*, 1988, 37, 9.
[198] J.C. Greenwood, Etched silicon vibration sensor. *J. Phys. Sci. Instrum.*, 17, 1985, 680–683.
[199] H.K. Trien, L. Ewe, W. Mokwa, M. Schwarz, B.J. Hosticka, Flexible silicon structures for a retina implant. IEEE Workshop MEMS 1998, January 25–29, Heidelberg, Germany, 515–519.
[200] S. Kolnsberg, K. Stangel, D. Hammerschmidt, M. Schwarz, B.J. Hosticka, L. Ewe, H.K. Trieu, W. Mokwa, CMOS micro transceivers in ophthalmology. World Microtech. Cong. Proceed. MICRO. tec, 2000, Expo 2000, Hannover, Germany.
[201] P. Mion, J. Kim, Feedback control of turbulence. *Appl. Mech. Rev.*, 47(5–6), part 1–3, 1994.
[202] W.H. Ko, The future of sensors and actuators systems. *Sensors and Actuators A*, 56, 1996, 193–197.
[203] M. Humayun, E. de Juan, G. Dagnielle, R. Greenberg, R. Propst, H. Phillips, Visual perception elicited by electrical simulation of the retina in blind humans. *Arch. Ophtalmol.*, 114, 1996, 40–46.
[204] A. Lal, R.M. White, Silicon microfabricated horns for power ultrasonics. Proceed. Transducers 95, Eurosensors IX, 8th Int. Conf. Solid St. Sensors and Actuators, June 25–29, Stockholm, Sweden, 1995, 405–408.
[205] S. Charles, R. Williams, T.L. Poteat, Micromachined structures in ophthalmic microsurgery. *Sensors and Actuators A*, 21–23, 1990, 263–266.

[206] A. Górecka-Drzazga, J. Dziuban, Fabrication of silicon microtips. Metal/nonmetal Microsystems: physics, technology, applications, Polanica Zdr. 11–14 Sept., 1994. *Proc. SPIE*, Vol. 2780, 380–383.
[207] A. Górecka-Drzazga, J. Dziuban, Technological aspects of FEA's fabrication. Proceedings of 2nd Int. Workshop on Vacuum Microelectronics of 12th Int. Vacuum Microelectronic. Conf., July 6–9, July 11–13, 1999, Darmstadt-Wrocław, 102–104.

Chapter 4

BONDING

Bonding is a set of technological procedures which permanently seal solid-state materials with smooth and flat surfaces, most often by chemical reactions. This method of material sealing is commonly used in microsystem technology for bonding of unprocessed flat wafers, deeply micromachined wafers, with movable micromechanical structures, whole wafers or particular chips or small details. Bonding is not only a simple packaging method of microsystem technology, as most often classified, it is the integral procedure of microsystems fabrication, playing a role in front-end as well as back-end technological procedures. The widest application in microsystem technology has been in the bonding of two silicon wafers and bonding of silicon wafer to glass substrate, as well as the bonding of chips and components made of these materials.

Bonding significantly broadens the technological capability of formation of 3-D microsystems coming from deep micromachining of silicon because:

- it allows the fabrication of multi-layer, sandwiched, micromechanical structures, in which each layer can contain three-dimensional structures,
- it allows the multi-chip production of microsystems with many micromechanical structures and/or microelectronic structures bonded to the single silicon or glass substrate platform,
- it is a very good technique for packaging of microsystems.

Bonding, regardless of its type, consists mainly of the three following steps:

- preparation of surface for bonding (cleaning, activation),
- alignment and pre-bonding of details,
- final formation of bonding.

These steps are closely related to, and depended on, the type of bonding procedure applied. Bonding is carried out in the range of high ($>700^{\circ}C$), medium (200–500°C) or low temperatures (20–200°C). It can be either a thermally activated process, without electric field excitation (fusion bonding, bonding through the low-melting glasses, eutectic bonding, HF and NaOH bonding, foil bonding) or supported by electric field excitation (anodic bonding). Bonding can be carried out in a clean air, in an inert/chemically active atmosphere or in vacuum.

Micromechanical sandwiched silicon structures are mainly fabricated by means of high-temperature bonding. Silicon-glass structures are produced by use of a wider range of medium-temperature bonding techniques. Among these are anodic bonding, so-called glass frit bonding using low-melting glasses, and eutectic bonding. Low-temperature bonding techniques, such as foil bonding, resist bonding and glueing of details, are employed more rarely. Mixed techniques of bonding are also very often applied.

All the above-mentioned techniques of bonding (sometimes called the micromechanical sealing of materials) will be discussed in this chapter, which is divided into two parts. In the first part, non-electric bonding techniques suitable for microsystem technology are presented. This material – based on the analysis of literature data and the author's experience – begins with the presentation of the methods of preparation of surfaces for bonding. Next, the mechanism of silicon to silicon fusion bonding for hydrophilic and hydrophobic surfaces is presented, as well as the influence of the SiO_2 interlayer on the process of fusion bonding and the quality of bonding (the transitory layers). Then, fusion bonding through other dielectric layers (Si_3N_4, SiC, diamond-like, etc.) is discussed, along with the main micromechanical applications of high-temperature fusion bonding. Finally, low-temperature fusion bonding, bonding through sodium silicates, and boron and phosphorus glasses, HF bonding, eutectic bonding and foil bonding are described.

In the second part of this chapter anodic bonding is discussed. To start with, the historical background and a general description of the method are presented, and then the characterization of the properties of glasses, leading to the description of anodic bonding. Next, the mechanism of anodic bonding is analyzed, including the phenomena occurring by the cathode and by the anode (at silicon and glass interface), formation of the depletion layer, the role of electrostatic pressure and the discussion of Baumann's and Schmidt's models of the process. Next, the transport of charges is discussed and the activation energy of anodic bonding is determined. The conditions of a good anodic bonding of silicon to different types of glasses are researched, as well as the quality, force and minimal conditions of bonding. Flexures, induced stresses, hardness after the process and the optimization of these parameters are analyzed. Multi-layer, sandwich bonding is discussed, along with the special techniques of silicon-glass and silicon to glass bonding through all the useful dielectric and metallic layers. In addition many examples of applications of anodic bonding, in the construction and fabrication of various silicon-glass microsystems, are presented.

4.1. SURFACE CLEANING AND ACTIVATION

The process of bonding of materials requires an adequately prepared surface of bonded wafers. A surface needs to be clean and dustless, because wafers must adhere to each other firmly. The most authoritative method of evaluating surface cleanness is the wetting angle of a water drop located on a surface (Fig. 4.1). The value of this angle θ is also a measure of the hydrophilic or hydrophobic character of a surface. The wetting factor β can be evaluated from the condition of equilibrium of forces on the boundary of three phases: solid state, liquid and gas (substrate, water drop, air):

$$\beta = \cos\theta = \frac{\gamma_{GS} - \gamma_{CS}}{\gamma_{GC}} \tag{4.1}$$

$$\gamma_{GS} - \gamma_{CS} - \gamma_{GC} \cdot \cos\theta = 0, \tag{4.2}$$

where: γ_{GS} = surface tension on the solid–gas boundary, γ_{CS} – surface tension on the solid–liquid boundary, γ_{GC} = surface tension on the liquid–gas boundary.

The value of wetting angle θ is smallest, and the wettability is highest, when γ_{GC} and γ_{CS} are as small as possible, while γ_{GS} is as large as possible. It is assumed that a surface is strongly hydrophilic for $\theta < 20°$, and hydrophobic for $\theta > 50°$. The state of the hydrophilic surface – wetting angle, cleanness, number of dust particles, etc. – all depend on the method of cleaning and drying of wafers. Therefore this pre-treatment is an essential key part of bonding, because – as will be shown in subsequent sections of this chapter – the state of surface limits bond ability of sealed wafers.

The preparation procedures of silicon and glass, especially the procedures of cleaning and activation (taking the specificity of materials into consideration) are quite similar. This is the reason why these procedures are presented as a unified description here, at the beginning of the chapter.

4.1.1. Silicon wafers and silicon wafers covered with SiO_2

The surface of silicon wafer can be either hydrophilic or hydrophobic. The state of surface, that is its hydrophilic or hydrophobic character, is obtained after utilization of a suitable cleaning procedure, which ensures a clean surface with the desired degree of affinity to water. Schulze [1] has given the wetting angles θ of silicon wafer surfaces after a few procedures typical for IC technology (Table 4.1) and the procedures of hydration of surface by washing (Table 4.2).

The authors of papers [2]–[6] recommend the single-stage hydration procedure: RCA1 washing (RCA1: boiling in $NH_4OH{:}H_2O_2{:}H_2O$ 1:1:5), followed

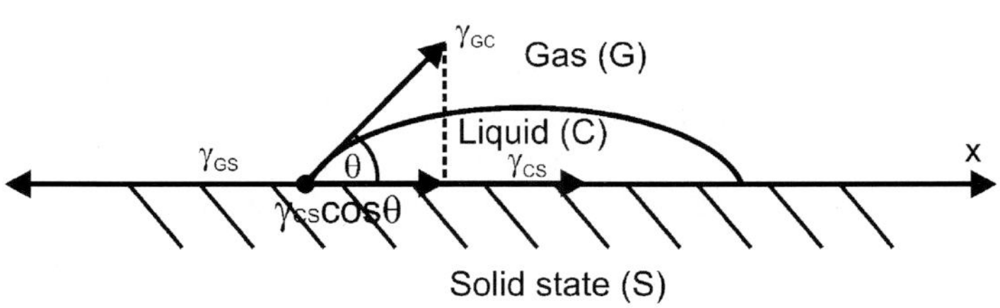

Fig. 4.1. A fluid drop at a solid-state surface.

Table 4.1. Wetting angles at the surface of silicon wafers treated in different ways

Surface type or interacting reagents	Wetting angle [°]
Washed in HMDS $[(CH_3)_3Si]_2NH$.	83.20
Washed in HMDS and TMAH for 50 s	71.30
Coated with AZ 1518	20.20
After short immersion in 5% HF	67.70
Si after developing of photoresist in TMAH (50 s)	51.60
Si after developing of photoresist in TMAH (50 s) and etching of Al for 4 min	21.80
Delivered substrates (typically)	17.60
After SCI procedure (H_2O_2 and NH_4OH)	5.50

Table 4.2. Procedures for hydrophilic surface preparation of silicon wafers [1]

Step	Chemical treatment	Temperature/time
1	Washing in DI water	10 min
2	CARO (30% H_2O_2/96% H_2SO_4)	120°C, 10 min
3	Washing in pure DI water	10 min
4	SCI (30% H_2O_2, 25% NH_4OH, H_2O)	80°C, 5 min
5	Washing in pure DI water	10 min
6	HF-immersion (5%)	3 min
7	Washing in pure DI water	10 min
8	SCII (30% H_2O_2, 37% HCl, H_2O)	70°C, 5 min
9	Washing in pure DI water	10 min
10	Drying in pure N_2	10 min

by drying in the nitrogen jet or RCA2 washing (RCA2: boiling in $HCl:H_2O_2:H_2O$ 1:1:6) and drying in N_2. They also advise the shortened "Piranha" washing (Piranha mixture: $H_2SO_4:H_2O_2$ (3:1)) at 70°C. Other hydrophilic procedures are as follows: boiling in 30% H_2SO_4 [6], HNO_3 [6–12], washing in NH_4OH [13, 14], plasma treatment in NH_3 [15], O_2 [2], in SF_6 or $SF_6 + O_2$ [16]*.

Hermasson [17] has presented fourteen procedures of treatment of silicon wafers and wetting angles θ at surfaces prepared in this way (Table 4.3, Fig. 4.2). The smallest wetting angle was attained in the RCA1 procedure [18], which is consistent with the works of Kissinger [19] and Bäcklund [20]. Bäcklund recommends "Piranha" washing or RCA1, and then boiling in HNO_3.

A bondable hydrophilic surface of silicon wafers can usually be obtained after complete washing and activation procedures [21–24]. In our own research it was noticed that an adequate procedure should start from degreasing and removing of other organic and inorganic impurities. It was also revealed that properly stored wafers, supplied by a reliable producer, can be treated† with

* Plasma treatment is the key procedure of low-temperature direct bonding discussed further in this book.

† This remark is valid only for anodic bonding procedure.

Table 4.3. Treatment procedures of silicon wafers

No.	Treatment	Temperature
1	RCA1, 10 min	100°C
2	Piranha: H_2SO_4:H_2O_2, 10 min	100°C
3	H_2SO_4:H_2O_2, RCA2, 10 min	96°C
4	RCA1 + RCA2 10 min each.	100°C
5	NH_3:H_2O:H_2O_2, HCl:H_2O_2:H_2O_2, each for 10 min	95°C
6	KOH	80°C
7	Wet oxidation	1100°C
8	Polyethylene glycol	room temperature
9	Non-treated, covered with intrinsic oxides	room temperature
10	Ethylenediamine pyrocatechol etched (EDP)	118°C
11	Isocyanide-propyl-dimethyl-chloromonosilane	room temperature
12	Etched in HNO_3:HF	32°C
13	Immersed in HF	room temperature
14	Dichloro-dimethyl-monosilane (DDS)	—

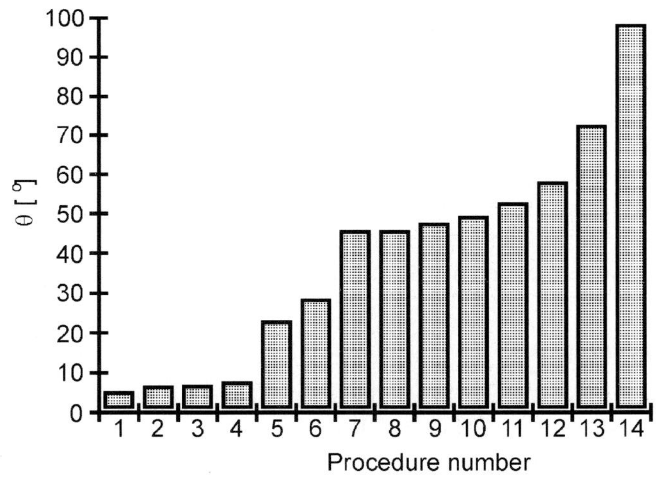

Fig. 4.2. Wetting angles for procedures from Table 4.3 [17].

the simplified washing and hydration procedures, that is: degreasing in hot acetone and trichloroethylene, then boiling for 30 minutes in 30% H_2O_2 or washing in the solution H_2SO_4:H_2O_2:H_2O (82.3:4.3:13.4) at 70°C, for 10 minutes, or washing in the solution NH_4OH:H_2O_2:H_2O (20:5:3) at 60°C for 30 min. Finally, wafers should be rinsed under megasonic DI water jet.*

The surface of the wafer covered with fresh thermal silicon oxide is hydrophobic but chemically and physically clean. Immediately after high-temperature oxidation the wafers can be boiled for a short time (10–20′) in 30% H_2O_2:H_2O.

* A "one tool" washing system for DI water megasonic cleaning is available: www.suss.de.

Clean wafers covered with layers which cannot be washed in aggressive solutions, especially with aluminum, can be prepared for bonding "in situ", utilizing softer procedures corresponding to the technological history of the wafer, e.g. RCA1 ($NH_4OH:H_2O_2:H_2O$), or boiling in 30% H_2O_2 for 30′.

It is commonly believed that bonding should be carried out immediately after cleaning and activation of wafers. Quenzer [9] says that the maximal delay after the preparation procedures cannot exceed 3 hours. It is also believed that the best durability of hydrophilic state of the treated surface is obtained in procedure RCA1 [25]. In our own works it has been pointed out that, after full procedures of washing and hydration, wafers can be stored in deionized water for many days without losing bonding ability. The loss of bond ability has been caused mainly by the development of bacterial flora and water dustiness.

Silicon wafers with hydrophobic surfaces can be also bonded [26–29]. The hydrophobic silicon surfaces can be easily obtained by washing the wafers in a weak aqueous solution of hydrogen fluoride acid, treatment by fumes of concentrated HF, or by plasma treatment in argon or hydrogen, or in a mixture of 4% H_2 in Ar [2]. Hydrogen atoms, as well as, in smaller quantities, OH groups and fluorine, occur at the hydrophobic silicon surfaces. Probably on such surfaces hydrocarbons are also adsorbed [30]. Hydrophobic wafers do not bond spontaneously (although in some sources spontaneous bonding was claimed), but need to be lightly pressed.

In many sources it is stated that bonding of wafers CMP (**C**hemical-**M**echanical-**P**olishing) polished is much simpler than bonding of micromechanical wafers prepared using standard methods [31–33]. This is particularly important for wafers covered with a thin layer of thermal oxide [33–35].

4.1.2. Glass substrates

Cleaning and activation of glass surfaces consist of the removal of organic impurities (degreasing), cleaning and hydration. Substrates delivered in packages, according to semiconductor requirements, can be cleaned in simplified procedures, that is washing in deionized water and boiling in 30% H_2O_2 for 30′. Other, "dirty" glass substrates need to be cleaned in the complete procedure:

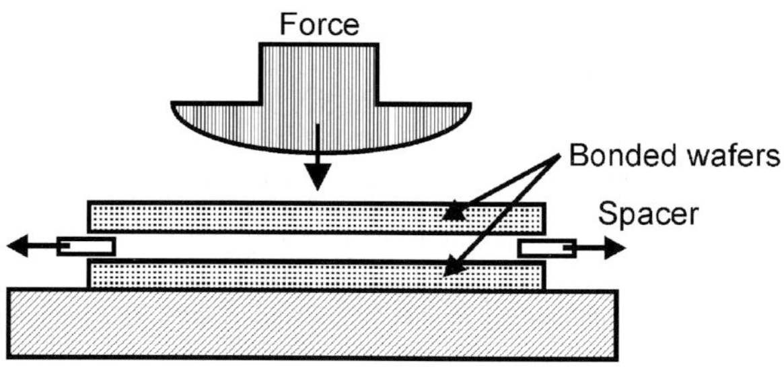

Fig. 4.3. Set-up for pre-bonding of wafers.

washing in warm detergents, degreasing in hot trichloroethylene or acetone, etching in 2% HF at 30°C, boiling in 2% K_2CrO_7 at 80°C and boiling in 30% H_2O_2. Rinsing of substrates in deionized water should follow each of these technological steps. The preparative procedures mentioned above reduce the wetting angle to approximately 10°, regardless of the state of wafer surface before cleaning. Further reduction of the wetting angle θ is attained after cleaning in glow discharge in argon under pressure of about 100 Pa and with voltage circa 5 kV [18]. The joined procedures of washing and cleaning in glow discharge* lead to a decrease in the value of θ to a few degrees (Fig. 4.4). In air, after taking the substrate out of vacuum, the state of surface worsens, while the wetting angle increases. Glass wafers stored for too long in air can lose their capacity for bonding, but they may be stored for hours in DI water.

4.2. HIGH-TEMPERATURE FUSION BONDING

The bonding of silicon wafers at high temperature without use of an external electric field is called silicon fusion bonding (SFB), silicon direct bonding (SDB), or silicon thermal bonding (STB).

In this technique the wafers have to be cleaned, and their surface should be activated before bonding. Substrates so prepared have a flat, smooth, clean and activated surface and can be brought into contact. Then a weak bonding occurs between wafers (so-called spontaneous bonding or self-bonding). The force of bonding increases and stabilizes at a high level during the high-temperature annealing of wafers.

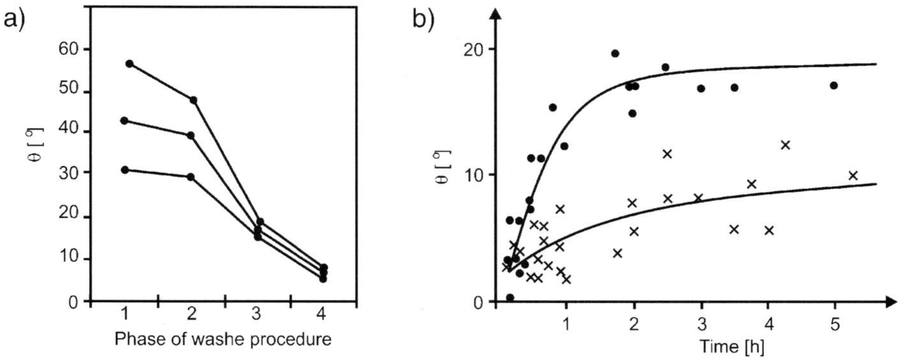

Fig. 4.4. Wetting angle of glass surface: a) after successive cleaning steps; 1 = after polishing and drying, 2 = after washing in organic solvents, 3 = after washing in $K_2Cr_2O_7$, 4 = after glow discharge, b) after removing from vacuum, in air, as a function of time [18].

---

* The new concept of dielectric-coupled atmospheric pressure scanned nitrogen/oxygen plasma purging of semiconductor and glass wafers has recently been developed by the Fraunhofer Institute (Germany) and Carl Suss Co. (Germany). $\theta = 2$–3° has been announced for glass/silicon surfaces, allowing direct ambient temperature bonding of these materials, see: www.suss.com.

4.2.1. Hydrophilic bonding

4.2.1.1. Silicon to silicon bonding (Si–Si)

The mechanism of high-temperature brand silicon wafer bonding has been described by many authors [1–3, 5, 6, 22, 35, 37–39]*. Two silicon wafers with mirrored hydrophilic surfaces brought into contact at room temperature attract one another and form a spontaneous bonding. Infrared spectroscopy, in the absorption spectrum of $1/\lambda$ from 1000 to 4000 cm^{-3}, exhibits absorption peaks attributed to Si–O, Si–H, O–H, Si–OH bonds in various configurations [18, 39].

This result indicates that, between bonded surfaces, an interlayer occurs, formed by: a thin layer of intrinsic oxide with thickness 0.4 nm, several monolayers of molecular water, a monolayer of Si–OH silanol bonds and numerous Si–H hydrogen bonds as well as free hydroxide OH^- groups. Some authors [18, 40, 41] claimed that Van der Waals' attraction forces cause the spontaneous bonding of wafers. Some others, including the author of this book, believe that this is a weak silanol and hydrogen bond spontaneously bonding the surfaces.

Immediately after contacting of two wafers, at first the spontaneous local bonding between contacted wafers occurs in one or several areas (points). The spontaneous bonding can be initiated in a chosen point at a wafer, in a so-called bonding precursor, for instance by local pressing of substrates. Then bonding propagates at the speed of a few centimeters per second, forming a wave [5, 14, 36], a propagation of a so-called wave of bonding can be observed. Smooth mirrored surfaces with high chemical and physical purity are characterized by rapid propagation of the wave of bonding. Pairs of the spontaneously bonded wafers can be handled; the sandwich will not come apart if treated carefully.

Spontaneously formed weak bonding of substrates must be reinforced by thermal annealing. A typical annealing process is as follows:

- slow placing-in of sandwich of wafers (0.5 mm/s) inside an oven, $t = 300°C$,
- increase of the temperature to 1100°C: gradient 5°C/min,
- annealing at 1100°C for 1 h, atmosphere N_2:O_2, 3:1, flow 10 l/min (for 4″ wafers),
- cooling-down to 300°C, gradient 5°C/min,
- slow removal of the substrate from an oven.

It is commonly believed that a strong bonding of silicon wafers after high-temperature annealing is formed by siloxane Si–O–Si bonds. The mechanism of formation of this sealing is multi-staged. Bonding obtained below 80°C is reversible [42] and can be applied in order to protect the silicon surfaces of special use. An increase in temperature results in the formation of strong, irreversible bonding.

Stengl and co-authors [38], and Weldon et al. [39], have proposed two models of multi-stage direct bonding complementing one another. The authors

* At least two world conferences organized by the Electrochemical Society Inc. are devoted to the subject of bonding: Semiconductor Wafer Bonding and SOI Technologies.

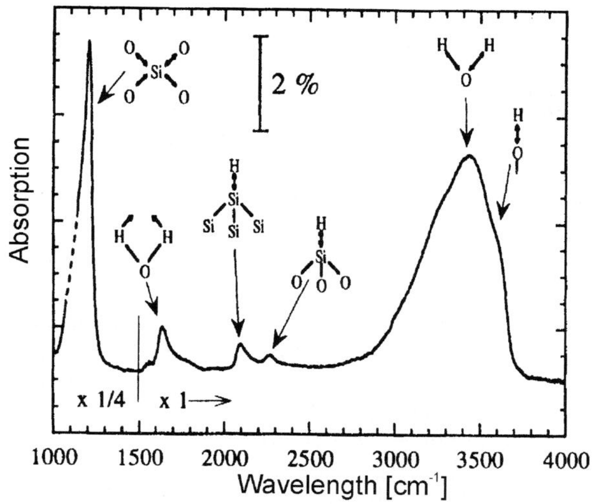

Fig. 4.5. Absorption of infrared radiation of two bonded silicon wafers with hydrophilic surfaces spontaneously bonded at a temperature of 80°C [39].

of paper [38] have elaborated a three-step process (Fig. 4.6). They assumed that, at room temperature, molecules chemically bonded to silicon (or only adhering to its surface) form a 0.7 nm-thick transient layer between the initially bonded wafers. An increase in temperature (first step, **20°C** $<$ ***t*** $\leq$ **200°C**) causes a transformation of molecular water trapped between the bonded wafers and formation of clusters of four water molecules. A long hydrogen bond evolved between the water molecules is formed. As a result of this transformation the distance between wafers is reduced to approximately 0.35 nm; the bonding energy γ increases to about 0.1–0.2 J/m^2. In the second step (**300°C** $\leq$ ***t*** $\leq$ **700°C**), as the temperature increases, the initial hydrogen bonds are replaced with siloxane Si–O–Si bonds, as an effect of polymerization of silanol, suggested to be the dominant mechanism in many works [22, 28, 35, 39]:

$$\equiv \text{Si–OH} + \text{HO–Si} \equiv \quad \text{gives} \quad \equiv \text{Si–O–Si} \equiv + \text{H}_2\text{O}, \tag{4.3}$$

where $\equiv$ represents bonds between surface silicon atoms and their bulk neighbors.

In the third step (**700°C** $\leq$ ***t*** $\leq$ **1100°C**), the siloxane bond formation continues, the released water forms the voids of non-bonded areas in the silicon–silicon transient layer. The voids disappear after prolonged annealing at 800°C or after a short time (for a few minutes) annealing at 1100°C. At this temperature water molecules disappear, silicon is oxidized locally, the additional oxygen bonds between sealed wafers are formed near-by bonds formed previously. Diffusion of atoms, and adhesive flow of oxides formed between silicon surfaces, fill the micro cavities and the nano-roughness, which also occurs on wafer surfaces. Surface bonding energy γ after annealing can exceed 2 J/m^2.

In Weldon's model [39] (Fig. 4.7) high-temperature bonding proceeds as

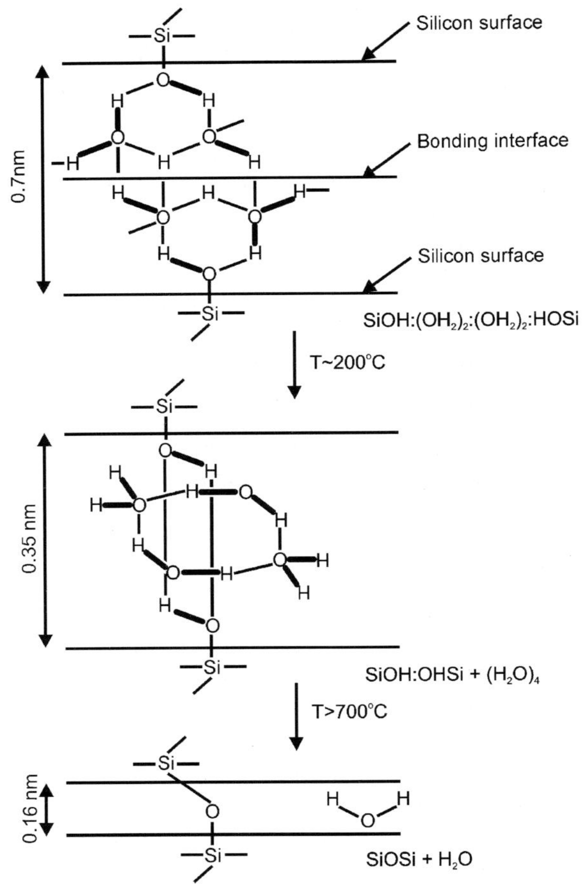

Fig. 4.6. Mechanism of silicon-to-silicon bonding; Stengl's model [38]. Illustration of transformation of a weak initial bond into a strong Si–O–Si bon.

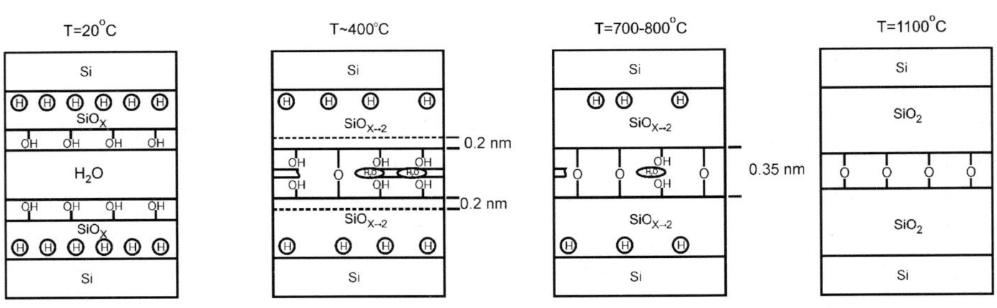

Fig. 4.7. Direct bonding according to Weldon's model [39].

follows: the first step (**20°C ≤ *t* ≤ 300°C**) begins with the loss of molecular water (up to 75%), which diffuses through the intrinsic oxide and forms an additional 0.4 nm-thick layer of silicon dioxide in the reaction shown below:

$$Si/SiO_2 + H_2O \rightarrow SiO_x/SiO_2 + H_2, \tag{4.4}$$

where Si/SiO_2 is a boundary between the pure silicon and initial intrinsic oxide, and x describes the non-stoichiometric compound formed in the reaction of internal silicon oxidation with oxygen deficiency ($O_2 \sim 2 \cdot 10^{15}$ atoms/cm^2).

The release of gaseous hydrogen is confirmed by spectroscopic examination of the composition of gaseous reaction products, which accumulate in cavities intentionally produced on the bonded surfaces [43].

In the second step (**300°C ≤ *t* ≤ 800°C**) further loss of water occurs, mainly by means of diffusion, which allows the surfaces to be brought nearer at a distance to 0.35 nm and the direct interaction of hydroxide groups, bonded with silicon.

Then the first siloxane bonds are formed, as described below:

$$Si{-}OH + HO{-}Si \rightarrow Si{-}O{-}Si + H_2O. \tag{4.5}$$

Released water causes further silicon oxidation according to the reaction given below:

$$Si/SiO_2 + H_2O \rightarrow SiO_x/SiO_2 + 2(O_3)Si{-}H, \tag{4.6}$$

where $2(O_3)Si{-}H$ refers to hydrogen bonded with a silicon atom, which is bonded with three atoms of oxygen coming from the oxide layer.

The excess water is kept in the micro gaps [39], in spite of relatively high temperatures, in the condensed state [43].

In the third technological step (**800°C ≤ *t* ≤ 1100°C**) the gap between surfaces completely "closes". Remaining hydroxide groups, bonded to silicon, form siloxane bonds. Hydroxide groups, produced during this process, which are a result of water molecule dissociation in high temperature, bond free silicon atoms at the surface. The process continues up to the complete consumption of available hydroxide groups.

$$Si{-}OH + OH{-}Si \rightarrow Si{-}O{-}Si + H^+ + OH^-. \tag{4.7}$$

Protons diffuse through silicon wafers; after neutralization they form gaseous hydrogen, which is evacuated.

4.2.1.2. *Silicon to silicon bonding through SiO_2 ($Si/SiO_2{-}Si$)*

Bonding of silicon wafers covered with insulating oxide ($Si/SiO_2{-}Si$ bonding) was developed for the first time in 1985–86, when Lasky [35] and Shimbo [6] fabricated SOI structures (***S****ilicon*-***O****n*-***I****nsulator*) without an external electrical field or pressure. High-temperature $Si/SiO_2{-}Si$ bonding gives the possibility of fabrication of a sandwich structure with a buried insulating oxide layer. Buried layers are mostly composed of high-temperature dry and/or wet silicon oxides or silica glazes PSG or BPSG. Wafers to be bonded must be properly washed and pass the hydration procedure. Usually only one of the wafers is covered with the insulating layer, although it is possible to bond two oxidized wafers. Properly prepared wafers easily form a spontaneous bond. The roughness of insulating layers, used in bonding of SOI wafers, influences the strength of bonding. This is why the quality of bonding of CMP wafers is higher, and the strength of bonding decreases for thicker, rougher oxides [22].

The mechanism of Si/SiO_2–Si bonding is generally similar to the course of Si–Si bonding [34, 35]. The most important difference between them is an increased diffusion of water and hydrogen in the layer of silicon dioxide and a viscous flow of relatively thick oxides, filling the micro cavities at the surface more easily. The surface of thermal oxides is smoother than the silicon one, and also absorbs smaller amounts of water (less by 40%). Moreover, micro trapping of molecular water does not occur in such an oxide layer in the temperature range 700°C to 800°C [39].

The kinetics of Si/SiO_2–Si bonding has been examined for the temperature range 20°C to 1400°C and the range of time from 10 seconds to 6 hours [14, 39, 44]. In the first phase of bonding, below 200°C, hydrogen bonds evolve either between OH^- groups adsorbed on the bonded surfaces, or between OH^- groups and oxygen atoms coming from the SiO_2 layer. Then in the second step, with the temperature increasing to about 1000°C, these hydrogen bonds are replaced with Si–O–Si oxygen bonds. In the third phase, when the temperature increases to over 1100°C the adhesive oxide flow (re-flow) results in the total bonding of surfaces. The surface energy γ of bonding between the wet oxides is similar to the surface energy of bonding of the dry oxides (Fig. 4.8).

Many researchers have concluded that bonding in an oxygen atmosphere [15, 35, 45, 47] is more effective than in a nitrogen atmosphere. However, according to paper [22] this is rather unlikely, because gaseous oxygen cannot diffuse laterally between two spontaneously bonded wafers at a distance similar

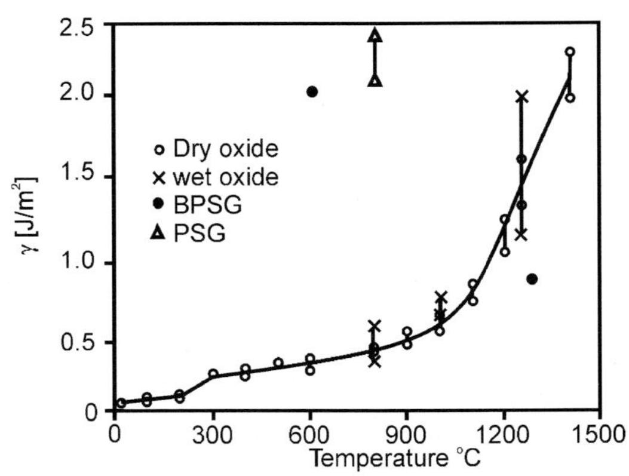

Fig. 4.8. Surface energy as a function of temperature of wafer-scale bonding covered with: 300 nm-thick wet oxides (of 3″ silicon wafers), dry oxides (wafers 3″), BPSG (4″), PSG (4″). Probes were annealed for 10 min in a nitrogen atmosphere [22].

to the wafer radius in the time and at temperatures applied in a typical process of bonding. It has been shown experimentally that there are no differences in the strength of bonding carried out in N_2 or O_2 atmospheres, although the number of bubbles on non-bonded areas is four times smaller in O_2 [46].

4.2.1.3. Voids

Non-bonded points and areas are called voids or bubbles. These two words are used somewhat interchangeably in literature sources. In this book the word voids mean empty non-bonded areas caused by mainly mechanical or physical factors. Bubbles have a more chemical origin. At least two types of non-bonded void can be distinguished: extrinsic voids, which evolve during the bonding of wafers at room temperature and do not recede after high-temperature annealing [22, 24, 27, 36] and intrinsic voids or bubbles [22], generated "in situ" during the annealing.

4.2.1.3.1. Extrinsic voids

Particles and other non-bondable materials at surfaces of bonded substrates or insufficient flatness of wafers generate extrinsic voids. Particles function as spacers and locally prevent ultimate surface contact, producing non-bonded areas in the form of empty cavities between two bonded substrates, with a specified diameter depending on the dimensions of particles and bonding strength. Dust particles lead to the formation of round-shaped voids [2], whose radius r_{void} is expressed by:

$$r_{\text{void}} = [1.2Et^3/\gamma]^{1/4} \cdot r^{1/2}, \tag{4.8}$$

where: E = Young's modulus, t = thickness of wafer, γ = surface bonding energy, r = dust radius.

The value of radius r_{void} depends on the procedure of wafer preparation and for hydrophilic washing can be given by:

$$r_{\text{void}} = 40r^{1/2}, \tag{4.9}$$

while for hydrophilic washing, followed by plasma treatment in oxygen-containing plasmas, it can be calculated from:

$$r_{\text{void}} = 11r^{1/2}. \tag{4.10}$$

High pulling forces, occurring between bonded substrates, deform the shape of particles. Therefore the dimensions of particles cannot be estimated on the basis of observation of void size. Moreover, voids usually have different structures: for example hard materials generate bigger, sharp-edged voids, while organic materials create smaller, ellipsoidal voids. Voids caused by surface imperfections and residual non-bondable layers (mechanical defects, residual resist and resins, metallic, dielectric layers, chemical impurities, etc.) are usually irregular (Fig. 4.9). Bigger forms of dust and other mechanical particles are usually easily visible before the process, so the bonded substrate can be cleaned.

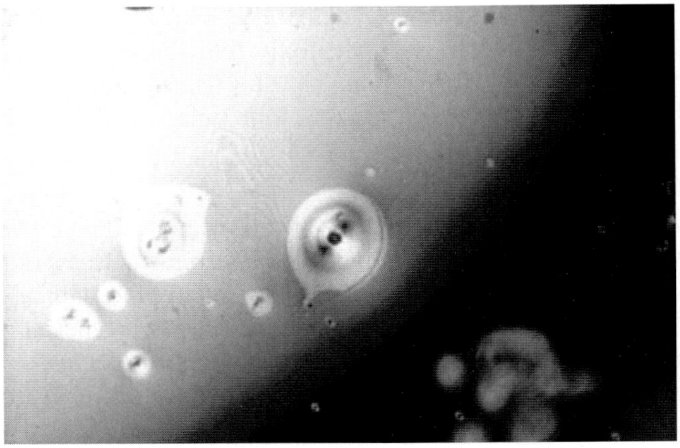

Fig. 4.9. Non-bonded areas: irregular shapes of dust or other small particles trapped between the wafers can be observed. The interference ring, which can serve for the determination of distance between non-bonded materials, is also visible.

The unaided eye cannot usually detect small-area imperfections or very thin layers of non-bondable materials. Any uses of testing devices can only worsen the state of the surface, so the process of washing, activation and bonding must be carried out as cleanly as possible. Washing and activation of wafers for anodic bonding usually needs class 100. Better conditions are necessary – preferably class 1 – for fusion bonding. This is very difficult at many laboratories and foundries not specializing in VLSI digital integrated circuit fabrication. Usually, laboratories at universities and small microsystem-oriented start-up companies do not use class 100. The solution is the use of megasonic cleaning and very careful, dust-free transport of wafers to the pre-bonding station. The most critical technical procedure is contacting of substrates to be bonded. Parkes [48] and Wilson [49] used the procedure of contacting wafers under the surface of very clean water. Wafers are transported from the washing station in covered quartz vessels and obtain ultimate contact immediately before bonding. In the author's practice the best results were obtained when wafers were positioned perpendicularly to the bottom of a vessel filled with filtered DI water. Stengl and co-authors [36] have proposed a spraying method of cleaning wafers before contacting, by jets of deionized water followed by a fast rotation. Water is sprayed between two wafers kept close to each other, and forms a water film. The wafers are rotated very quickly, and Corriolis forces remove water. A stream of water cleans surfaces. This method is called a local class 1 clean room, and has been implemented in industrial conditions.

Bonded wafers are usually wavy, not perfectly flat. What is more, the internal stress of substrates, caused mainly by thermal mismatch of sandwich of thin-film layers covering substrates, results in a concave or convex deflection, folding or warping of wafers. Deformed wafers can be bonded if the external force or pulling force appearing during bonding is sufficient to match surfaces. A higher value of surface energy of bonding γ ensures better matching of wavy surfaces [31]:

$$\frac{h}{l^2} < \sqrt{\frac{\gamma}{Et^2}}, \tag{4.12}$$

where: h = amplitude of wave, l = length of wave, E = Young's modulus, t = thickness of one of the wafers.

So, the higher the bonding energy γ is, the more wavy substrates can be bonded. For example, 625 μm-thick, 6″ wafers can be bonded, assuming that Young's modulus equals 168 GPa (see Table 2.4) $\gamma = 1\ \mathrm{J/m^2}$ and waving is characterized by: $h \approx 0.5$ μm, $l \approx 0.32$ cm. The boundary values of h–l pairs of spontaneous bond without influence of an external force at room temperature ($\gamma \approx 0.1\ \mathrm{J/m^2}$) are given in Table 4.4*.

The matt substrates, e.g. the non-polished, rough silicon wafers, do not bond. Such substrates before bonding should be covered with a bondable layer and/or polished. Abe [50] has revealed a strong dependence of roughness density on the formation of bubbles. The smoothness of the silicon substrate surface, typical for standard IC processes, can in some cases be insufficient for obtaining bonding without voids. The use of special polishing decreases the number of voids.

Many researchers recommend chemical–mechanical polishing (CMP) of surfaces prior to bonding [2, 32–34]. This ensures an excellent mirrored surface, because the surface layer penetration – the so-called Hörtz interaction or radius R_s – by polishing media, usually water–glass, is very small:

$$R_s = \frac{3}{4}\Phi\left\{\frac{p}{2K_pE}\right\}^{2/3}, \tag{4.13}$$

where: Φ = diameter of polishing powder particle, p = pressure corresponding to the force applied to a polishing tool, K_p = constant, determining the packing density of particles of polishing powder at the surface. The polishing medium used in CMP is water–glass. Silicon dioxide nano-crystals work as a "polishing powder". They are very small, their diameter Φ is equal to 100 nm, so, for K_p =

Table 4.4. The bond ability of wavy brand silicon wafers, data for spontaneous bonding at room temperature

Amplitude h [nm]	Wavelength l [mm]	Amplitude h [nm]	Wavelength l [mm]
10	0.25	40	2.5
14	0.5	200	5
17.5	0.75	400	7.5
20	1	800	10

* Equation (4.13) is not applied for anodic bonding, because matching of surfaces is determined by the pulling force caused by a strong electrostatic field. See other sections of this book.

0.5 and $p = 1.5$ MPa the penetration radius R_s equals 0.03 nm (0.03% Φ) can be obtained. Therefore, a surface suitably prepared in the CMP process is practically free of cracks and scratches. The CMP process can be utilized for pre-bonding preparation of unprocessed or microengineered wafers; very often this is the only method of preparing bonding wafers with deeply micromachined patterns [33].

Trapping of gases coming from the gaseous atmosphere surrounding the bonded substrates can also form extrinsic bubbles. This happens if the spontaneous bonding wave starts simultaneously from two or more points, or if substrates are waved in such a way that the trapping of gas bubble may occur. To prevent trapping of gases the bonded substrates are mechanically deformed in their central parts by a special tool, so the bonding wave starts from the desired point.

Some of the extrinsic bubbles are caused by organic particles trapped between two bonded substrates. Such impurities are subject to chemical reactions in the presence of the atmospheric oxygen trapped in a void surrounding the impurity, or generated during annealing. The reaction products significantly enlarge the surface of non-bonded area. The use of an external electrostatic or electric force can partly prevent the formation of large-area bubbles [47], but the local increase in gas pressure inside a bubble can affect residual destruction of the substrates.

4.2.1.3.2. Intrinsic bubbles

In contrast to the extrinsic bubbles that evolve in all types of brand silicon (Si–Si) or silicon covered with oxides (Si–SiO_2 and SiO_2–SiO_2) bonding, the intrinsic bubbles occur only for SiO_2–SiO_2 bonding*. The intrinsic bubbles are generated during the annealing of bonded substrates at about 200°C and they disappear when the temperature increases to 700–1000°C [6, 22, 27, 28, 51, 52]. Although this phenomenon occurs on both hydrophilic and hydrophobic surfaces (see section 4.2.2.1), it is more often observed for substrates immersed in HF [27, 51].

Intrinsic bubbles are not formed during low-temperature annealing at micromachined wafers with deep cavities [19, 53], which proves that such bubbles are generated by gases trapped in the area of interlayer. Two models of this effect have been proposed. In the first one it is assumed that water presented between wafers forms the bubbles [6, 28]. Water molecules adsorbed on silicon dioxide layers are released during the formation of surface siloxane bonds. The formation of bubbles begins at about 200°C to 300°C, when molecular water accumulated in micro/nano cavities existing in oxide layers is released, generating the bubbles. At higher temperatures this water dissociates and forms additional oxides and the bubbles disappear.

Harendt [53] has investigated the composition of a residual gaseous atmosphere existing inside deeply etched cavities by means of the method of mass

---

* This concerns Si/SiO_2–Si bonding. The bubbles also evolve during Si/Si_3N_4–Si_3N_4/Si bonding.

Schulze [1] has examined the bonded surfaces under the acoustic microscope for the frequencies $f = 200$ MHz and 400 MHz, detecting small and very small bubbles. Because of the small penetration depth of an acoustic wave (≈ 30 μm and ≈ 70 μm, respectively), the observed wafers need to be prepared prior to measurement (abraded and polished at a proper angle). Acoustic methods are often used to characterize the quality of bonding of wafers with metallic layers non-transferrable to infrareds.

Magic mirror method

In the method of magic mirror a light beam illuminates a mirrored, polished surface of one of the bonded substrates under small misalignment from the perpendicular axe. Reflected light is projected onto a screen or a photosensitive plate. Small convexities formed around voids and bubbles are visible at the screen as black contrast. This method is compatible with the acoustic method.

Thinning

Thinning and then etching of one of the bonded wafers until bubbles are detected is a destructive but precise method of estimation of voids and bubbles. This procedure is particularly useful when at least one of the wafers is oxidized. Bubbles undetectable for the IR method can be detected by means of thinning.

4.2.1.4. Transitory layers

Transitory layers do not appear between properly aligned (in a crystallographic sense), hydrophilic bonded surfaces of two brand silicon wafers. According to Stengl [38], Weldon [35] and Mitani [31, 53] the transitory layer, corresponding to siloxane bonds, has a thickness ranging from 0.16 to 0.4 nm and does not limit electric conductivity perpendicular to bonded surfaces. However, Schulze, who has studied electric parameters of the bonded brand silicon substrates, observed that voltage-current characteristics were linear for hydrophobic bonding (see section 4.2.2) and non-linear for hydrophilic bonding. This can be a proof of the presence of a transitory layer, which limits electric conductivity perpendicular to bonded surfaces [1]. Bengston states that the non-linearity of *I-V* characteristic is caused by the depleted layer remaining on the border of two bonded surfaces [2]. This layer has to be an effect of modulation of surface potential φ_s by the surface states, which are generated in thin interlayer oxides, formed on the bonded surfaces during the bonding procedure. Shimbo [6], and Furukawa [54] describe the high lattice compatibility in a direction perpendicular to the wafer surface, obtained after substrates anneal at 1000–1100°C. A silicon lattice is continued across the bonding, although in the closest vicinity of surface many dislocations occur.

In the transitory area of misaligned wafers, or wafers with different crystallographic orientation of surfaces, for instance (100) and (110), an amorphous layer of $SiO_{0.54}$, 0.5 to 0.45 nm-thick [54–57], was determined by means of the SIMS method [58]. Thickness and continuity of the silicon-oxide-like transitory

layer depend on the degree of disorientation of two bonded wafers and on the method of crystallization. The layer formed between two silicon wafers made of mono-crystal obtained by float-zone crystal growth (FZ) disappears after high-temperature heating for 2 hours at 1100°C. The continuous amorphous silicon-oxide-like layer has been observed in sandwiches of bonded substrates annealed at 1100°C made of Czochralski's mono-crystals (CZ), for disorientation higher than 3°. The layer does not disappear, even after prolonged heating at a temperature higher than 1100°C (several days, 1150°C). Relatively thick (12 nm) islands of amorphous areas remain after such a procedure [29, 55].

Between two bonded silicon wafers, if at least one of them is covered with a thermal oxide layer, a detectable (using TEM) transitory layer on the SiO_2–SiO_2 or Si–SiO_2 border [14, 55] does not appear. The SiO_2–SiO_2 surface of adhesion is not detectable even in images with very high resolution. However, a few small ($\phi < 2.5$ nm) defects can be distinguished, such as microscopic voids, perhaps caused by adsorbed water, which remain in numerous microcavities at a temperature of 700–800°C.

4.2.1.5. *Strength of bonding – tests*

Razor blade method

Maszara proposed a simple and elegant method for measuring the strength of fusion bonding [14]*. A thin metal blade (razor blade) is inserted between the bonded surfaces. The blade functions as a wedge and causes the formation of micro fractures in a limited area. Propagation of fractures is observed in infrareds, the value of surface bonding energy γ is calculated from knowledge of a sample and blade dimensions and elastic coefficients of silicon.

The relation between the thickness of blade y, length of fracture propagation L and surface energy γ is given by:

$$\gamma = \frac{3}{8}\frac{Et^3y^2}{L^4}, \tag{4.17}$$

where: E = Young's modulus, t = thickness of wafers.

The accuracy of this method, taking into account that the length of fracture propagation can be measured with an accuracy of ± 0.5 mm, equals approximately 20%. The surface energy determined in such a way is an average value for the area in which the blade is inserted.

Tensile/shear tests

Tensile and shear tests are good engineers' methods of bonding quality evaluation. These methods of determination of bond strength are based on examination of the force causing the cracking of test structures. The strength unit obtained by means of the measurement of tensile strength is N/m^2, while the razor blade method provides results of surface energy measurements in J/m^2. Mathematical correlation of results is not possible. For instance, the strength of an Si–Si bond

---

* This method can be used for tests of bonding quality of other bonding processes.

of test structures after an RCA hydrophilic procedure and annealing at 1000°C equals about 2.5 J/m^2, while its tensile strength is in the range of 20–60 GN/m^2. In some cases, when the strength of bond is significantly higher than of the bulk silicon, the tensile tests cannot be applied because the tested structure breaks in the bulk, not showing the true value of the bonding force.

4.2.2. Hydrophobic bonding

The spontaneous bonding of two hydrophobic silicon surfaces is much weaker than that observed for hydrophilic surfaces. At room temperature bonding energy is below 0.1 J/m^2; double and triple hydrogen bonds predominate. At a higher temperature (400°C) triple hydrogen bonds disappear, surface bonding energy γ increases to approximately 0.4 J/m^2, and Si–H direct bonds are formed. At a temperature of 600°C Si–H direct bonds are replaced with Si–Si direct bonds, while γ exceeds 2 J/m^2. At 700°C, γ equals 2.5 J/m^2. This is comparable to the maximal value of γ obtained for hydrophilic bonding at 1100°C after a prolonged annealing of substrates. Above 800°C all hydrogen bonds are replaced by Si–Si bonds; γ reaches 4 J/m^2. Infrared absorption spectroscopy confirms this mechanism [44] (Fig. 4.10).

The course of the chemical reaction is presented below:

$$\text{Si–H} + \text{Si–H} \rightarrow \text{Si–Si} + H_2. \tag{4.18}$$

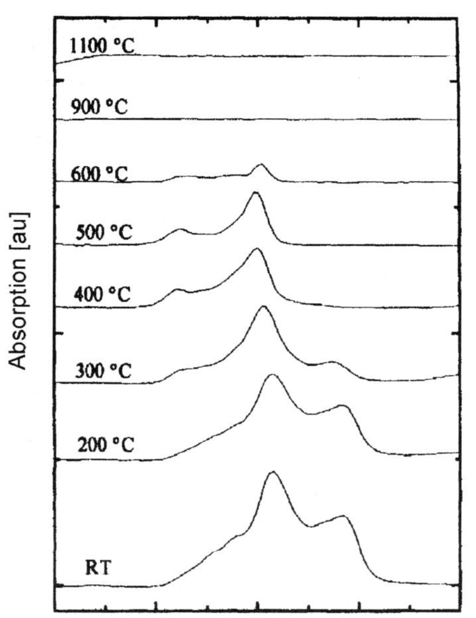

Fig. 4.10. Spectrum of the absorption of infrared radiation – hydrophobic bonding [44].

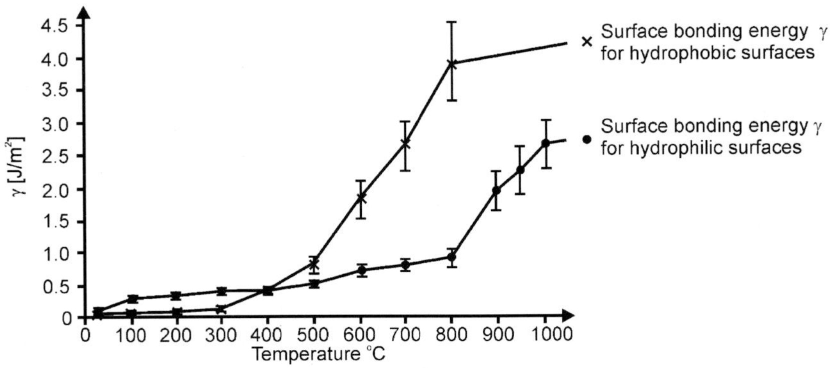

Fig. 4.11. Surface bonding energy γ as a function of temperature: hydrophilic and hydrophobic silicon fusion bonding [1].

At 400°C the surface bonding energy γ of hydrophobic bonding is significantly lower than the bonding energy γ of hydrophilic bonding. The maximal surface bonding energy γ of hydrophobic bonding is higher than the surface bonding energy γ of hydrophilic bonding; the bond is stronger than bulk silicon (Fig. 4.11).

4.2.3. Bonding of silicon to other materials

At present, fusion bonding of silicon substrates covered with silicon nitrides and carbides, diamond-like layers, sapphire or quartz, is not so well developed as fusion bonding of brand silicon. However, such materials are finding ever-wider applications in microsystem technology [59].

Nitride bonding, e.g. bonding of silicon substrates covered with a thin layer of silicon nitride (Si/Si_xN_y–Si_xN_y/Si bonding) LPCVD deposited from dichlorosilane in the presence of ammonia, under a pressure of 30 Pa, on hot (800°C) silicon substrates, has been presented in paper [60]. After deposition the Si_3N_4 layer substrates were treated with oxygen plasma (O_2, 200 W, 100 Pa, 30 min), washed (RCA-1), and then annealed in steam for 30 minutes at 900°C. Fusion bonding was obtained after the optimal annealing of the pre-bonded substrates for 4 hours at 1100°C, in nitrogen. Surface bonding energy γ did not exceed 1.5 J/m^2. It is believed that the mechanism of bonding [33, 60] of oxygen-enriched silicon wafers covered with Si_xN_y layer is similar to the fusion bonding of surfaces covered with silicon oxide. X-ray photoemission spectroscopy (XPS) confirms that sealing occurs only between oxynitride $Si_xN_yO_z$ layers formed on the Si_xN_y surface.

The authors of paper [21] studied the bonding of silicon wafers covered with 1 μm-thick Si_xN_y layers deposited by means of the LPCVD method and polished using the CMP method. After polishing, the substrates were washed in a solution of H_2SO_4:H_2O_2:H_2O, 1:1:5 at 80°C, and next immersed in a vapor of concentrated HNO_3, boiled in 70% HNO_3 at 90°C and finally washed in deionized water. The speed of spontaneous bonding wave formation equaled 3 cm/min. After heating at 1000°C, for 24 h, the surface bond energy γ amounted to $\gamma \sim 2.8$ J/m^2. This method was successfully used in the fabrication process of micromechanical bolometers.

High-quality bonding of quartz substrates has been described in paper [61]. Surface bonding energy of wafers hydrophilized in RCA1 and RCA2 procedures for 10′ and heated at 316°C equaled 2.0 ± 0.9 J/m^2. Patterned plates with 18 nm-deep cavities etched in quartz were bonded after 1 hour annealing in nitrogen, at 500°C.

The Smart Cut* fusion bonding of 4″, 327 μm thick silicon wafers and 4″, 525 μm-thick quartz substrates, heavily doped with boron ($N_A > 1 \times 10^{19}$ cm^{-3}) has been presented in [62]. Such a pair of wafers did not bow during the process of bonding, either at the annealing temperature of 150°C, or at Smart Cut process temperature 270°C, when the quartz layer implanted with hydrogen with a dose of 5×10^{16} cm^{-2} at 160 keV was separated.

The SOI substrates with a diamond-like layer have been discussed in paper [32], while the fusion bonding of diamond wafers covered with silicon carbide with silicon covered with thermal oxides (αC/SiC–SiO_2/Si) has been presented in paper [65]. The fusion bonding of (100) and (111) silicon wafers with 410 μm-thick, R-cut sapphire substrates at 400°C has been described in publication [66]. After cooling the silicon wafers cracked in the directions corresponding to ⟨100⟩ crystallographic axes. Some small regions evolved and they were heated at 800°C for 2 hours in air. The attained strength of bonding was equal to the strength of monolithic silicon.

The fusion bonding of GaAs/InP to silicon [67], along with the smart cut technique, become the standard processes of fabrication of heterojunction structures and lasers made of these materials [68, 69]. It is believed that the bonding of $A_{III}B_V$ to silicon, performed at 650°C, will be commonly applied in integrated optoelectronic circuits.

Excellent fusion bonding of GaAs wafers to sapphire, with surface bonding energy γ reaching 3 J/m^2, has been obtained for (100) GaAs hydrophilic surfaces, bonded to R-cut sapphire substrates at a temperature of 500°C in a hydrogen atmosphere. No cracks of bonded plates after cooling were observed, even in liquid nitrogen [66].

$A_{III}B_V$ materials (and materials similar to them) are not applied on a wide scale in silicon micromechanics. However, the development of MEOMS will lead to the integration of micromechanical and optoelectronic devices. Then the fusion bonding of non-silicon materials will be used more often.

4.2.4. Application of fusion bonding in microsystem technology – some chosen examples

Fusion bonding serves as a technique of assembling varying types of three-dimensional constructions for sensors, actuators and other more complex

---

* "Smart Cut" and "Unibond" are registered names of (respectively): (i) the process of separating thin layers by means of hydrogen implantation and high-temperature heating, (ii) SOI substrates fabricated using this method [63, 64].

devices. Mass-produced pressure sensors; accelerometers; pumps; valves; even small, post-stamp-size rocket propellers for the nano-satellite technique, have been produced using this method of silicon sealing. The number of applications of fusion bonding in silicon microsystem technology increases with the development of this domain of the technique.

4.2.4.1. Prefabricated wafers for membrane pressure sensors

The minimal dimension k_{min} of the chip of the classical pressure sensor with a square, anisotropically wet etched membrane (Fig. 4.12a, at left) is given by:

$$k_{min} = a_0 + 2u + 2\frac{d-m}{\sqrt{2}} + 2r, \tag{4.19}$$

where: $d =$ thickness of wafer, $m =$ thickness of membrane, $u =$ underetching of (111) plane, $a_0 =$ length of the edge of membrane, $r =$ minimal width of the frame.

For large membrane $a_0 = 1000$ μm and: $m = 20$ μm, $d = 380$ μm, $u = 3$ μm, $r =$ 200 μm, which are typical values of parameters, characterizing – for example – the structure of a piezoresistive pressure sensor for 100 kPa, the parameter k_{min} equals 1520 μm. For small membrane $a_0 = 250$, $k_{min} = 759$ μm. These results mean that the edge of a larger chip is almost one and a half times wider than the edge of a bigger membrane, and two and a half times wider for a smaller one.

Similar calculations performed for a structure with "inverted" geometry (Fig. 4.12a, at right) for $a_0 = 250$ μm, show that the parameter k_{min} equals 350 μm, that is almost two times less than possible in a classical structure. From reasons discussed earlier in this book, "inverted" geometry cannot be wet

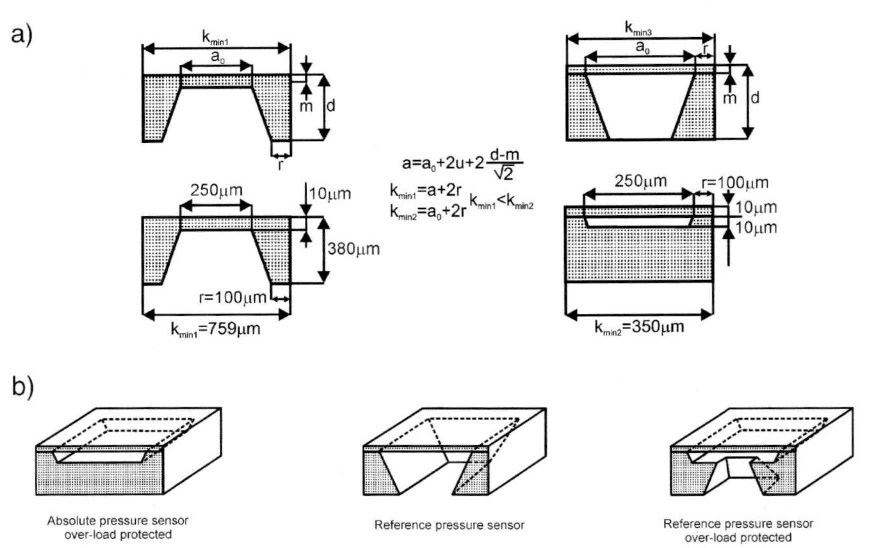

Fig. 4.12. Structures of pressure sensors: a) comparison of classical and "inverted" geometry, b) possible "inverted" constructions.

anisotropically etched through the etch window formed in a mask covering the (100) silicon wafer. Such geometry (Fig. 4.12b) may, in comparison, be easily obtained by at least two methods of fabrication applying fusion bonding of micromachined silicon substrates (Fig. 4.13) [70–72].

In the first of the described fabrication methods two (100) substrates are used. Initially, shallow cavities, with planar dimensions corresponding to membranes, are selectively etched in the first silicon wafer in KOH. The depth of cavities is selected in such a way that membranes to be produced can touch the bottom while deflected by a maximal permissible pressure. Next, a silicon wafer is fusion bonded to the micromachined substrate. Finally, membranes are formed by mechanical grinding and CMP polished.

In the second method, shallow cavities are etched in a similar way in the first of (100) silicon wafers. Next, a sufficient dose of protons at the precisely determined depth is uniformly implanted in the second substrate. Then, the implanted side of this substrate is bonded to the first, micromachined wafer. Next, the high-temperature annealing "Smart Cuts" a thin layer of silicon in

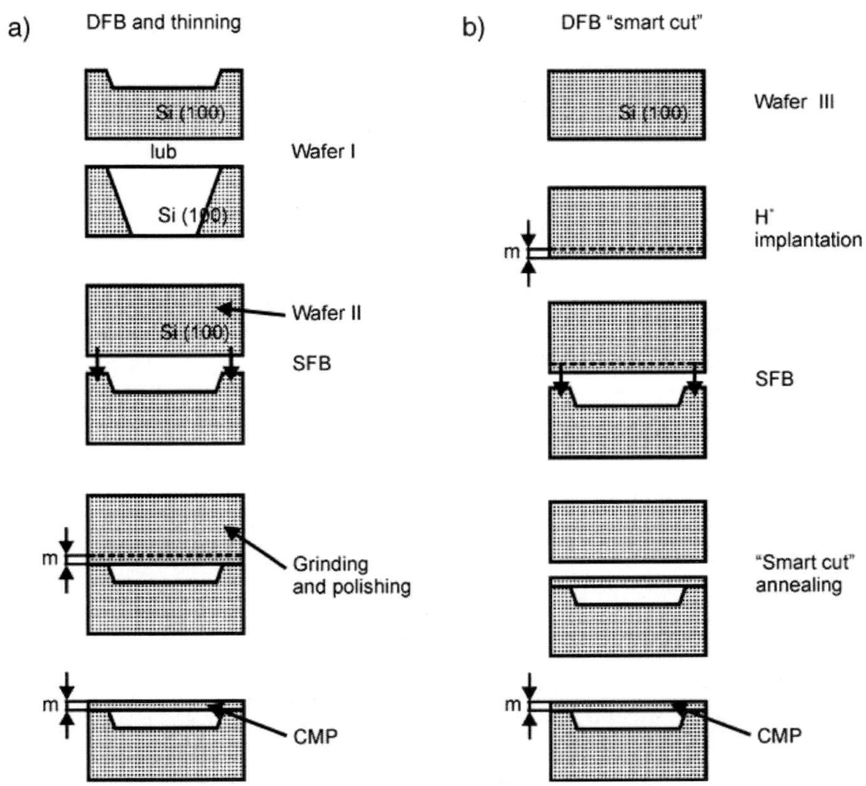

Fig. 4.13. Methods of fabrication of a pressure sensor structure with laminated membrane: a) thinning technique, b) "Smart Cut" technique [61, 70–72].

the plane of implantation, forming membranes. The surface is finally CMP polished.

The methods discussed above were applied in the manufacture of so-called prefabricated substrates with laminated membranes, which were developed in 1993–1995 [73–78]. These wafers [77], many inches in diameter, are technologically identical to classical, non-processed wafers. They have the thickness, diameter, smoothness and crystallographic quality of surface meeting general process requirements. Prefabricated wafers technologically satisfy the requirements of IC and CMOS production lines, which significantly simplify the fabrication of micromechanical sensors. Another advantage of these wafers is a very high repeatability of membrane thickness on the substrate scale and in the production series [75, 76]. Structures of pressure sensors obtained in such a procedure are overload-protected and/or much smaller than the classical structures. This is significant in the construction of sensors designed for the lower ranges of pressure, as well as in the fabrication of subminiature pressure sensors assembled in catheters used in prenatal examination or medical examination of small children.

Classical silicon membranes of pressure sensors are often etched in the SOI wafers, which are produced by fusion bonding of wafers covered with silicon dioxide. The etch front stops at the oxide layer; the thickness of the membrane corresponds to the thickness of the thin top silicon layer of the SOI substrate. This method of membrane thickness control, called the oxide etch-stop, can easily replace other popular etch-stop methods, but is more expensive because of the high price of SOI substrates [77, 78].

4.2.4.2. Sensors, actuators and microsystems

Fusion bonding is used in the technology of many different pressure and acceleration sensors. For example, a barometric sensor (barometric altimeter) with excellent measuring accuracy (± 0.01 kPa), designed for application in civil aviation and in the air force, has been described in paper [79]. In this sensor the force of pressure applied on the membrane stiffens the vibratory beam, which changes the self-resonance of the beam. An electronic phase-frequency system causes the renewed resonance of the beam: the balancing signal is a measure of pressure. This extremely complex three-dimensional structure is fabricated by means of the hydrophilic fusion bonding of only its silicon components (Fig. 4.14).

A differential, silicon, capacitive pressure sensor, consisting of three micromechanical structures, etched anisotropically using the wet method and bonded utilizing fusion bonding, was presented in paper [80] (Fig. 4.15). A thin, supported silicon membrane, located between two perforated immovable silicon reference electrodes, was applied in this sensor.

A simple, capacitive pressure sensor produced by use of fusion bonding in three-step technological procedures, applying photolithography, was discussed in paper [81] (Fig. 4.16).

A capacitive accelerometer made on an SOI wafer, with a movable mass in

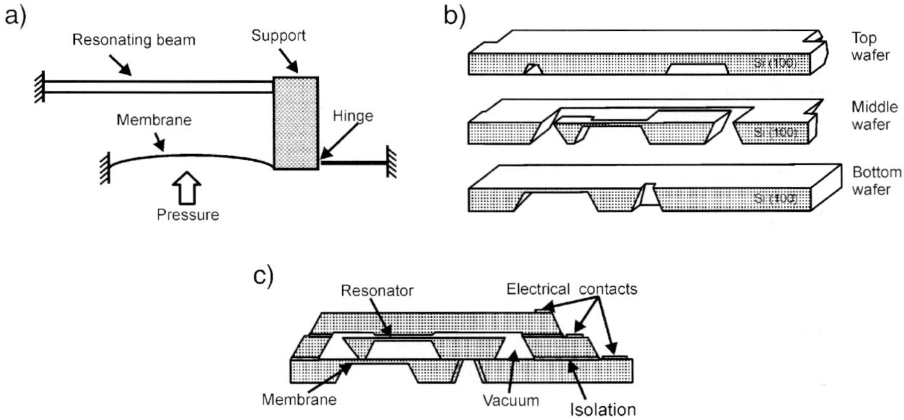

Fig. 4.14. Pressure sensor for aviation (made by Sextans Avionique): a) working principle, b) structures before bonding, c) cross-section [79].

the form of a flat silicon electrode, hung 1 μm above the surface of the wafer, was described in paper [82]. The distance between electrodes can easily be determined technologically, because it corresponds to the thickness of the SiO_2 layer of the SOI substrate. In paper [83] Kleasen with co-authors presented vibration detectors and angular resonators (Fig. 4.17), which were made of two silicon plates, bonded using the SFB method.

A piezoresistive pressure sensor with membrane fabricated on SOI substrate by means of the oxide-stop method was presented in paper [84] (Fig. 4.18). In the fabrication of this sensor, the self-compensation system with a double Wheatstone's bridge was applied.

Direct bonding of silicon is very often used in the technology of valves, pumps and flow controllers for analytical microsystems [84–89]. The construction of a valve exemplifying this group of microsystems is shown in Fig. 4.19. The valve contains a bossed membrane, etched anisotropically using the wet method with the application of the compensation of under-etching of convex corners. The membrane is thermally moved by diffused heating resistors, utilizing the bimetal effect in the Al–Si bi-layer, which shuts or opens the gas flow through the valve-seat.

A microminiature, silicon-quartz, iodic discharge lamp has been discussed in paper [90]. These integrated lamps with dimensions ~2 mm × 2 mm × 2.4 mm were produced on 100 mm silicon wafers. Their fabrication process included tri-layer fusion bonding performed under high pressure (10 MPa) in a halide atmosphere at a temperature of 1300°C. Light obtained in the lamps was very intensive: 2650 lumens for supplying power of 42 W.

A complete modular jet-propulsion device for nanosatellites, made of silicon with use of fusion bonding (Fig. 4.20), has been presented in paper [91]. Its driving block consists of filters, proportional dosage valve, PZT actuators, pressure transducer and heat exchanger with temperature sensor and jet nozzle.

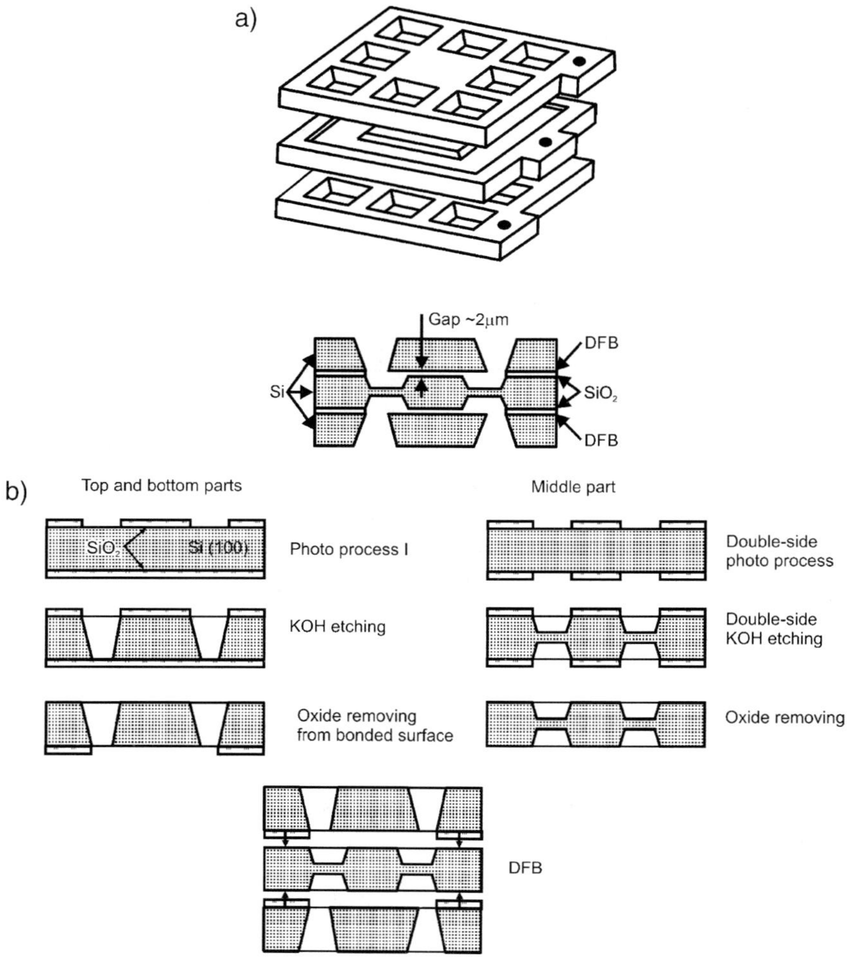

Fig. 4.15. Differential pressure sensor: a) construction and cross-section after DFB, b) scheme of fabrication [80].

Silicon components were fabricated applying different micromachining techniques. The jet flow of a compressed gas is controlled precisely by the microprocessor system coupled with a module.

A rocket micromotor elaborated by Ayon with co-workers from MIT (MA, USA) was described in paper [92]. This 18 mm × 13.5 mm × 3 mm device contains six fusion-bonded wafers. The fabrication process includes ten processes of photolithography and eighteen procedures of anisotropic dry and wet etching micromachining of silicon. The intensive internal system with ETOH liquid medium under a pressure of 30 MPa ($\sim$300 atm) for cooling of the combustion chamber is used. The heat exchange factor amounts to over 200 W per 1 mm^2 of cooled surface. The fuel (methane and oxygen) is kept under high pressure.

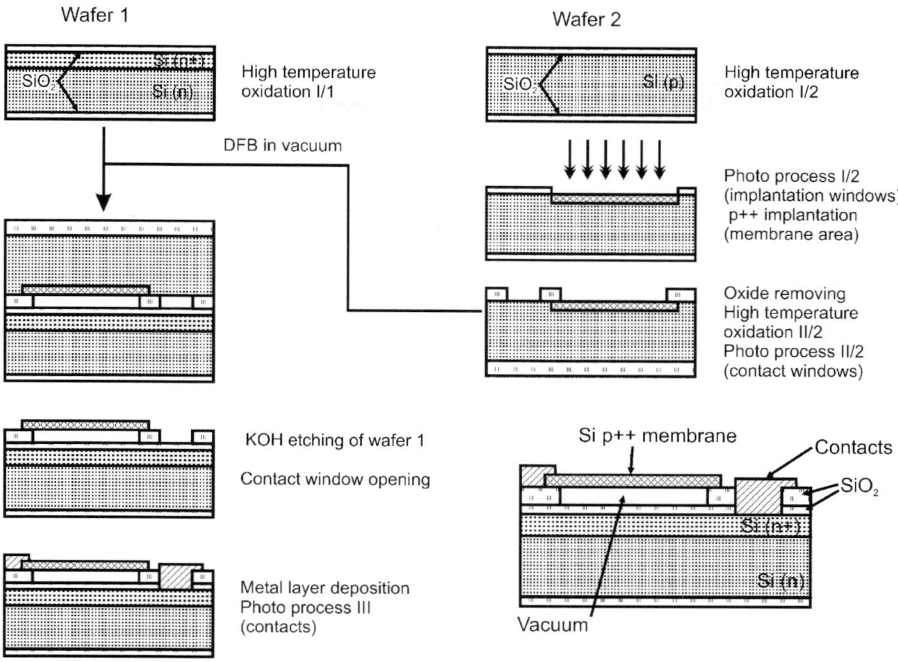

Fig. 4.16. Capacitive pressure sensor – construction and fabrication procedure [81].

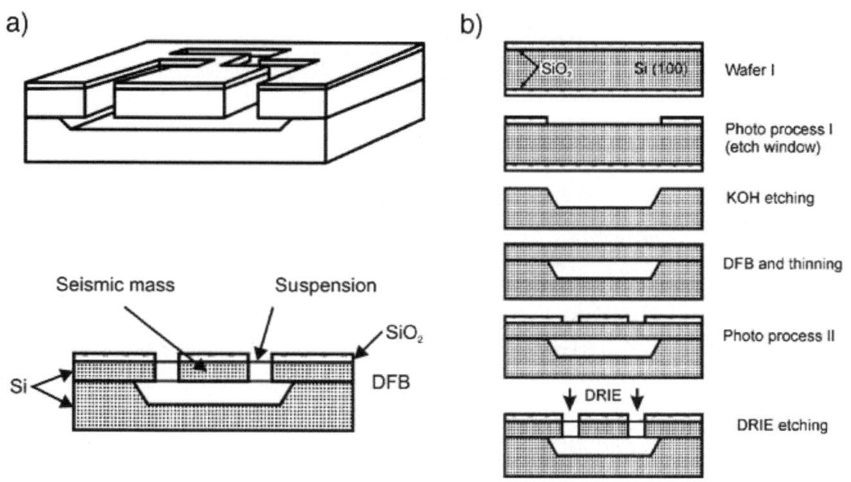

Fig. 4.17. Vibrating structures: a) construction and cross-section, b) scheme of fabrication [83].

The temperature of outlet gases exceeds 1725°C. 1 N thrust has been attained for pressure of 1.2 MPa (~12 atm), which corresponded to a power of 750 W. The ratio of motor mass to thrust equals 1:85. For the maximal pressure of 30.0 MPa (~300 atm), the expected parameters are as follows: thrust equaling

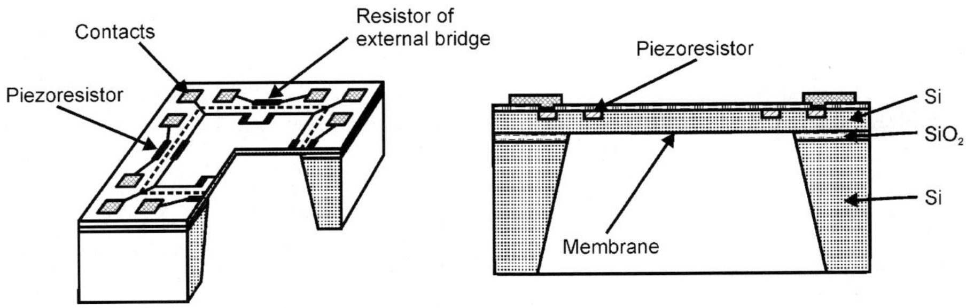

Fig. 4.18. Piezoresistive pressure sensor on SOI substrate: construction and cross-section [84].

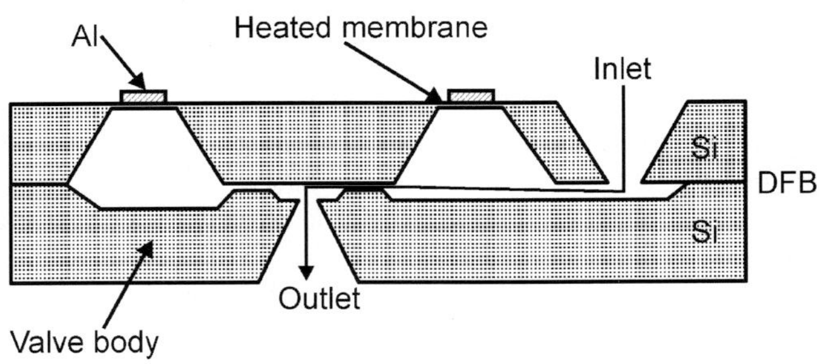

Fig. 4.19. Silicon microvalve [86, 88].

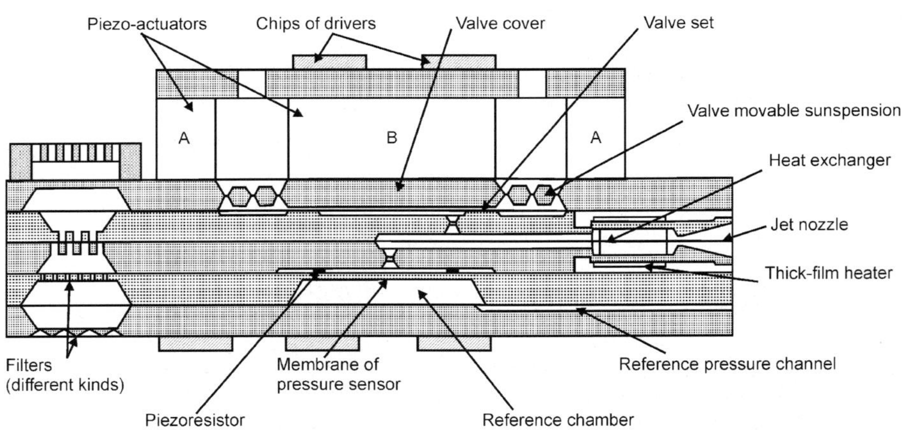

Fig. 4.20. Cross-section of the jet-propulsion engine [91].

15 N, power of 20 kW and the ratio mass/thrust 1:1000. Fuel and coolant were delivered to the motor by steel tubes with Kovar (Fe–Ni–Co alloy) flanges, which are bonded to silicon through a glass frit (Fig. 4.21).

The other example of the sophisticated microsystem in which a fabrication process fusion bonding was used is the atomic force microscope, integrated to an autonomic silicon structure, which is able to penetrate the area of

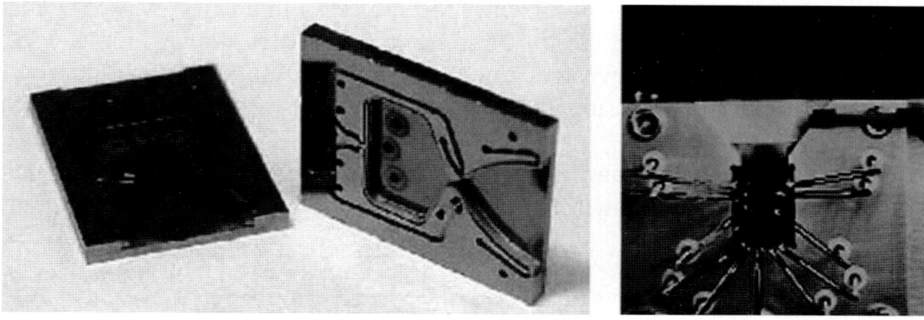

Fig. 4.21. Silicon rocket propulsion engine [92].

3 μm × 3 μm and to investigate the nano-properties of the investigated surface [93, 94].

The unique vibrating pressure sensor structure has been described in paper [95]. In this device a silicon resonator is suspended onto two prisms made of thick silicon membrane. The pressure of gas deflects the membrane; a small movement of the membrane produces tensile stresses in the resonator, changing its resonant frequency of vibrations. The components of the sensor have been wet and dry and fusion bonded (1100°C, 4 h) after wet high-temperature oxidation.

Silicon galvanic mirrors for laser television have been presented in paper [96].

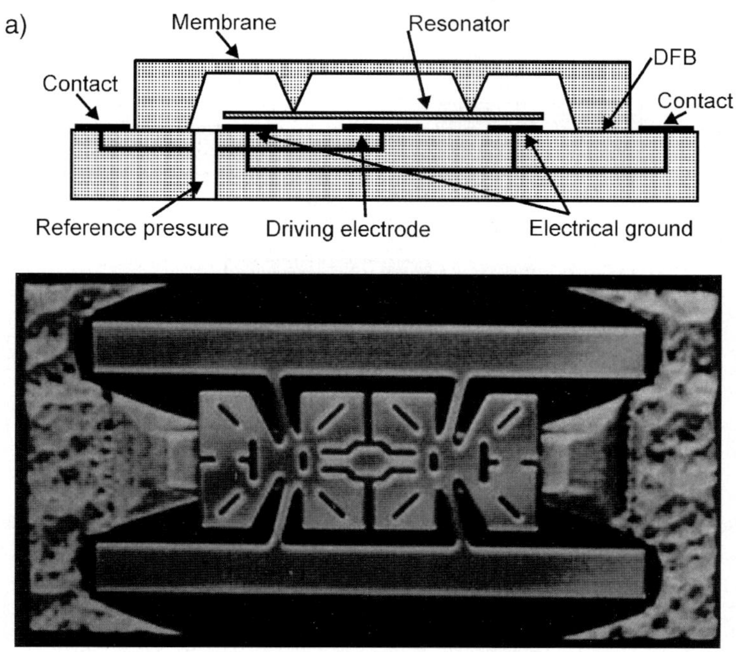

Fig. 4.22. Pressure sensor: a) construction, b) view from the resonator side [95].

This revolutionary technique is currently the subject of intensive research*. The fabrication process utilizing fusion bonding and CMP is shown in Fig. 4.23. Varying types of exciting micromachines: micromotors, clocks, flying butterfly-like objects, made on SOI substrates were described by Minotti and Ferreira [97].

Many other fusion bonding technological applications can be found in the domain of microchemical microsystems. As an example, in paper [98] liquid

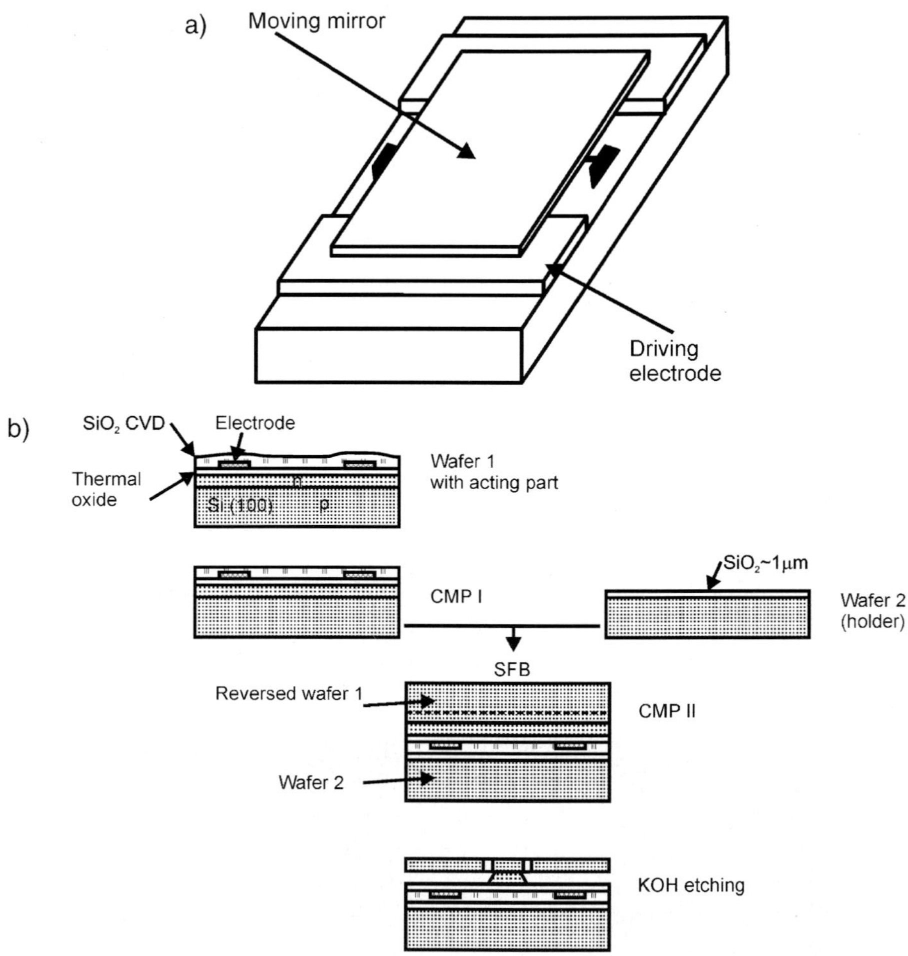

Fig. 4.23. Scanning mirror for laser television applications: a) construction, b) method of fabrication [96].

* In order to develop the potential market possibilities of laser television, a joint-venture of companies Schneider Rudfunkwerke A. G. TǦ rkheim and Daimler Benz A. G. Stuttgart has been formed under the name "Laser Display Technology".

chromatography planar columns were presented. Channels of hexagonal columns were formed by deep dry etching of a pattern in a silicon wafer, and fusion bonding to another silicon wafer. CO_2 sensors with an IR filter made by fusion bonding were described in paper [99].

4.3. LOW-TEMPERATURE BONDING

High-temperature fusion bonding of silicon wafers needs a final formation of bond by curing of the sandwiched substrates at high temperature – about 1000°C. This is a critical step for materials having low decomposition temperature or thermally mismatched (large difference of thermal expansion coefficients). Such materials must be bonded at a reasonably low temperature. Additionally, high-temperature annealing can shift doping profiles, and may introduce defects and/or destruction of dielectric isolating layers or metallic contact thin-film layers. Thus, it cannot be used widely in microsystem technology, either as a back-end technique or as the final chip assembly method.

Low-temperature bonding (LTB) is a sealing process in which the definitive formation of bond can be obtained at room temperature or, after annealing in the low temperature range, below 300°C. There are many types of LTB processes, some of them are modified hydrophilic and hydrophobic fusion bonding, others utilize a sol-gel or low-melting glass interlayer, thin layers of resins or foil, and last but not least, eutectic soldering of materials. Bonding at room temperature can be obtained in so-called HF and NaOH bonding.

4.3.1. Modified hydrophilic bonding

An additional activation of hydrophilic bondable surfaces by plasma treatment in oxygen [100–107], made after typical washing and wet activation of substrates, can lower the temperature of the final bond curing well below 300°C. Such a hydrophilic fusion bonding process can be classified among the low temperature bonding methods.

At room temperature the surface bonding energy γ of oxygen/argon plasma-treated silicon is higher than γ of conventionally washed (RCA) silicon (Fig. 4.24a). γ equals from 1 to 1.5 J/m^2, so strong sealing at room temperature results from the increased initial concentration of OH groups at bonded surfaces, which provides a higher number of siloxane bonds formed at lower temperatures [100]. This seems to be one of two dominant mechanisms of low-temperature hydrophilic bonding. On the other hand, increasing of OH groups leads to the formation of bubbles.

Room temperature spontaneous adhesion of silicon wafers is – as commonly agreed [37–39, 99, 101] – an effect of bridging of the bonded surfaces by water clusters. These water clusters must be removed from the bonded interface, to allow forming of direct silicon-to-silicon or siloxane bonds. Removal of surface water in high-temperature direct bonding is obtained by residual high-temperature oxidation of the bonded surface of silicon during annealing. Farrens and

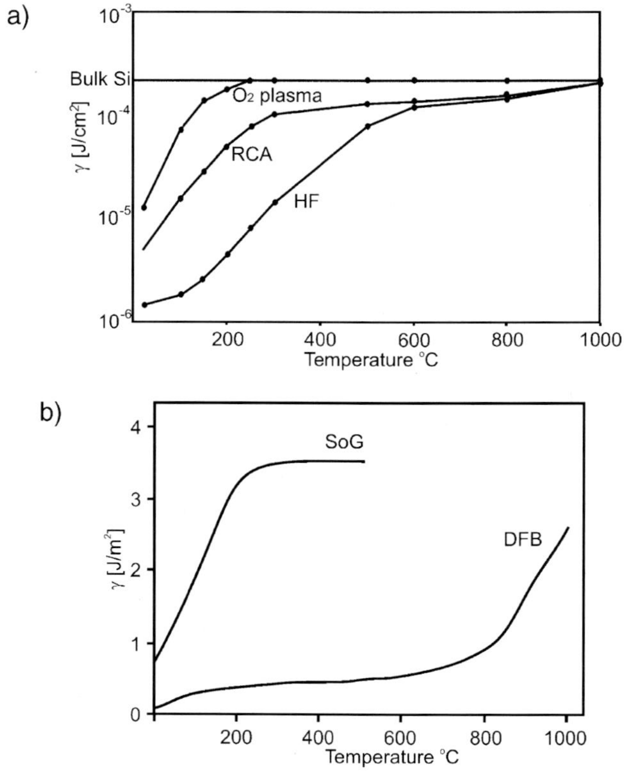

Fig. 4.24. Bonding energy γ as a function of temperature annealing: a) comparison of hydrophobic, hydrophilic (RCA procedure) and oxygen plasma-treated LTB, b) comparison of fusion bonding and spin-on deposited sodium silicate LTB (SoG) [100, 109].

Tong [37, 101], as well as Bengtsson and Amirfeiz [102], explained that plasma treatment removes molecular water from bonded surfaces; as a result, covalent direct Si–Si bonds can be formed, even at room temperature, as observed for high-temperature direct bonding. Lean and De Voe [100] compared activation energy E_a of high-temperature hydrophilic and hydrophobic bonding to low-temperature O_2 plasma hydrophilic bonding, finding the lowest activation energy for plasma-treated surfaces at about 300°C. Data collected in Table 4.5 show representatively the differences between the here three bonding methods discussed above. Hydrophobic surfaces express the lowest E_a below 150°C, hydrophilic at 300–800°C, plasma-treated at 300°C. A lower E_a can be interpreted as a higher chemical affinity of bonded surfaces at the desired temperature, documenting the highest bond ability of the plasma-treated silicon surfaces.

Oxygen plasma treating leads to the formation of fixed charge and interface states in a silicon-oxide sandwich [101]. An increase in interface states (traps) can be reversed by forming gas annealing in N_2 or H_2 at 450°C, for 10 minutes.

Table 4.5. Bonding activation energy in eV for different procedures [100]

Hydrophobic direct	Hydrophilic direct	Plasma LTB	Temperature range °C
0.047	×	0.205	RT-150
	0.159	×	RT-300
0.252	×	×	150–800
×	×	0.087	150–300
×	0.041	×	300–800
0.164	0.193	×	>800
×	×	0.003	300

Such a treatment decreases by half the final bonding surface energy γ, reducing by 70–96% fixed charges and electrically active states.

A new method of pre-bonding plasma treatment, "nanoPREP technology", was proposed by the Carl Süss Company (Germany) [103]. In the method, wafers to be bonded are treated with an ambient pressure plasma jet. The active plasma is limited to a narrow but long area. A nitrogen/oxygen atmospheric pressure plasma "strip" is generated by use of a dielectric barrier discharge (DBD) at 20 kHz and scanned over all wafer sizes (up to 300 mm) substrate or small chips. A very hydrophilic surface of silicon, silicon oxides, germanium, Ga As and borosilicate glasses is ensured after short ($\sim$2 min) processing.

4.3.2. Boron, amorphous layer and fluorine bonding

In this method the surfaces to be bonded, after typical silicon wafer washing, are plasma-treated in boron-containing plasmas, to achieve close contact to form a spontaneous bond, and annealed. The bonding energy γ of the spontaneous bond at 20°C is low, about 40 mJ/m^2, γ grows to 0.9 J/m^2 at 150°C and finally reaches 2.5 J/m^2 at 350°C. The course of the chemical reaction during bonding is:

$$\mathrm{Si{-}B{:}H + H{:}B{-}Si = Si{-}B{-}B{-}Si + H_2}, \qquad (4.19)$$

in which SiB represents a strong silicon–boron bond and B:H weaker hydrogen–boron bond. During annealing the hydrogen bonds are replaced by direct silicon–boron–silicon bonds, and hydrogen is released. The key factor of high γ at low temperature is the release of hydrogen from the bonding interface. This may be obtained for varying pre-bonding treatments, e.g. implantation, B_2H_6 or Ar plasma treatment or sputter deposition of α-Si followed by a HF dip.

An example of the technological procedure of boron bonding, described in paper [108], is as follows:

Wafers were washed (RCA1 procedure), rinsed in 1% HF in order to remove all residual oxides, then treated with B_2H_6 /He/Ar plasma for 30 s to 5 min, for plasma self-polarization DC potential 100 V. After this treatment the wafers were rinsed again in 1% HF, dried and kept in low vacuum $\sim$700 Pa. Next the wafers contacted (γ for HF dipped wafers is only 10–20 mJ/m^2) and annealed in air at 350°C.

The modified method applied bonding of silicon wafers, covered with amorphous layers, produced by Ar ion bombardment in the RF excited plasma (13.54 MHz), at approximately 3–10 Pa, by 15 s to 20 min, for DC potential 400 V. After plasma treatment the wafers were dipped in 1% HF and contacted, then stored for 24 hours in low vacuum. Final formation of bond was made at 400°C, the maximal value of γ was comparable to the bulk silicon fracture energy.

So, the most advantageous features of boron bonding are:

- very high bonding energy γ – reaching the fracture energy of bulk silicon
- perfect sealing of silicon wafers.

Fluorine-enhanced bonding

Strong, covalent silicon-to-silicon bonds can be formed between two silicon wafers covered with native silicon dioxide formed by washing of wafers in a HNO_3:HF solution, as reported in paper [109], next brought to contact at room temperature and annealed at 100°C. γ reaches 2 J/m^2, which is higher than the bonding energy of the hydrophilic bond of two RCA1-treated silicon wafers annealed in 700°C. This can be obtained because:

- fluorine in the chemically grown oxide forms strong bonds,
- by-pass products of the chemical reaction can be absorbed by nanocavities formed at the surface of the oxide,
- the surface of oxide very slightly etched by HNO_3:HF solution is perfectly smooth.

4.3.3. Bonding through low-temperature melting glass

In this method of bonding of silicon wafers the glass interlayer is deposited onto the surface of one of the sealed wafers. Spontaneous bonding of wafers is usually supported by mechanical pressing of the wafers.

The following kinds of low-temperature melting glass interlayer are used:

- glass frit (sintered gasket) made of Corning 7570 lead glass deposited selectively by means of a screen process, fusion temperature ≥ 415°C [110],
- spin-on deposited liquid sodium silicate or aluminum phosphate glass, fusion temperature 200°C (sodium silicate) or 350°C (aluminum phosphate) [9, 26, 111, 112],
- ion sputtered boron glasses [7–9, 113, 114], fusion temperature 450°C.*

The one most often applied in microsystem technology is sol-gel bonding

* LTB bonding of silicon substrates doped by phosphorus or through boro-phosphoro glasses deposited onto oxidized silicon wafers is not possible. The temperature of formation of sealing increases to 900°C.

through the sodium silicate or phosphorus glass. Both methods can be used in a standard laboratory environment, as they tolerate class 1000 conditions.

4.3.3.1. Sol-gel bonding

Wafers are coated with a spin-on deposited highly concentrated aqueous solution of sodium silicate (N_2O:SiO_2 as 1:3). This material has good wetting properties and forms a homogeneous layer at hydrophilic surfaces of wafers. Due to the elasticity of sol all of the smaller imperfections or dust particles on the surface are filled up; the surface of a layer is very smooth and hydrophilic. The bonding wave appears immediately after contacting wafers and propagates quickly.

Sol consists of partially polymerized silicate ions and H_2O (up to 20%). The sol layer loses water during heating. Thus, the sol-gel method is preferred for sealing of wafers covered with SiO_2 because the silicon dioxide layer absorbs the water from the sol, which results in strong siloxane bond formation and perfect sealing. According to Quenzer [9], the SiO_2 layer is indispensable to perform void-free low-temperature bonding, but good sealing can be obtained for oxidized silicon ($\gamma \sim 3.2$ J/m^2), oxidized silicon to silicon covered with silicon nitrides ($\gamma \sim 2.1$ J/m^2), and for brand silicon ($\gamma \sim 2.0$ J/m^2). Bonding of two surfaces covered with silicon nitrides is possible, but the bond is weak ($\gamma \sim 1$ J/m^2) and voids evolve.

The energy of bonding γ increases with increase in temperature (γ at ambient temperature equals 1 J/m^2) and may reach 3.2 J/m^2. The maximal temperature of bond formation should not be above 200°C because at higher temperature the gel SiO_2 layer peels off. Wafers need to be annealed for 24 hours [109]. After bond formation the wafers may be annealed for a short time – a few minutes – at 350°C in order to stabilize sealing. Deng and others found [111] that interaction between spin speed and temperature of annealing is the most important factor influencing the energy of bonding; the refractive index of the interlayer depends strongly on its thickness (2.2 to 1.1 for 83.7 nm to 7.2 nm). Roughened wafers can be covered with thicker layers; in that case bond formation should be made at 150°C for 30 min, then at 200°C for a few hours. Higher temperatures cannot be applied. Liquid glass should not be stored for more than 2–3 months, because it suffers degradation under the influence of atmospheric CO_2 [9]. Another method of treatment is curing of spun silicate at 90°C for 1 hour or left overnight at 25°C, which leads to a very high value of surface bonding energy – about 2.7 J/m^2.

The biggest disadvantage of the method lies in the spin-on deposition of an interlayer. Corriolis force scatters liquid glass at walls of the equipment, the glass quickly dries there, large particles of the glass dust re-deposit at the bonded surface, causing extrinsic voids. The spin-on equipment must be stored clean and dust-free.

The silicate based sol-gel bonding of glass, ceramic, sapphire, NaYaG, $LiNa_3$,

Table 4.6. Combinations of surfaces to be bonded through boron glass

Bottom wafers	Top wafers			
	p-type Si	Wet Si/SiO_2	Dry Si/SiO_2	Si/Si_3N_4
p-type Si	+	+	+	+
Wet Si/SiO_2	+	+	+	+
Dry Si/SiO_2	+	+	+	+
Si/Si_3N_4 CVD	+	+	+	+
Si/SiO_2 covered with Al 0.5 μm	−	+	+	−

+, good bonding; −, bad bonding.

CaF_2 and MgF_2 substrates was reported by the Schott Glass Company*. A silicate solution is applied between two bonded wafers, components are brought into contact, and stored undisturbed for 6–12 hours. Aligning of wafers is allowed for 2 minutes after contacting. The final bond is obtained after 1 week curing or 120°C short-time annealing. The bonded sandwich has to be cycled at 77 K–200°C to release stresses. The bonding interlayer thickness varied from 100 nm to 2.2 μm. Helium leak-proof sealing was obtained.

4.3.3.2. Boron glass bonding

Bonding of silicon wafers with use of a boron glass interlayer 1 to 5 μm thick has been discussed in papers [7–9], [114]. Washed and hydrophilic wafers are covered with a thin glass layer by magnetron sputtering and bonded directly after this deposition. The sandwich of wafers, is pressed (a few kPa) and annealed at 450–475°C for 10–30 minutes [8]. Boron glass fuses at the temperature of 450°C and planarizes the small geometric imperfections of the surface. Various combinations of bondable surfaces are possible (Table 4.6). Bonding of brand silicon, silicon covered with silicon oxides, nitrides or metallic layers can also be performed.

The sealing temperature in the presence of small quantities of P_2O_5 and addition of BPO_4 is – according to the phase diagram of the ternary system SiO_2–B_2O_3–P_2O_5 [114] – significantly higher. BPO_4 melts at temperatures a few hundred degrees centigrade higher than B_2O_3. P_2O_5 melts at even higher temperatures than BPO_4. Therefore, the process of bonding needs to be carried out at increased temperature†.

The resistance of boron glass bonding to the impact of water, alcohols and esters is lower than the resistance of the sol-gel interlayer. What is more, boron glass shows a tendency to form crystals of boron acid under the influence of

* Method presented at Photonic West, 25–30 Jan. 2003, San José, CA, USA, see web.site or: M. Strzelecki, L. Gilroy, N. Wychoff, R. O'Maldey, L. Gilroy, D. Schimmel, *Sticking together*, SPIE Magazine of Photonics Technologies & Application, May 2003.

† The phosphor-silicon glass (PSG) can also be used in bonding of silicon wafers covered with thermal oxides [111]. The fusion temperature of PSG exceeds 1000°C and bonding is performed at a temperature of 1100°C after 30 minutes of heating. The quality of bonding is excellent. However, this is not a low-temperature process.

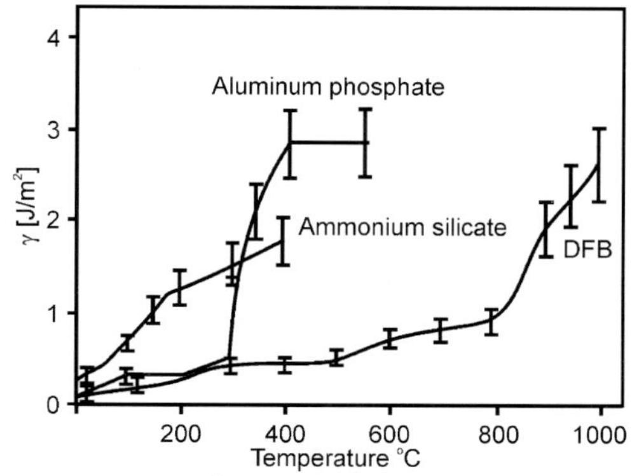

Fig. 4.25. Bonding energy γ of silicon wafers as a function of temperature of heating: aluminum phosphate, ammonium silicate and DFB [105].

ambient atmosphere. Thus, wafers covered with boron glass should not be stored too long at increased temperatures.

Other methods of LTB silicon wafer bonding through a low-melting interlayer reported in the scientific literature play a secondary role in microsystem technology. The LTB process with an interlayer of sodium-poor glass deposited from a mixture of powder and isopropyl alcohol was described in paper [110]. Tetraethyl orthosilicate (TEOS) bonding was shown in papers [6] and [112], and ammonium silicate and aluminum phosphate bonding in paper [103]. The bonding energy of wafers bonded through an ammonium silicate layer was $\gamma \sim 1$ J/m^2, sufficient to ensure sewing of the bonded sandwich. A very high surface bonding energy γ was obtained for aluminum phosphate [103] (Fig. 4.25), but unfortunately many voids evolved.

4.3.4. Room temperature bonding

The simplest procedure of very low temperature (20–180°C) sealing of wafers and/or details is so-called adhesive polymer bonding. A thin layer of adhesive is spread by use of the spin-on method or is used in the form of a foil*. Sealing is attained by pressing the wafers against each other and heating at a suitable, reasonably low temperature. Small repeatability, thermal instability, gassing and moisture penetration are the factors that bring the reliability of this method of bonding into question. The process of adhesive bonding is not compatible with microelectronic procedures, and cannot be treated as a true bonding technique (no irreversible chemical bonds are formed!). However, there are several application examples of this method in micromechanical device fabrication which seem to be very attractive for fluidic microsystems fabrication.

* Foil "bonding" has been successfully used, together with glass–silicon anodic bonding, to form spectacular gas-injecting systems for micro-gas-chromatographs and other micro-fluidic devices.

Typical spin-on adhesive sealing of wafers is based mostly on SU8 negative resists, which, when UV-exposed, have low volume shrinkage and high resistance against wet chemicals. The silicon–glass substrate sealing procedure starts with Piranha washing followed by dehydration at 200°C for 40 minutes. The SU8 layer is spin-on deposited on the first silicon substrate, soft-baked at 95°C for a few minutes, UV-exposed in order to form the desired pattern and post-baked at 90°C for 10 minutes; then it is developed (unexposed SU8 can be removed in propylene glycol monomethyl). The second, glass substrate is spin-coated with SU8, the resist layer is pre-baked, substrates are brought into contact and SU8 layers are joined together under the influence of low pressure. Following this, the SU8 interlayer is blanked exposure through the glass substrate and the SU8 interlayer is finally hard baked at 150°C. Razor tests show a surface energy of sealing of 0.42–0.56 J/m^2 [115].

PDMS (polydimethyl siloxane) sealing needs plasma treatment of spin-coated layers and immediate contacting of surface, followed by low-temperature annealing. PDMS viscosity is reduced to spin thinner films by xylene, a small amount of photoinitiator of polymerization DMAP (dimethoxy phenyl acetophenon) is added if pattern formation is needed (420 nm light). Silicon or glass substrates are Piranha washed, rinsed in DI water, and spin-dried in pure N_2. Next the PDMS layer is spin-coated, dried and O_2 plasma-treated at about 20 Pa for 20 seconds. Substrates are brought into contact under a pressure of 0.7–1.4 kPa and annealed at 50°C for 10 minutes. Bonding is achieved by a hydrolysis cure of the plasma-damaged PDMS surface which becomes hydrophilic after plasma treatment [115]. PDMS sealing is suitable for contact bonding of PDMS layers itself, glasses and oxidized silicon.

Very strong (pull tests force 19 MPa!) BCB (bisbenzocyclobutene) of silicon and/or glass sealing can be provided by covering one of the sealed substrates with adhesion promoter AP 8000 (Dow Chemical Company) followed by spin-coating with BCB 3022-46 from the same producer, and pre-curing in vacuum for a few minutes at 70°C. Final forming of sealing needs annealing at 180°C for 4 hours under a small pulling force. Patterned bonding can be achieved after etching of substrates heated to 200°C covered with BCB in CF_4/O_2 plasma at 0.1 Pa for 30 minutes [116]. Such a method was used for hydrophobic valve fabrication with channels covered with strongly hydrophobic octafluorocyclobutane (C_4F_8).

Low-temperature strong (pull tests force 4.3 MPa) sealing of silicon and glass using a spin-on deposited Teflon®-like amorphous fluorocarbon polymer cured at 160°C was reported by Oh and co-workers in paper [117]. The method was used for assembly of a microfluidic system in which excellent chemical resistance was among the important factors.

HF bonding

HF bonding is defined as bonding at room temperature, in which aqueous solutions of hydrofluoric acid are applied to firmly bond materials typically used in silicon micromechanics (such as: silicon, glass, quartz and their combinations). The procedure of bonding, described in papers [118] and [119] was relatively simple (Fig. 4.26). Wafers were washed (hydrophilic procedures) and etched in a 1% aqueous solution of hydrofluoric acid (1% HF) for 1 minute. Next the wafers came in close contact, and a drop of a weak HF solution (0.5–1% HF) was introduced into the gap between the wafers. Wafers were pressed against each other for at least a dozen or so hours by a force equivalent to a pressure of 0.04 to 1.3 MPa.

The mechanism of HF bonding has not yet been explained. On the basis of the results of SIMS examination of the interlayer formed between bonded surfaces it can be inferred that SiO_x complexes can appear on bonded surfaces. SiO_x complexes contain hydrogen and fluorine that form the interlayer (Fig. 4.27). The thickness of the interlayer decreases with increase of pressure applied to the bonded wafers: for 1.3 MPa it is smaller than 5 nm. The strength of bonding depends on the concentration of HF solution and on the applied pressure (Fig. 4.28). The strength of Si–Si, Si–glass bonding equals 4 to 10 MPa and is sufficient for cutting or sewing of the sandwich.

NaOH bonding

The other method of room temperature bonding is sealing of wafers by use of concentrated NaOH. This method was successfully used by its inventors (Becker et al. [120]) for low-temperature bonding of two quartz wafers and of quartz to Pyrex glass. In the process, several drops of a 30% aqueous solution of NaOH covered clean, hydrophilic surfaces of wafers. Next the wafers were

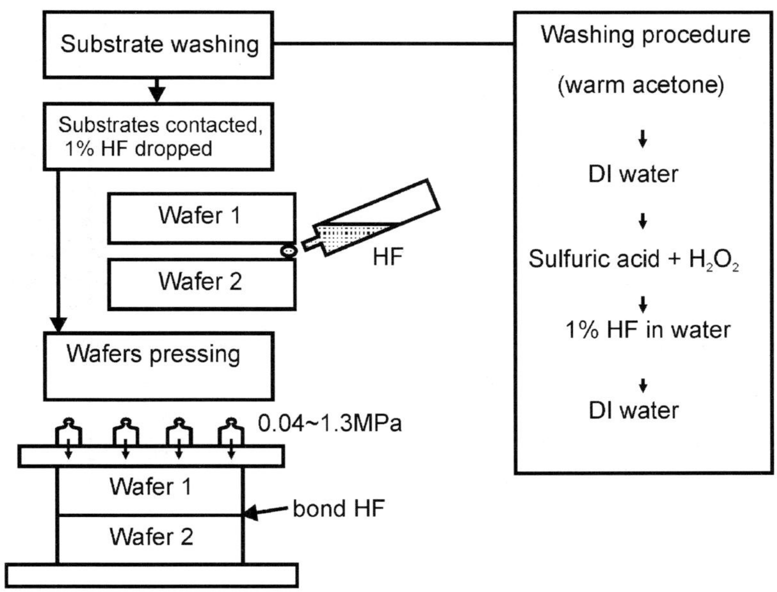

Fig. 4.26. Procedure of HF bonding [118, 119].

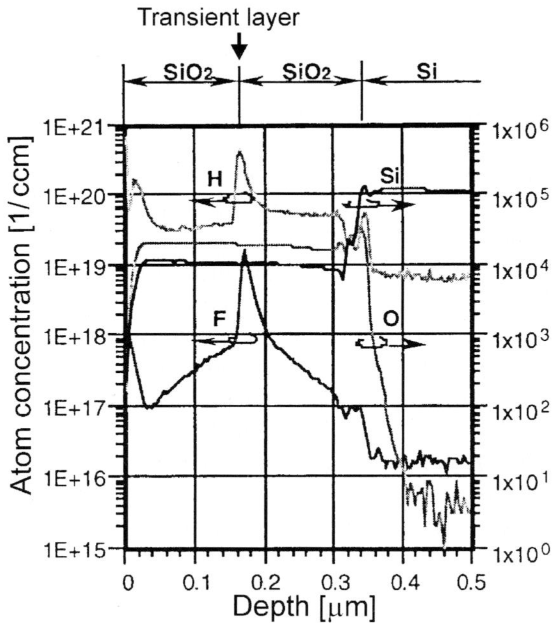

Fig. 4.27. Profiles of H, F, O and Si in the interlayer (bonding HF Si/SiO_2–SiO_i–Si) [118].

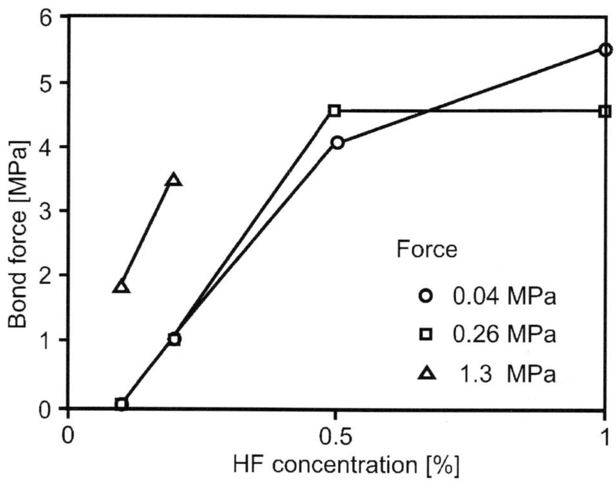

Fig. 4.28. Strength of bonding determined by the method of destructive pulling test as a function of HF concentration for different external force contacting silicon wafers [119].

pressed against each other to form a thin, continuous NaOH layer between them, and annealed at 500°C for 1 hour pressed at approximately 0.1 kPa. The destructive pull test force exceeded 7.7 MPa.

4.3.5. Eutectic bonding

The Au–Si eutectic soldering of silicon structures to metal housing is an industrial process which has been applied for many years in the packaging of silicon microelectronic chips. The application of this process in microsystem technology seems to be very natural. Gold with silicon form a eutectic alloy with composition 19 ± 0.5 at% Si in Au at a temperature of 363°C (Fig. 4.29).

The eutectic alloy of Au/Si covers a substrate on which the silicon wafer or structure has to be bonded. Bonded parts are covered with a thin-film layer of gold. The bonded parts are localized at the substrate and a sandwich is heated to a temperature a little higher than 363°C. Then, at the eutectic point, the liquid phase of the eutectic gold–silicon alloy is formed, gold dissolves in silicon, the eutectic point shifts to a higher temperature and the solid-state phase of the gold–silicon bond is formed. An example of the bonding procedure of silicon wafers through gold thin-film layers was shown in paper [121]. Samples were washed, then native oxides were dry etched away (PERIE or RIE). Immediately after dry etch, a 0.6 μm-thick gold layer was deposited onto the bonded parts by use of magnetron sputtering at circa 10^{-3} Pa. The gold-coated surfaces were contacted, and the samples were pressed lightly against each other. A eutectic bond was formed after 4 hours annealing at 400°C in vacuum.

Usually the temperature of the true sealing process exceeds the eutectic point.

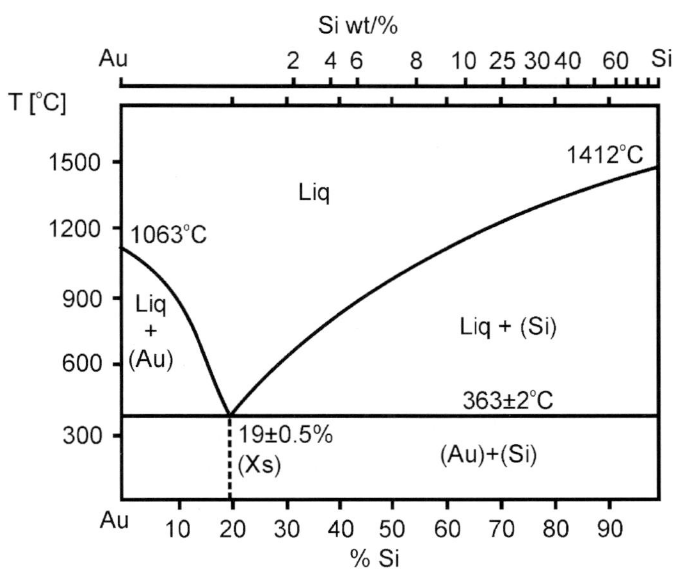

Fig. 4.29. Si–Au phase diagram.

This involves silicon dissolution in gold and stops the uncontrolled diffusion of gold to silicon. At higher sealing temperatures texturization of the bonded surfaces occurs, and fibrous microstructures with eutectic composition are produced (19 at% Si in Au) [122], which makes obtaining continuous, uniform bonding difficult or impossible. In order to avoid texturization it is recommended to perform a very short period (about 10 minutes) of Si–Au eutectic bonding at 365°C.

Low-temperature eutectic bonding has reasonable but difficult technological properties. The main advantages of Si–Au eutectic bonding are its technological compatibility with IC processes and a low temperature of process that does not destroy the metallization. On the other hand, the biggest drawback of this method of bonding is weak wetting through gold of residual and technological oxides, which occur on the silicon surface. Intrinsic oxides, which evolve during bonding in the air, need to be "broken" using the method of mashing, in order to ensure direct contact between silicon and gold. This mechanical operation is quite "brutal" and can destroy the fragile, three-dimensional micromechanical constructions fabricated on bonded wafers. In addition the method of mashing causes "congealing" of bonded details in random positions.

Si–Au–Si bonding is unstable in time and the failure of vacuum-tight connections can occur. Moreover, gold can reduce the lifetime of charge carriers in silicon. Si/SiO_2–Si bonding with use of Cr, Ti and gold layers, that adhere to SiO_2, requires complex preparation. The process of bonding needs to be performed in vacuum (similar to some Si–Au–Si bonding). Difficult access to bonded details during bonding is another unfavorable feature of eutectic bonding, which makes its wider application in microsystem technology impossible. As well as silicon to silicon gold eutectic bonding another eutectic sealing has been described.

Si–Si eutectic bonding with a Si–Cr/Au interlayer at 455°C to 520°C for pressure of $4–5 \times 10^{-4}$ Pa has been discussed in paper [123]. The authors of work [123] announced, that they have obtained a strength of eutectic bonding comparable to fusion bonding. This appears quite probable, taking into consideration the very small dimensions bonded by those parts.

Nd-YAH laser light (1064 nm)-induced Al/Au residual eutectic bonding of Pyrex-like glass and silicon substrates was presented in paper [124]. A metal thin-film layer was deposited on silicon and glass bonded surfaces. Substrates were contacted, scanned, heated with focused laser light, through the glass substrate metallic interlayers to the eutectic point, forming a residual bond.

Eutectic bonding of silicon wafers covered with thermal oxides (Si–Si, Si/SiO_2–Si) with use of Ti (30 nm) and Au (120 nm) bi-layers was presented in paper [122]. Bonding was carried out at 520°C. During the process the phenomenon of sintering (also observed in the Si–SiO_2–Al array) enriched Ti layers with silicon and led to the local discontinuity of the oxide layer. An Au–Si eutectic alloy, as well as an Au–$TiSi_2$–Si alloy, were formed. A fibrous microstructures did not evolve, and bonding remained continuous.

A successful low-temperature Si–Si bonding by an Al interlayer* was also discussed in paper [122]. In this method one of the silicon wafers with hydrophobic surfaces was covered with a magnetron-sputtered 1 μm-thick Al layer. Wafers were contacted and annealed at 650°C, very good bonding was obtained. The usefulness of Si–Al eutectic bonding in bonding of silicon wafers is questionable, because at the temperature of successful bonding (650°C) a degradation of aluminum metallization occurs. What is more, this process is not compatible with IC technology, though it can be applied in the initial steps of the formation of three-dimensional micromechanical structures.

Si–Si bonding by an Al–Au interlayer was discussed in paper [125]. A pressure of 45 MPa was applied at a temperature of 350°C for 60 minutes. Next, samples were annealed at 155°C for 500 to 1000 hours. Bonding was formed by an Au_4Al alloy interlayer, its stability was ensured by excess of gold. Silicon to silicon bonding with use of Al–Au interlayers can be very useful in micromechanics, due to very low temperature of bonding (<350°C).

Silicon to silicon bonding with use of an Al–Ge alloy (30 at% Ge, eutectic phase 424°C), designed for MEMS application, was presented in paper [126]. Si–Si bonding through a layer of silicides PtSi was described in paper [127]. The bond was formed by sputtering of Pt on one of the bonded substrates, contacting of bonded surface and annealing at 350°C for a short time followed by annealing at 700°C for 2 hours. The bonding obtained was excellent, with very good electric conductivity across the bond. Low-temperature Au–In–Au eutectic bonding was shown in paper [128]. It is possible that bonding with the use of layers of Al–Ge, PtSi, $TiSi_2$ will be applied on a wider scale in future.

4.3.6. Application of low-temperature bonding in microsystem technology – some chosen examples

The low-temperature direct vacuum bonding of silicon to silicon was applied in the fabrication process of the microminiature accelerometer and gyroscope made by NASA's Jet Propulsion Laboratory†. Glass frit bonding was used for assembly of a commercial pressure sensor from Motorola Co. [129]. Direct room temperature bonding after rigorous RCA cleaning of two glass parts of a fluidic chip (more than 2 years of perfect work, as reported) was shown in paper [130]. Foil bonding and glueing were used to build a gas-injection device for the first integrated gas chromatograph‡ and micro pump with valve [131] (Fig. 4.30). SU-8 bonding, which seems to be the most popular method of low-temperature sealing of varying materials used in gas/fluidic device fabrication, has been used several times in microsystem technology. Chosen examples are Fabry-Perrot resonators [132], micronozzles [115], chemical and biochemical reactors and analyzers [133, 134].

---

* Eutectic alloy Al–Si is formed at a temperature of 577°C for 12 at% Si.
† www.nasatech.com.
‡ www.mtigc.com., this solution will be discussed in more detail in the following chapter.

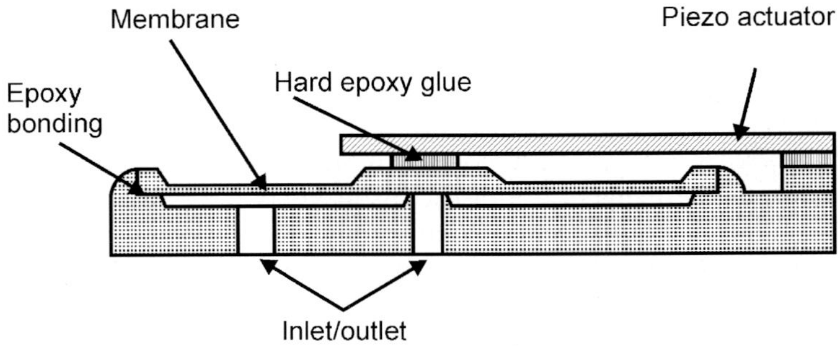

Fig. 4.30 VAMP micropump [131].

The HF bonding of a device for capillary electrophoresis (μCE) was described in paper [119]. The chip dimensions were 47 mm × 22 mm × 2 mm. It was made of IWAKI 7740 glass and TOSHIBA-T4040 synthetic quartz. Canals for distribution of analytes and buffer, as well as a detection chamber, were etched in glass and quartz substrate, through a sandwich Cr–Au mask (20/200 nm-thick), in a strong HF (46%) at 30°C. Glass and quartz wafers were bonded by use of a 1% aqueous solution of HF and a pressure of 0.04 N/cm^2, at room temperature. In the described device, fabricated by the development laboratory of SHIMADZU Co., good separation of amino acids was obtained (Arg, Met, Gly) (Fig. 4.31). A similar device, NaOH bonded, was presented in paper [120].

A three-dimensional silicon construction, consisting of three details with dimensions equaling 200 μm × 350 μm × 860 μm, assembled in vacuum by means of this method, was described in paper [125]. Au–In–Au eutectic bonding was used to assemble a thermoelectric cooler [128] and a micro-optical device [134].

The eutectic Si–Au/Al–Si method has been applied to bond the elements of a detector of infrared radiation [125] (Fig. 4.32).

Si/Au eutectic bonding with a Ti layer, carried out under a pressure of about 10^{-5} Pa at a temperature of 380°C for 3 minutes, has been utilized in technology of a microgripper (Fig. 4.33) [135]. The drive of grips is provided by Ni–Ti–Cu layer of shape memory alloy (SMA). Dimensions of the device are 1000 μm × 200 μm × 380 μm, thickness of silicon beams equals approximately 12.5 μm, the SMA layer is about 5 μm thick. Maximal spacing of microgripper equals 110 μm, while pressing force amounts to 40 mN.

4.4. ANODIC (ELECTROSTATIC) BONDING

Electric field-enhanced, low-temperature (below 500°C) bonding of metals or semiconductors to inorganic dielectrics, especially of silicon to glass, is called anodic bonding, electrostatic bonding or electrostatic welding.

Anodic bonding was invented – as the secondary result of research work on conductance of glass – by Pommerantz in 1966 and patented in 1968 in USA

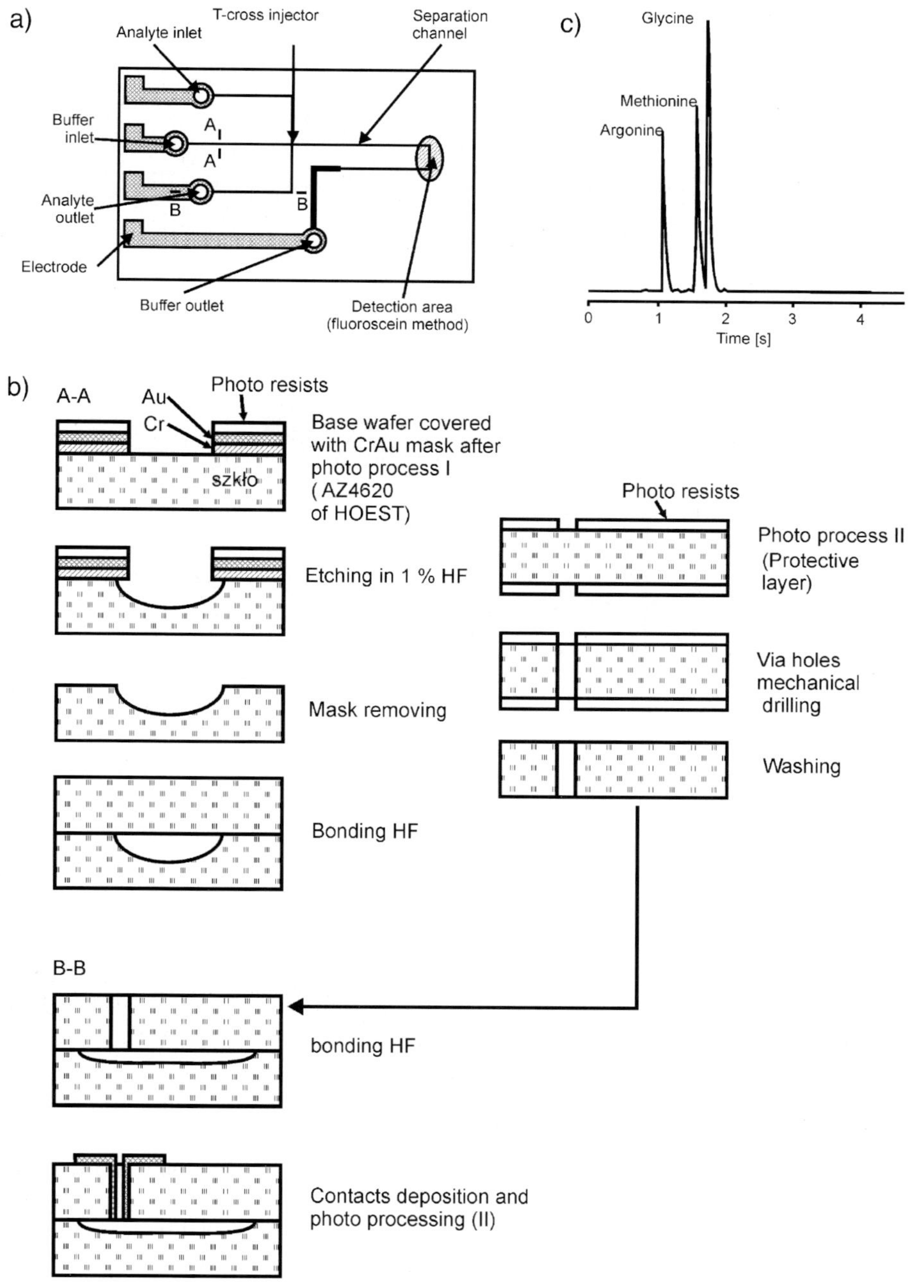

Fig. 4.31. μCE SHIMADZU liquid chromatograph [119]: a) lay-out, b) fabrication method, c) example of separation of amino acids – carrier of 50 nM triborate, pH = 8.6, detection: fluoroscein (FITC), argon laser 488 nm, 8 mW.

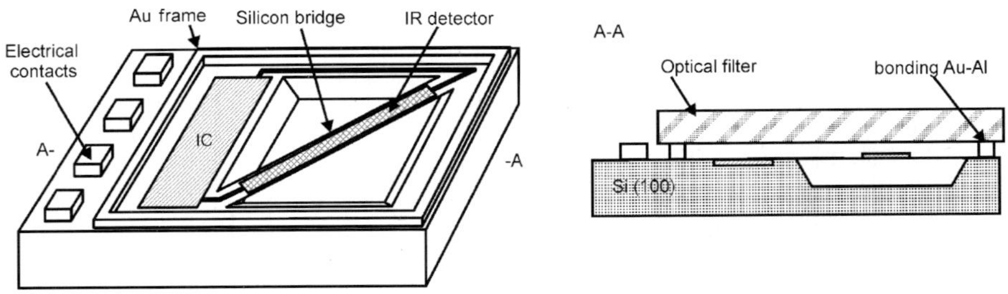

Fig. 4.32. Radiation detector: frame fabricated by means of the method of Al–Si bonding [125].

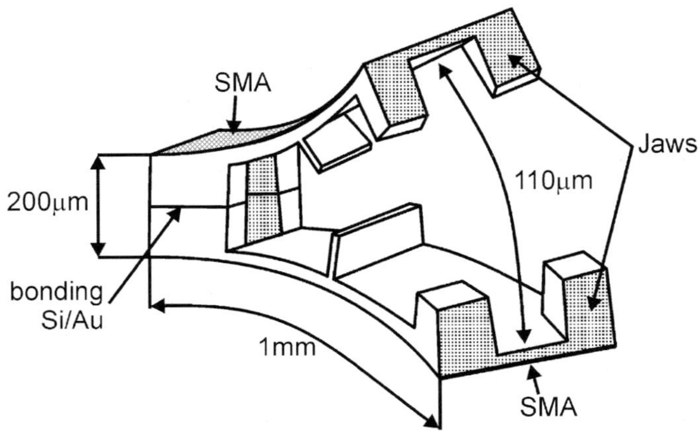

Fig. 4.33. Microgripper [135].

patent no. 3397278. The patent claims from 1968 were broadened by Pommerantz, Wallis and Dorsey in 1969 in USA patent no. 3417459. The first, most often cited work of Wallis and Pommerantz, that concerns anodic bonding, was written in 1969 [136]. Partially new patent claims were filed in 1974 by the company General Electric in USA patent no. 378321, in 1981 by the company Boening in USA patent no. 4294602 and in 1990 by the National Research Development Corporation in UK patent no. 9020908.1.

Electrostatic bonding, according to the patent claims of Pommerantz, consisted in bonding of inorganic, ion-conductive insulator (glass, quartz, sapphire, alundum ceramics) to electron conductive metals and semiconductors (including silicon, germanium, GaAs). Suitably prepared, smooth surfaces of materials were brought into contact and then heated to a temperature below the softening point of insulator in order to increase its conductance. The insulator was electrically polarized negatively (cathode) in relation to the metal or semiconductor (anode).

Pommerantz, Wallis and Dorsey prematurely put in their patents different combinations of bonded materials and conditions of anodic bonding, which resulted in patenting of various technological absurdities, e.g. anodic bonding of precious metals (Pt, Pd) to insulators with ionic conductance, which, as has been verified in further research, is not possible. They did not recognize that, among others, the very important condition of sealing of materials by the anodic bonding method is forming of a strong oxide bond between anode and insulator. Some of the proposed combinations, for example anodic bonding of gallium arsenide to glass, or sapphire to silicon, have never been technologically developed sufficiently to achieve wider applications in microelectronic or micromechanical technology. Pommerantz properly predicted that anodic bonding would be applied in electronic technology on a wide scale. In his first patent he proposed – among other things – to use anodic bonding for hybrid, multichip assembly of semiconductor devices to glass substrates, as a method of fabrication of electric connections between structures and metallic thin-film paths on glass and as a method of air-tight sealing and encapsulation of structures (Fig. 4.34).

Although anodic bonding was invented many years ago, and at present silicon to glass anodic bonding is, besides wet etching of silicon, one of the most important processes in silicon microsystem technology, the nature of this process has not yet been fully understood and explained. The literature of this subject provides dispersed, often contradictory information, without the crucial know-how.

For years Pyrex 7740 glass of Corning has been the preferred material, widely used for anodic bonding of silicon and glass. In the 1990s new types of glass for anodic bonding were fabricated (SD-1/2 of Hoya, Borofloat 33 of Schott). The main goal of the author is to present state-of-art details on the anodic bonding phenomenon, as well as to give a detailed technological, experimental, analysis of the applicability of old and new glass. Parts of the experimental material obtained by the author are presented in this book for the first time, although the author's research on anodic bonding has been carried out for years.

At the beginning of the next section types of glass for silicon–glass anodic bonding are discussed. Next, the mechanism of anodic bonding is examined including the division into phenomena occurring near the cathode and anode

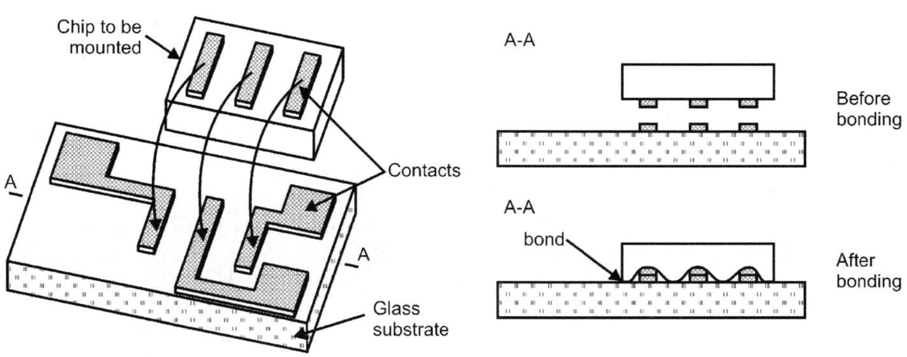

Fig. 4.34. Application of anodic bonding according to Pommerantz (1966).

with an emphasis on the depletion layer. Then, the role of electrostatic pressure, which supports the ultimate contact of bonded surfaces, is described, along with chemical reactions, models of bonding, charge flow and activation energy of the process of charge transport. So-called conditions of good bonding of silicon and glass plates are presented, the influence of the selection of process parameters on the quality and strength of bonding is analyzed, as well as unwanted after-bonding wafer deformations and stress generation. Wafer-scale silicon to glass sandwich bonding, the multi-layer and selective bonding and bonding of small details is discussed. Examples of microsystems – taken mainly from the author's experience – are presented.

Silicon–glass anodic bonding – introductory information

The basic equipment required to perform silicon to glass anodic bonding includes a heated conductive table with a flat, smooth surface; a high-voltage DC power supply; DC current meter; and a touch-type needle polarization electrode (Fig. 4.35).

Usually the silicon wafer is placed on the hot table first, with the glass ones on top. Wafers adhere to each other and come into close contact due to the Van der Waals forces. Immediately after contact, Newton's rings and interferential colors formed at the wafers interface can be seen through the glass wafer (Fig. 4.36a/1). The silicon–glass sandwich is heated to bonding temperature (300–500°C). Next, the sandwich is polarized by high voltage (500–2000 V): the silicon wafer, placed on the table, is polarized positively, while the glass wafer, located on silicon, is polarized negatively. The electrical conductance of hot glass is high so, under the influence of the strong electric field, the ion current of movable positive charges – mainly sodium – flows in glass. As a result the sodium depletion layer reach in negative charge evolves near the silicon–glass boundary, forming a kind of flat capacitor polarized by high voltage, whose plates are a few dozen nanometers apart. A drop in polarization voltage occurs only at the silicon–glass boundary. The result is the formation of a very strong electrostatic attraction force pulling the bonded materials against each other. Surfaces of silicon and glass are brought to ultimate contact, the distance

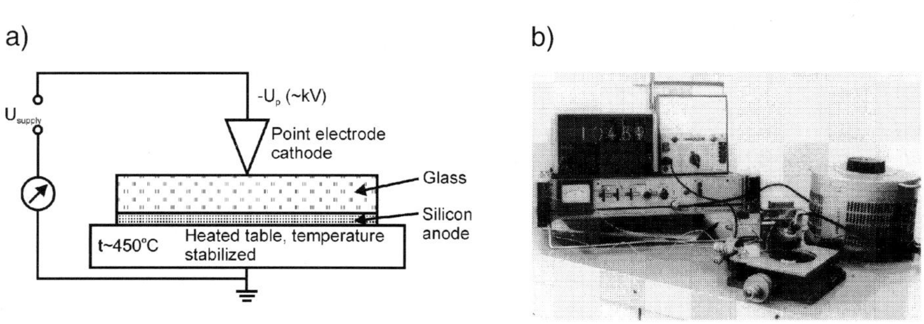

Fig. 4.35. The simplest version of stand for anodic bonding: a) scheme, b) view.

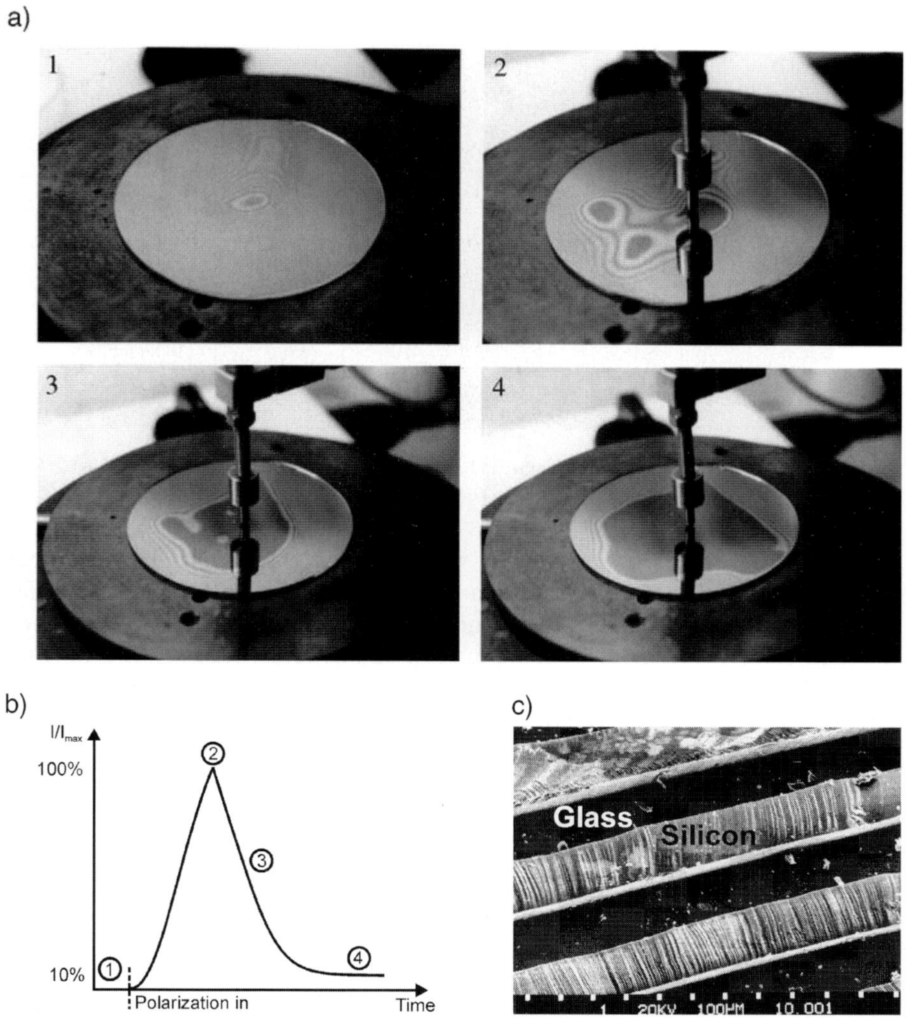

Fig. 4.36. The course of bonding (n-type, 3″ silicon; Pyrex Corning 7740, 3″ glass): a) successive phases of the process, b) typical normalized bonding current as a function of time for constant polarization, c) sample after destructive pull mechanical tests; broken bulk silicon strips can be seen.

between surfaces becomes so small that direct siloxane bonds are formed between the sealed materials. The permanent bonding of the sealed materials can be observed through the glass wafer as the color of bonded surfaces changes from metallic-silver to dark blue-grayish.

Bonding proceeds in the form of a so-called wave of bonding: it propagates from the area of initial bonding (bonding precursor) at a speed reaching a few centimeters per second (Fig. 4.36a/2/3/4). Since it is not a rule that the bonding precursor corresponds to the region of the electrode point pressure, it can

Fig. 4.37. Fracturing of thermally mismatched silicon and Palex sodium glass ($\alpha_g = 6 \cdot 10^{-6}\,K^{-1}$) sandwich, bonded anodically, observed after cooling down of the sample from 400°C to 20°C.

actually begin from any area on the wafers (Fig. 4.36a/2). During the process of bonding an electric current flows in the circuit (Fig. 4.36b). Immediately after bringing the materials into contact and applying the polarization the current is the highest; as the wave of bonding extends the current decreases as a function of time. It is commonly believed that the process of bonding is finished when the value of current reaches approximately 10% of its maximal value.

The strength of correctly performed bonding of silicon to glass is at least equal to the tear resistance of silicon. Attempts to separate bonded wafers usually result in destruction of bulk silicon or glass without any influence on the bonded interface (Fig. 4.36c).

4.4.1. Glass for anodic bonding

Any type of plain glass can be anodically bonded to silicon. After the bonding procedure the sandwich of silicon and glass has to be cooled down from about 400–500°C to room temperature. Whenever the coefficients of thermal expansion of silicon α_{Si} and of glass α_g do not match, cooling causes stresses, destroying the sample (Fig. 4.37).

The coefficient of linear thermal expansion of silicon α_{Si} is a nonlinear function of temperature. In the range of very low temperatures, that is from 30 K to 100 K, it is negative, then positive, growing nonlinearly as a function of temperature (Fig. 4.38)*.

Nonlinear, but small thermal expansion of silicon is the most important reason for difficult matching of this material to other materials commonly used in microelectronic technology, including plain glass (Fig. 4.39). Therefore anodic bonding of silicon to glass needs a special type of glass. The ideal glass for this

* Numerical data and diagrams presented here were developed out in the 1960s and 1970s.

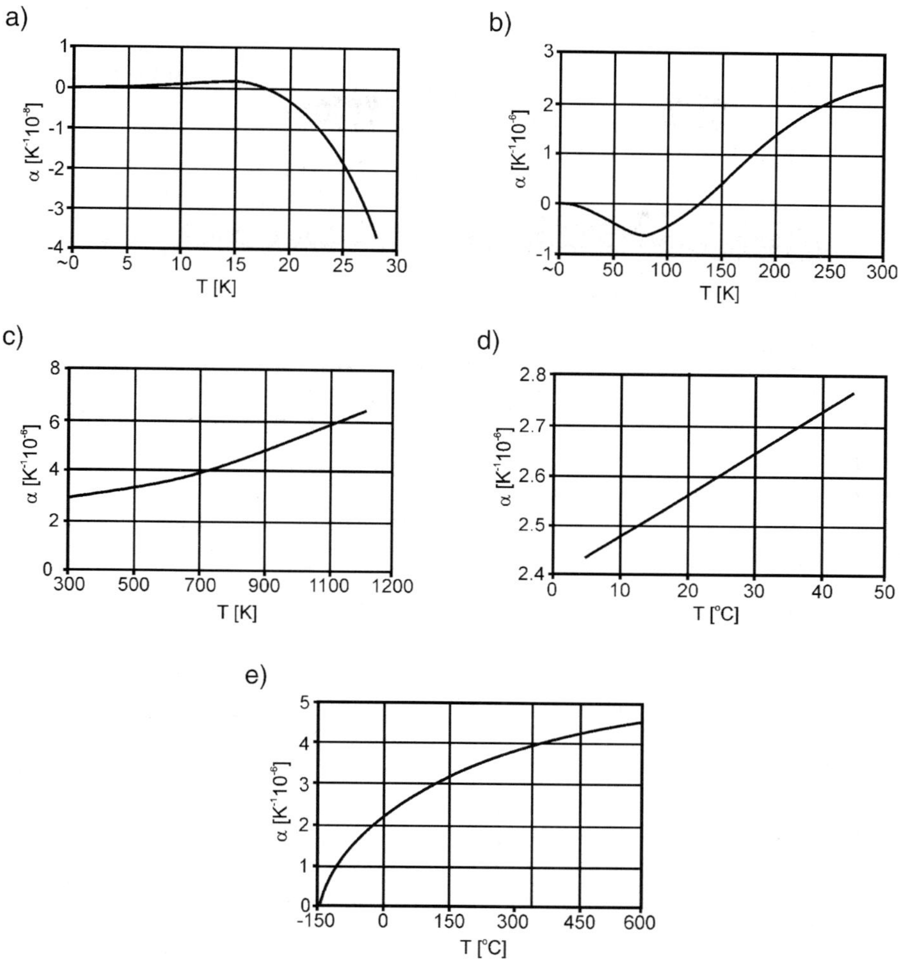

Fig. 4.38. Coefficient of linear thermal expansion of silicon as a function of temperature [139–141]: a) below 30 K, measured using the volume dilatometer, b) to 300 K, monocrystalline silicon, c) 300 K to 1200 K, powdered material, d) 277.2 K to 319.2 K (5°C to 45°C), monocrystalline silicon [141], e) −150°C to +600°C, monocrystalline silicon.

process should possess very good and temporally stable mechanical parameters, high thermal shock resistance, a thermal expansion coefficient matching that of silicon, suitable electrical conductivity and high resistance to electric breakdown at elevated temperature, as well as low price and good availability.

From the early beginnings of investigations into the silicon–glass anodic bonding process, it has been observed that the Pyrex* borosilicate glasses almost

---

* Pyrex® glass was invented by Sullivan and Taylor in 1915 as a glass resistant to thermal shocks for use in marine lighthouses and reflectors, and Addis's lamps of warships during the First World War. The composition of this glass was classified up to the late 1920s.

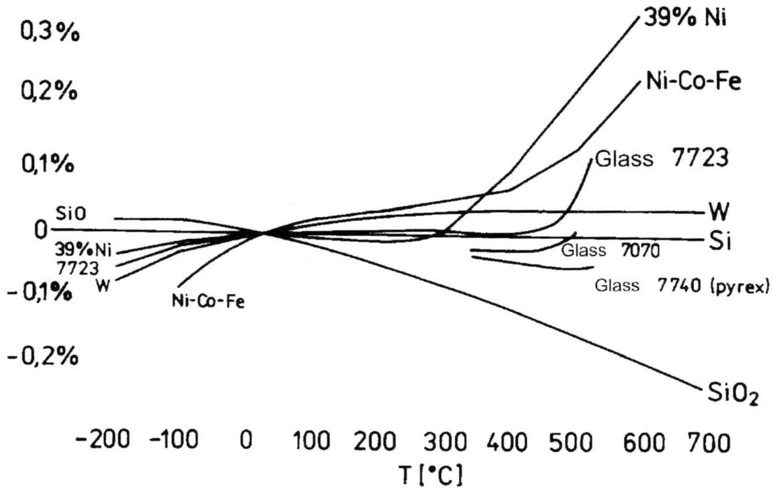

Fig. 4.39. Relative difference of thermal expansion of several standard materials utilized in microelectronics in relation to thermal expansion of monocrystalline silicon.

ideally fulfilled the conditions listed above. This glass has been produced for many years, in many plants (appearing under many trade names, such as Hysil, Monax, Duran, Termisil), being applied mainly in laboratory glass chemical equipment and kitchenware.

The chemical composition of Pyrex-like glasses for anodic bonding does not differ significantly from the 1915 prototype: SiO_2 80.5%, B_2O_3 12.9%, Al_2O_3 2.2%, Na_2O 3.8% and small amounts of MgO, CaO, BaO (Table 4.7).

At the moment the main glasses for anodic bonding are: Corning (Pyrex) 7740, Corning 7070, Borofloat 33 and SD1 and SD2, whose chemical composition has not been published. The basic physical parameters are collected in Table 4.8.

The coefficients of linear temperature expansion α_g of all types of glasses suitable for anodic bonding are very similar in the temperature range 20–300°C and "match" the α_{Si} of monocrystalline silicon. However, the curves of unit elongation $\Delta l/l$ of particular types of glass and silicon as a function of temperature are different (Figs 4.40 and 4.41).

Below 240°C the thermal expansion of 7740 and 7070 glasses is higher than the thermal expansion of silicon ($\alpha_g > \alpha_{Si}$), while above this temperature the relation is reversed ($\alpha_{Si} > \alpha_g$). In the whole 50–500°C temperature range SD1 glass expands less than silicon ($\alpha_{SD1} < \alpha_{Si}$), while SD2 glass expands almost ideally like silicon. Expensive, but ideally matching to silicon, SD2 glass seems to be an ideal material for silicon–glass anodic bonding, but it is not. The usefulness of glass depends not only on the compatibility of coefficients of thermal expansion of glass and silicon, but also on other properties.

Corning 1729 glass has few alkaline ions (<0.05%), which means that it is useful in the fabrication of optical sensors working at elevated temperatures

Table 4.7. Chemical composition of Pyrex and Pyrex-like glasses [133. 142–144])

Name of glass, producer, country, year	Component content in mass%									
	SiO_2	B_2O_2	$H_2O_3 + Fe_2O_3$	CaO	MgO	Na_2O	K_2O	Al_2O_3	Li_2O	ZnO
Pyrex 1929	80.75	12.0	2.2	0.3	—	4.1	0.1	0.4	—	
Pyrex 1945	80.02	11.31	2.71	0.76	—	4.74	0.35	—	—	
Pyrex 1964	80.60	12.60	2.24	0.1	0.05	4.15	—	0.1		
Pyrex 7740. currently	80.5	12.9	2.2	—	—	3.8	0.4	—	—	
Pyrex 7070. currently	70	28	1.1	—	—	—	0.5	—	1.2	
Hysil, U.K.	80.60	12.20	2.70	0.12	—	4.15	—	—	—	
Monax, U.K.	74.66	13.44	3.89	0.75	0.49	5.89	0.79	—	—	
Duran 50 D	79.69	10.29	3.10	0.77	0.87	5.20	—	—	—	
Termisil, Pl.	81.25	11.51	2.42	1.02	—	3.7	0.1			
Borofloat 40. D	73–78	9–12	1–4	1–3	—	1–5	1–5	—	—	1–2
Borofloat 33. D	80	13	2.5	—	—	3.5	1	—	—	—

Table 4.8. Chosen properties of glasses for anodic bonding

Parameter	Name of glass†						
	Pyrex 1915‡	Corning (Pyrex) 7740	Corning 7070	Borofloat 33	SD1 Hoya	SD2 Hoya	Corning 1729
Density [g/cm^3]	2.23	2.23	2.13	2.23	2.59	2.60	2.56
Transformation temperature [°C]	560	560 (528)	—	530	725	721	855
Annealing point [°C]	590	590 (510)	456	560	673	669	799
Softening point [°C]	820	821	—	815	788	789	—
α_g [10^{-6} K^{-1}]	3.2 (20–300°C)	3.23* (20–300°C)	3.20* (20–300°C)	3.3 ± 0.1 (20–300°C)	3.05* (30–100°C) 3.25* (30–450°C)	3.20* (30–100°C) 3.41 (30–450°C)	3.34*
Young's modulus [GPa]	69	62.75	51.0	63	90.1	88.6	80.8

* The thermal expansion coefficient curves are presented in Figs 4.40 and 4.41.
† Pyrex is a registered trademark of the Corning company, Tempax and Borofloat 33 are registered trademarks of the Schott company, SD1 and SD2 are registered trademarks of the Hoya company.
‡ Given for comparison, not utilized.

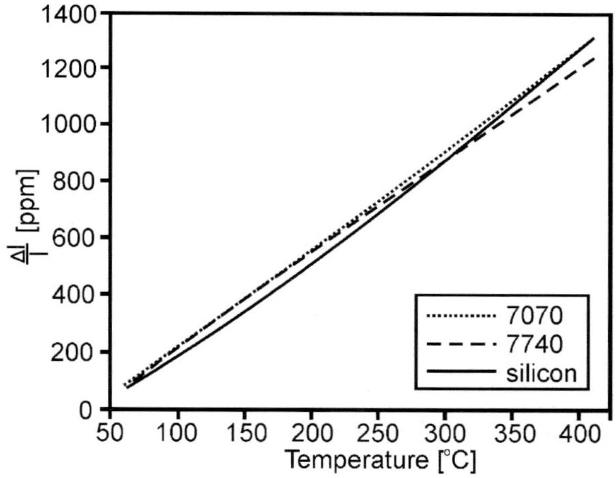

Fig. 4.40. Unit elongation of Corning 7740 and Corning 7070 and silicon as a function of temperature [142].

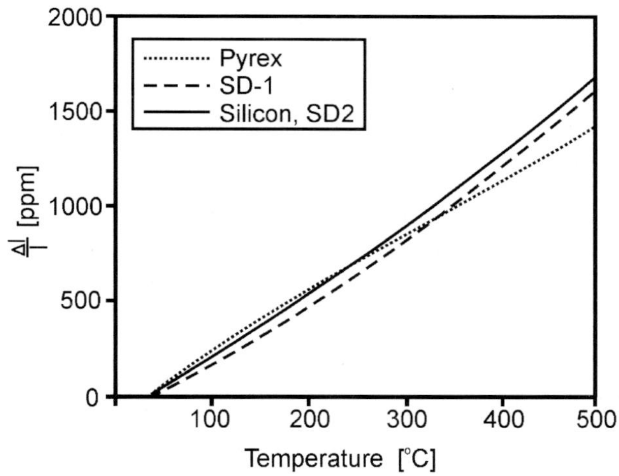

Fig. 4.41. Unit elongation of SD1, SD2, Corning 7740 and silicon (for comparison), as a function of temperature [142].

[145]. The high resistivity of Corning 7070 glass is a significant advantage during anodic bonding in vacuum, when an increased voltage is required. Corning 7740 glass, which is the most popular, ensures the smallest warping and waving of bonded sandwich, SD2 Hoya glass enables the lowest thermal shock stresses after bonding, and the sealing process can be performed at a higher temperature. Borofloat 33 is as "a process easy glass" with a reasonably low price and good product ability [146].

Another important issue is the availability and price of wafers.* The ratio of the prices of glasses of similar quality can equal 10:5:1 (SD2:Corning 7740:Borofloat 33).

The statement of drawbacks and advantages of different types of glass for anodic bonding, allowing selection of glasses from the point of view of their specific applications, is not easy to prepare. In addition, besides a few well-known types of glass, many glass-making companies introduce their own products: Desay [147], Owens ES1 [148].

Parameters of bonding, especially force of sealing and type and quantity of mechanical stresses built in the silicon–glass bi-layer and evolving during anodic bonding, depend chiefly on selection of the parameters of process (voltage, time, type of electrode), temperature curing prior to, and after, bonding [149] and the state of the bonded surfaces [150]. Complete matching of thermal expansion coefficients of these materials before anodic bonding is less relevant (Table 4.9)†.

Glass for anodic bonding is most often fabricated in the form of round wafers with diameters equaling 2″, 3″ or 4″ (bigger diameters are used sporadically), thickness from 0.7 to 3 mm, flat parallel surfaces with coding mark (22.4 mm). Diameter tolerance in a batch equals $\pm$0.5 mm, thickness tolerance is $\pm$0.05 mm (Schott) or $\pm$0.01 mm (Corning). TTV (Total Thickness Variation) equals less than 20 μm, waving is less than 20 μm. The surface of wafers has to be smooth, mirror-like (3 nm), while glass needs to be clear, without any inclusions or voids. Although surface scratches are inadmissible, minor defects of surface, concentrated in nature, are tolerated. Their number in a delivery cannot exceed 4/m^2 for major defects (larger than 2.5 mm) or 2/m^2 for minor ones (smaller than 2.5 mm). Defects less than 0.5 mm in diameter have to be placed at a distance of at least 50 mm. Single inclusions, if not bigger than 0.3 mm, are acceptable. Glasses for anodic bonding are delivered in the same packages as silicon wafers. Single glass wafers are protected by small bags made of dustless paper or plastic foil (Fig. 4.42).

Glass for special applications can be prepared in the form of squares or rectangles, that may be mechanically machined in order to drill holes. The most sophisticated glass substrates are fabricated for anodic bonding of micromachined silicon wafers with hundreds of thin membranes for piezoresistive pressure sensors. The fabrication process of these sensors utilizes glass substrates with an array of hundreds of holes (Fig. 4.43). Most often, the internal diameter of holes (ID) equals 1 mm or 0.8 mm $\pm$50 μm. Module error cannot exceed $\pm$50 μm, any defects at hole edges are acceptable. Furthermore, wafers with holes have to meet all the requirements for the glass wafers used in silicon to glass anodic bonding.

Glass wafers with holes designed for pressure sensors are very expensive; the

---

* About prices; original substrates offered by local sellers can be several-fold more expensive than those offered by producers, large quantity is usually cheaper.

† Some of the producers of glass wafers for microsystem technology: Bullen Ultrasonic Ltd 4613 Canden Road, Eaton OH.45320 USA. Schott Desag AG Hütterstraße 1, D31073 Grüneplan, D, Plan Optik GmbH, Unter den Eichen, D56479 Elsoff, D.

Table 4.9. Selected properties of glasses for anodic bonding

Type of glass	Price	Availability	*R* surface resistance (at the temperature of bonding)	Voltage applied in bonding *U*	Mechanical resistance	Matching with silicon	Preferred application	Comments
7070	high	easy	high $\log R = 8$	medium	small	poor	vacuum bonding	
7740	medium	easy	small $\log R = 5$	low	medium	good	universal, mass production	applied very widely
SD1	high	difficult	high	high	high	poor	vacuum bonding	end-user-certificate
SD2	very high	difficult	small $\log R = 9$	high	high	excellent	space equipment, big changes of temperature	
Borofloat 33	medium/ low	OK	—	low	medium	good	universal	becoming popular

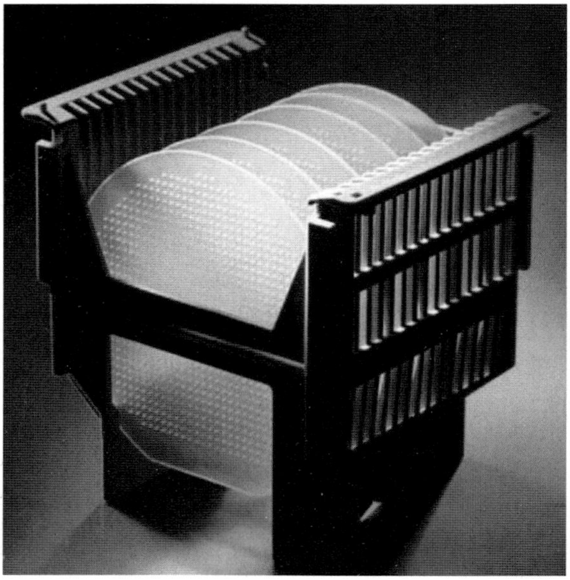

Fig. 4.42. 4″ Corning 7740 glass wafers in the transporting package.

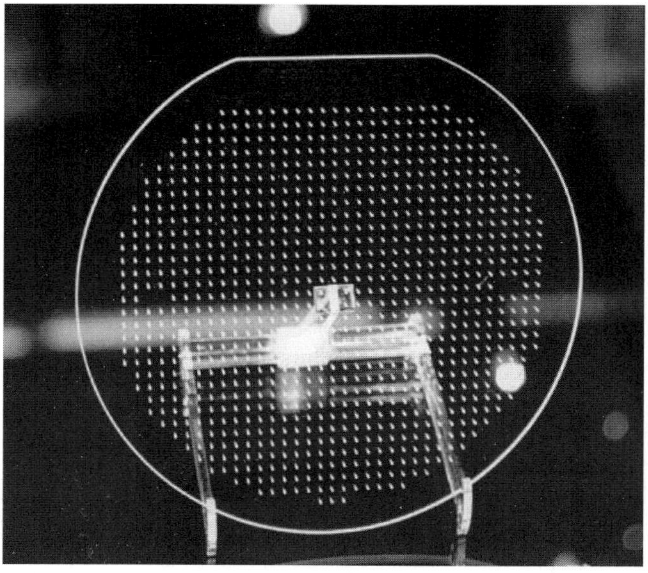

Fig. 4.43. 3″ glass wafers with array of holes.

price of 3″ wafer glass, 1 mm thick, with array of holes in a net 2.4 mm × 2.4 mm is at least five times higher than the price of the non-machined wafer. The cost of glass with holes is the main price-generation factor of the production of silicon piezoresistive pressure sensors.

4.4.2. Mechanism of bonding

The resistivity of glasses decreases with increase in temperature, which is related to the increase in ionic current (Fig. 4.44). Under the influence of electric field cations, the active Na^+ ions particularly drift quite freely in glass toward the cathode. At the same time anions, which are practically motionless, remain almost fixed in the glass matrix [151]. An anode made of silicon functions as a blocking electrode, so the injection of positively charged silicon into glass cannot take place. Additionally, in borosilicate glass, that consists mainly of two solid phases of SiO_2 and B_2O_3, silicon is placed stably in a glass matrix and does not drift under the influence of an electric field. Hence, it is not possible to compensate the drift of Na^+ positive charge. At the silicon–glass interface depleted in sodium and reached in anions a thin layer is formed [152–164] (Fig. 4.45). The phenomena occurring during anodic bonding can be divided into two main groups: near the cathode, which are strongly dependent on the type and shape of cathode, and at the silicon–glass boundary which are, first of all, the result of a depletion layer forming and chemical reactions occurring at the silicon–glass interface.

The mechanisms of silicon to glass anodic bonding discussed elsewhere are based on local oxidation of silicon anode. In some of them it is assumed that O^- and OH^- ions localized near the silicon–glass interface in the depletion layer can drift toward the anode under electric field excitation and reach the

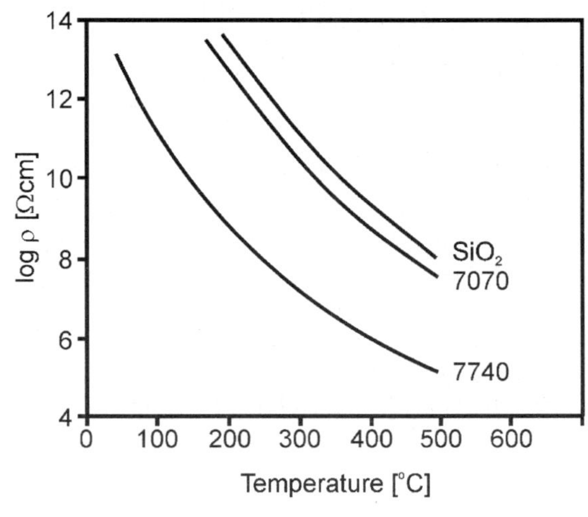

Fig. 4.44. Bulk resistivity of Corning glasses as a function of temperature (for comparison a curve obtained for SiO_2 is presented) [144].

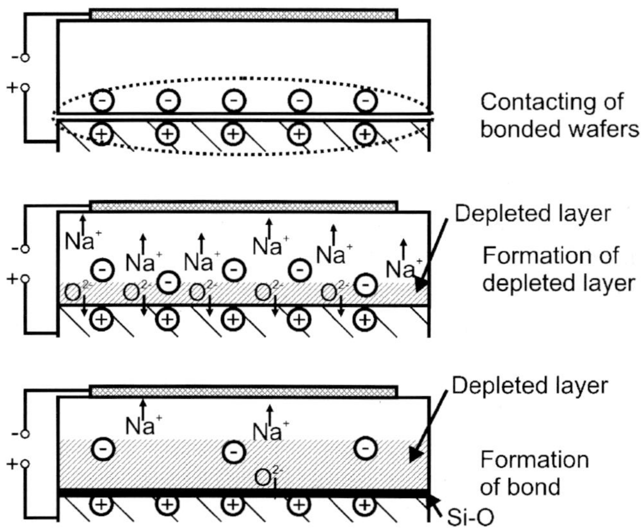

Fig. 4.45. Ions flow in the silicon/glass/cathode assembly for three phases of bonding [159].

glass surface. They oxidize silicon (because bonded materials are contacted very closely – note the role of the electrostatic pulling force). As a result, strong siloxane bonds between glass and silicon are formed [160]. In the other, the role of the electrostatic force is assumed to be similar, but siloxane bond formation is assumed to be a result of the so-called "water pump" phenomenon. This consists in the decomposition of molecular water trapped on the boundary surfaces of silicon and glass. Hydroxide groups take a part in oxidation of surface silicon and forming of siloxane bonds between sealed materials. Released hydrogen ions drift toward the cathode together with the sodium ions.

Despite the nature of anodic bonding, it is a commonly accepted fact that during bonding the stream of sodium ions flows through the glass and reaches the surface of glass in the region near the cathode, where it is neutralized. In this region sodium reacts with oxygen taken from the glass matrix or from the air, forming Na_2O. This process changes the glass composition near the cathode. An increase of a few percent of Na_2O concentration in glass in the area near the cathode is possible (Fig. 4.46) [153]. Another possibility of neutralization of sodium ions is the formation of sodium hydroxide (NaOH) at the expense of molecular water taken from the glass surface or/and at direct chemical reactions between sodium and the cathode material. The physicolchemical nature of anodic bonding will be discussed in detail in later parts of this book. Here we will concentrate on the secondary but unpleasant effects of sodium ion drift.

4.4.2.1. Cathode

The local near-cathode concentration of sodium ions – and negative effects introduced by sodium – depends on the density of ionic current, being a result

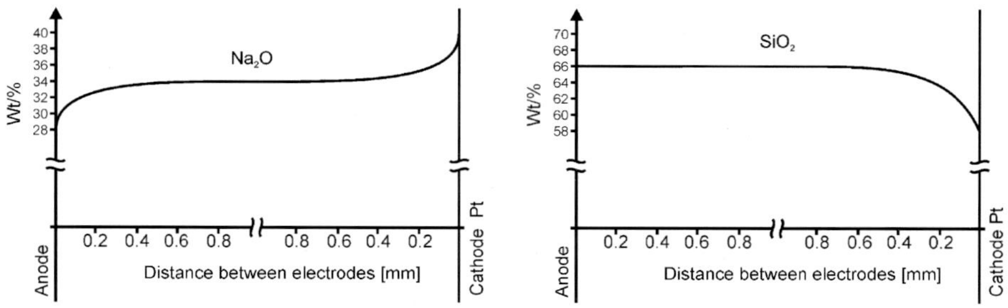

Fig. 4.46. Distribution of the concentration of Na_2O and SiO_2 in glass. The utilized Pt cathode was deposited in vacuum, glass was annealed at a temperature of 500°C, the polarization voltage of cathode equaled 500 V [153].

of the type of cathode [165]. Since the type and shape of cathode influence phenomena occurring in its area, it is very important to select the right cathode for bonding. The cathodes most often applied in the processes of bonding of materials are flat metal, glass–metal and tip ones.

Flat metal cathode

Let us assume that the flat cathode is smooth and adheres tightly to smooth glass (any thin-film glass metal cathode meets these conditions). Electrical field lines are parallel to the cathode, the field is uniformly distributed across a bonded glass wafer. The distribution of ionic current in glass is uniform. Non-uniformity of current flow, caused by microscopic inequalities of glass structure and changes in composition, inclusions and voids, can be neglected in this case. Once these ions reach the cathode they are neutralized. If the cathode is integrally connected with glass and made of precious materials (Pd, Pt) not reacting with sodium or its products, the cathode is not subjected to corrosion and uniform enrichment in Na_2O near the cathode can be noticed.

In technological practice permanent thin-film cathodes are rarely employed. Most often applied are removed inox-metal flat cathodes, made in the form of stainless-steel plate with a polished working surface. However, there are always microinequalities on the surface, that cause only local contact of cathode and glass. The distribution of the electrical field is disturbed, not uniform, and the density of ionic current Na^+ can be locally increased (Fig. 4.47). At the very beginning of the sealing process, sodium and its compounds attack at several points the cathode and, at the elevated temperature of anodic bonding, cause formation of the net of local tip cathodes. At the tip cathodes the electrical field is enhanced by orders of magnitude. Local tunnel emission of electrons to glass is possible, along with the channeled flow of ionic current Na^+ near the tip. The structure of glass is destroyed at several points. Glass becomes turbid, and a white Na_2O sediment precipitates on its surface. After cooling the sediment passes to the hydrated forms (sodium hydroxide) and acid sodium carbonates. The process described here can destroy the quality of the surface of a bonded

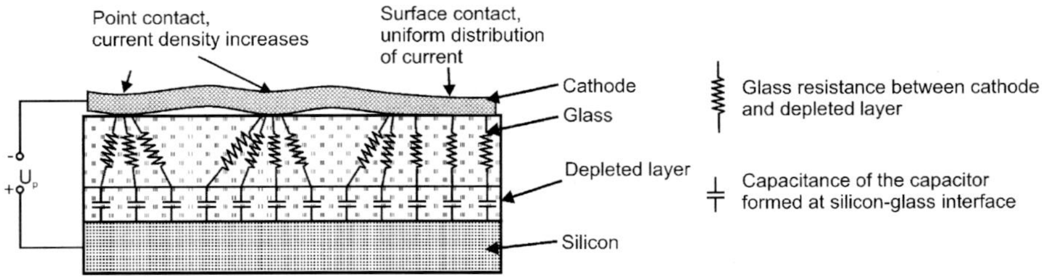

Fig. 4.47. The real flat metal cathode contact and its impact on ionic current flow in glass [159].

glass wafer. The planar cathode has to be polished prior to the anodic bonding process.

Metal–glass cathode

The effect of micro tips formation does not affect the near-cathode area in flat metal/glass cathodes. Such a cathode comprises a sandwich of the inter-cathode, made of a flat-parallel borosilicate glass wafer with mirrored surface placed onto the glass substrate to be bonded with silicon and a typical metal plate cathode. In this configuration only the metal cathode is polarized. Because the surface of inter-cathode glass wafer is optically polished (roughness below 3 nm) two glass wafers adhere to each other very tightly. This ensures a very uniform distribution of electrical field across wafers; what is more, both glass wafers are chemically inert, so forming of micro tips does not evolve on the boundary of bonded glass and inter-cathode.

Under these conditions sodium propagates almost ideally uniformly, and the possible non-uniformities of propagation caused by defects of the glass structure are not important (Fig. 4.48). The drift of sodium ions in an electric field, at elevated temperature of the process of anodic bonding, occurs in both glass wafers. However, defects appear only on the side of the metal cathode, that is on the top surface of the inter-cathode. Insignificant turbidity of the bonded glass can be observed. This is spread quite uniformly and can be removed by boiling in deionized water for 20 minutes.

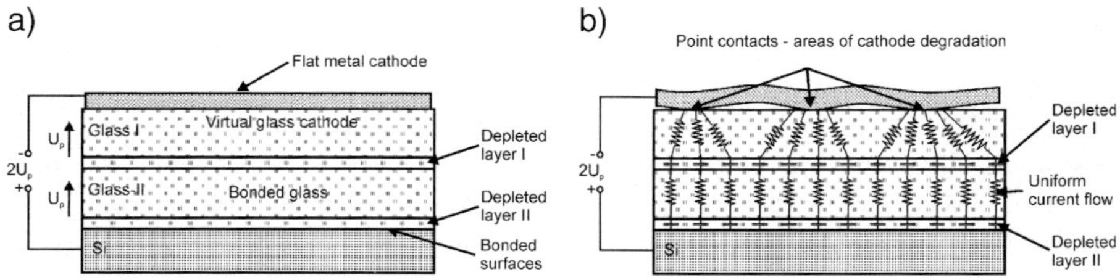

Fig.4.48. Sandwiched metal glas cathode: a) configuration, b) ionic current flow.

Tip cathode

Successful bonding of silicon and glass plates is obtained with use of the commonly used tip cathode. The name is misleading, because cathodes are made of stainless-steel bars with diameter equaling 2–3 mm, with a round and polished end ($r \sim 1$ mm). A tip cathode ($r < 0.1$ mm) under typical conditions of anodic bonding (400°C, 500 V) disturbs the electrical field (Fig. 4.49a). The electrical field near the cathode becomes so high that local avalanche brakedowns are possible. The local concentration of Na^+ ions is high enough to cause the formation of an "ionic current channel" in glass, under the cathode. This channel enables air to penetrate the glass, which results in decomposition and destruction of glass in the area of several diameters of electrode. A large number of sodium ions physically reaches the surface of the glass, Na_2O sediments, acid sodium carbonates and hydrated sodium hydroxide can be observed around the tip cathode (Fig. 4.49b).

A track of the point electrode appears on the glass surface even after a short-time flow of ionic current. The part of the glass wafer which is touched by the tip cathode during the anodic bonding process is no longer technologically useful. This is not burdensome at the step-and-repeat technique of fabrication

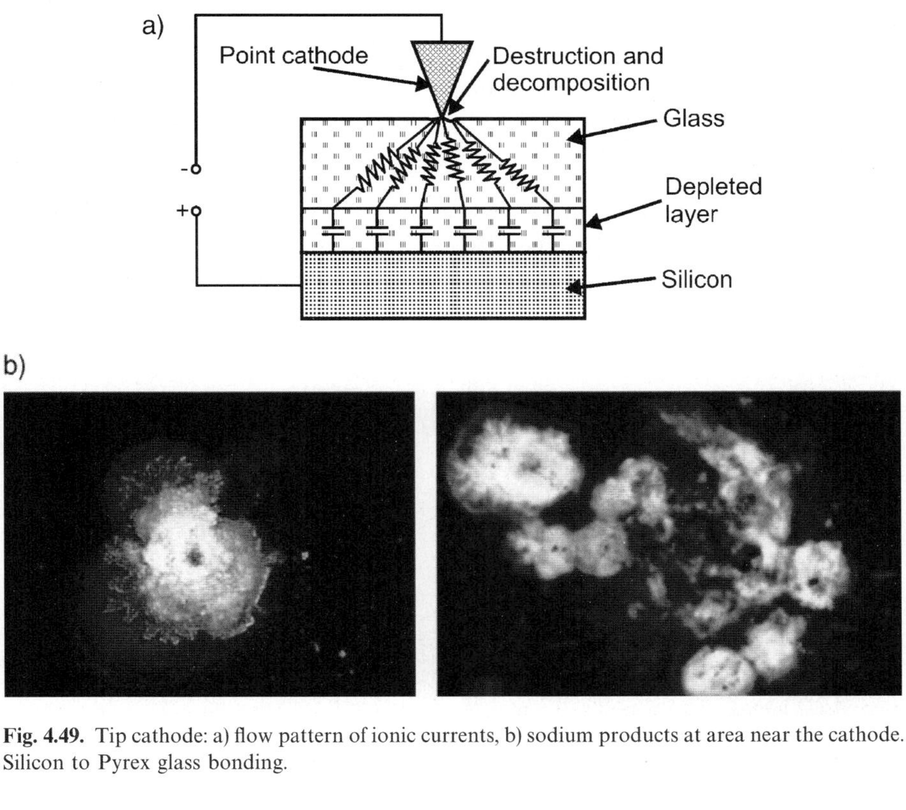

Fig. 4.49. Tip cathode: a) flow pattern of ionic currents, b) sodium products at area near the cathode. Silicon to Pyrex glass bonding.

of small-area micromechanical sensors, but becomes a serious problem in large-area wafer-scale microsystem technology. In this case a sandwich planar glass–metal cathode should be preferred.

Selection of cathodes

The shape and type of cathodes should be selected to meet the requirements of the process of bonding. A flat metal thin-film electrode, deposited using vacuum methods, ensures the best uniformity of electrical field near the cathode and does not practically cause any local defects. After bonding, the material of cathode has to be removed from the glass surface, which can be difficult to perform, because the cathode material can adhere strongly to glass after bonding.

Flat metal or graphite electrodes need to be polished prior to use. Nevertheless, use of a perfectly polished flat cathode does not ensure a clear glass surface, without defects, after anodic bonding.

A point electrode causes the formation of defects on a glass surface, and a significant change of its composition in the area of a dozen or so diameters. In the region of electrode adhesion local electrical breakdown may occur, which makes high-voltage bonding impossible to perform. An ion current channel is formed, as well as a large number of sodium chemical by-products.

The assembly with an intermediate electrode is devoid of the drawbacks of flat, metal electrodes. In this assembly the intermediate glass electrode adheres firmly to the surface of the proper glass (similar to the sputtered electrode); possible damage, resulting from the surface phenomena occur only on the upper surface of the intermediate electrode (on the side of the flat metal cathode) (Fig. 4.50). The parameters of bonding need to be selected carefully; elevated values of polarization voltage should be applied.

4.4.2.2. Anode – depleted layer

Drift of positive ions toward the cathode in hot glasses, under the influence of an electric field, was researched by many authors in the 1970s. It has been

a) b)

Fig. 4.50. Damage to metal cathode (a) and different types of cathodes (b).

confirmed that the chemical composition and physical properties of the depletion layer after anodic bonding are more similar to quartz (silicon dioxide) than to bulk borosilicate glass; that the change of composition of glass in the depletion layer is permanent and does not recede after disconnection of polarization or cooling [151–155, 158, 162].

The almost complete escape of cations from the 0.1 μm-thick area of glass for a non-blocking flat iron–chrome (FeCr) anode was observed by Gossile [162] (Fig. 4.51). Sodium ions left the depletion area most rapidly, while Ca^+, K^+, Al^+ etc. ions drifted more slowly in hot glass. The tested glass near the anode contained mainly silicon dioxide (SiO_2) or aluminum oxides (Al_2O_3). Small quantities of iron and chromium oxides (Fe_2O_3 and Cr_2O_3) from the non-blocking anode were also observed (Fig. 4.52).

Many researchers have found different chemical resistivities of the depletion layer and the bulk borosilicate glass. In a strong aqueous solution of HF depletion is etched more quickly than the borosilicate glass [151], but significantly more slowly, almost ten times, in weak solutions of HF [154, 155]. The etch rate of the depletion layer equals a few hundreds of nanometers per minute, a typical value observed for quartz.

Carlson and co-authors researched the polarization phenomena in hot glasses by means of the method of reflection infrared spectroscopy [152, 154]. The

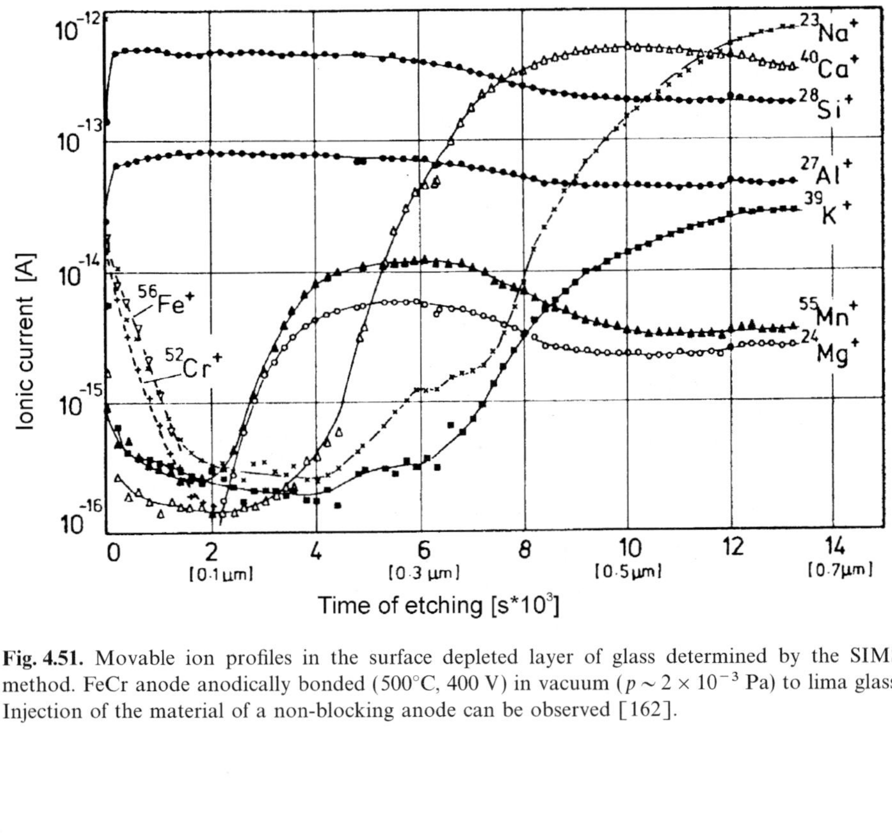

Fig. 4.51. Movable ion profiles in the surface depleted layer of glass determined by the SIMS method. FeCr anode anodically bonded (500°C, 400 V) in vacuum ($p \sim 2 \times 10^{-3}$ Pa) to lima glass. Injection of the material of a non-blocking anode can be observed [162].

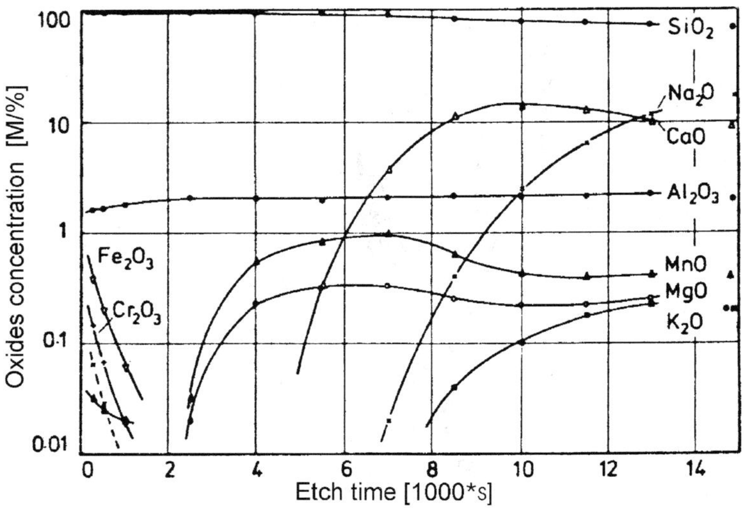

Fig. 4.52. Composition of glass after anodic bonding of FeCr to glass [162].

spectral characteristics of the depletion layer formed in glass and of bulk sample, of fused quartz, obtained by them, were almost similar (Fig. 4.53).

The electrical resistance of the depletion layer is a few orders of magnitude higher than the electrical resistance of bulk borosilicate glass. For example, the electrical resistance of a depletion layer formed in Corning 7740 glass, after a successful anodic bonding procedure, can be increased by 10 000 times (10^4) in relation to the electrical resistance of this glass [154] (Fig. 4.54).

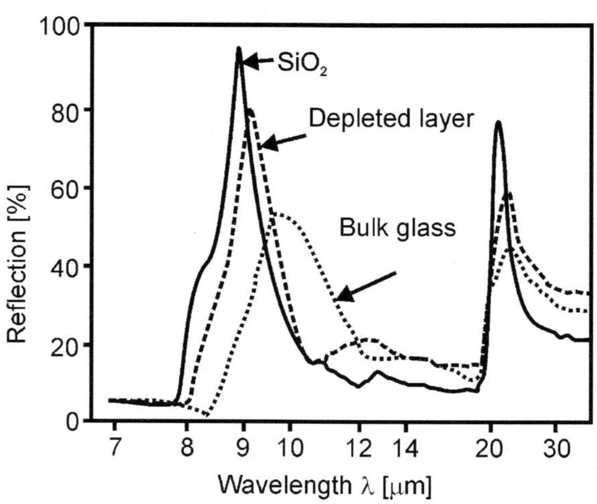

Fig. 4.53. Spectrum characteristics of glasses before bonding, depletion layer, fused quartz [154].

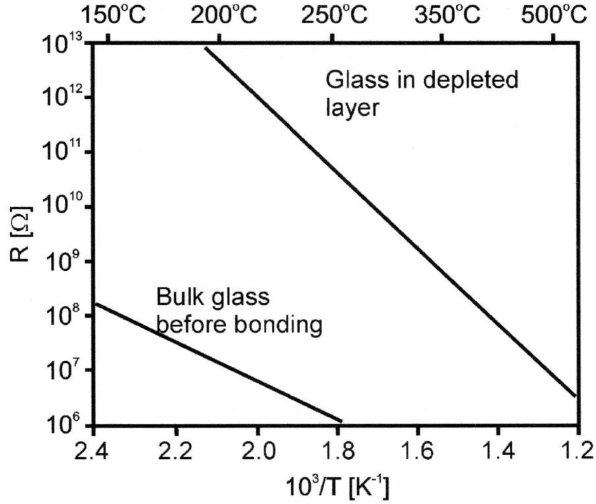

Fig. 4.54. The electrical resistance of bulk Corning 7740 glass before bonding (R_N) and in the depletion layer after bonding (R_D), as a function of temperature [154]. R_N was measured using the low-frequency AC method, R_D was measured after bonding, for cathodic polarization $U = 500$ V, thickness of the depletion layer $d_z = 10$ μm.

4.4.2.3. *Electrostatic pressure, alignment of surfaces*

The force F, which clamps glass to silicon during anodic bonding, can be expressed analogically to the force of a flat condenser with capacity C and area of electrodes S, with a dielectric made of glass (Fig. 4.55). Anode (silicon) is the bottom filler sheet of the condenser, while the top filler sheet – before the formation of a depletion layer – is formed by a flat cathode, negatively polarized in relation to silicon and at a distance of d from the anode. Force F is given by:

$$F = -\frac{1}{2} U^2 \frac{C}{d}, \tag{4.21}$$

where: U = polarization voltage, d = distance from anode.

Electrostatic pressure can be expressed by the following equation:

$$P = -\frac{F}{S}, \tag{4.22}$$

while for the flat condenser, whose capacity C is given by:

$$C = \frac{\varepsilon_0 \cdot \varepsilon_r}{d} \cdot S \tag{4.23}$$

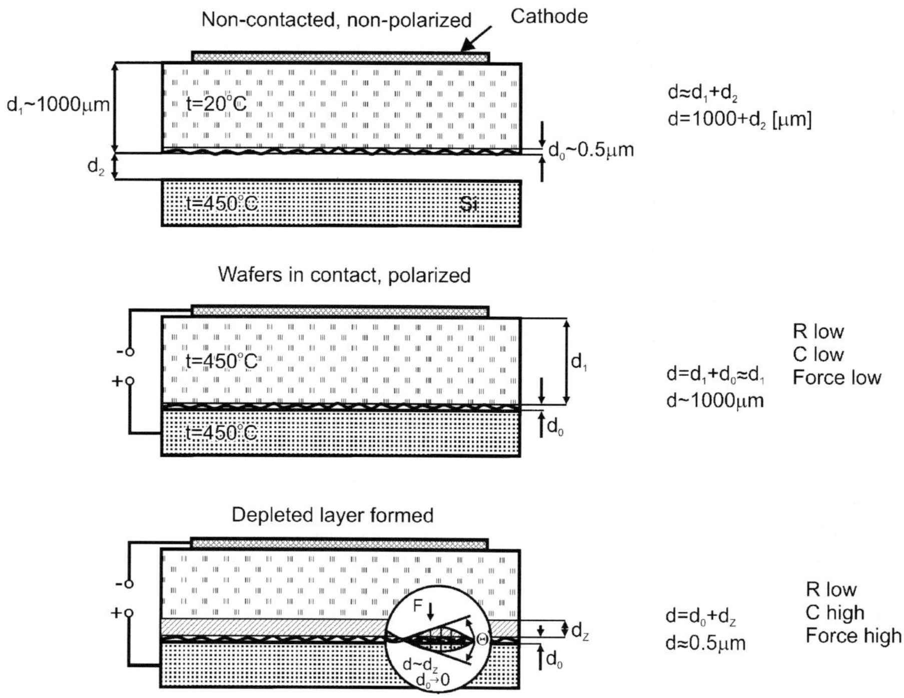

Fig. 4.55. Three steps of bonding, forming of flat condenser.

and

$$P = \frac{1}{2}\frac{\varepsilon_0 \cdot \varepsilon_r}{d^2} \cdot U^2, \tag{4.24}$$

or

$$P = \frac{1}{2}\varepsilon \cdot E_p^2, \tag{4.25}$$

where: E_p = electric field intensity.

Immediately after the polarization voltage is applied the electrostatic pressure is low, because the drop of voltage U occurs on the thick ($d_1 \sim 1000$–3000 μm) dielectric glass layer and large air gap, corresponding to the substrate's maximal waviness or height of large particles at surfaces. Electrostatic pressure P strongly depends on the effective air gap (Fig. 4.56); for larger gaps, over 1 μm, its value becomes insignificant.

After the formation of a negatively charged depletion layer in highly conductive glass, a drop of almost full polarizing voltage U occurs mainly at the narrow air-gap between the smooth surfaces of silicon and glass ($d_0 \ll 0.5$ μm).

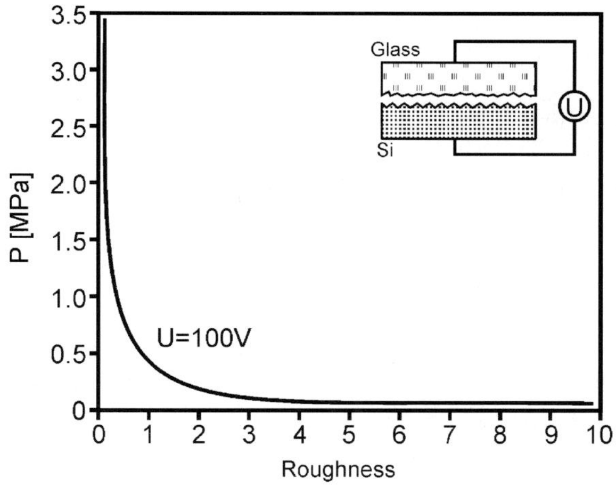

Fig. 4.56. Electrostatic pressure P as a function of the roughness of surface approximated by a model presented in Fig. 4.57 [150].

Electrostatic pressure increases. In some, statistical points (bond precursors) surfaces of bonded substrates are dropped to the ultimate contact so that the air-gap becomes equal to parts of a nanometer. A drop of polarizing voltage U occurs only on the very thin ($d_z < 20$ nm), highly resistive, quartz-like depletion layer. Electrostatic pressure and pulling force, that clamp glass to silicon, become very high (Table 4.10), and the bonding wave starts to propagate across the bonded surfaces. The pulling force is so high that smaller surface obstacles (hard dust particles, steps of thin-film layer) as well as waviness of wafers, are tolerated much better in comparison to other bonding methods – especially to direct bonding. Surfaces have a tendency to align to each other. The alignment of bonded surfaces, caused by electrostatic pressure, has been investigated in paper [150]. Surfaces were approximated using the deformation model illustrated in Fig. 4.57. It was observed that the elastic deformation of the glass

Table 4.10. Electrostatic pressure for varying voltage U, air gap d_0 and d_z [1]

Voltage U [V]	d_0 [μm]	Thickness of depletion layer d_z [nm]	Electrostatic pressure P [MPa]
400	0	16	27 700
600	0	20	34 200
600	0.5	<20	6.4
1000	1	—	4 (air capacitor)

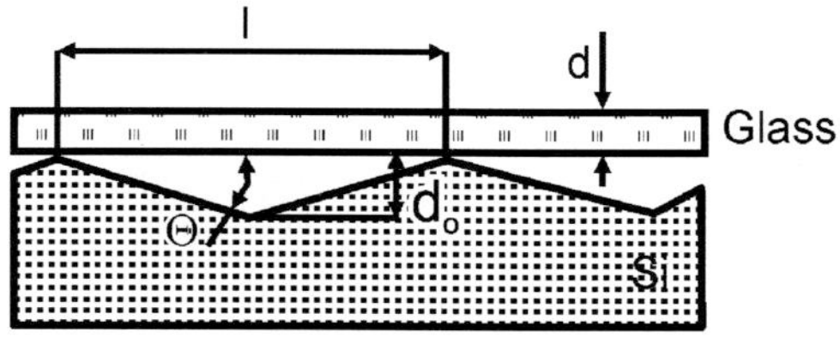

Fig. 4.57. Model of elastically deformed surfaces [150].

layer with thickness *d*, allowing complete contact of bonded surfaces, can be obtained for polarization voltage *U* given by the following formula:

$$U \geq \sqrt{\frac{h\nu d^3}{80E \cdot l^4}}, \tag{4.26}$$

where: ν = Poisson's ratio, h = height of waviness, l = length of wave, and E = Young's modulus.

For $h = 2$ μm, $l = 20$ μm and $d = 1$ mm, the elastic deformation of Pyrex glass matching perfectly bonded wafers is obtained for $U = 150$ V only.

The polarization voltage *U* sufficient for elastic deformation of glass bonded to silicon covered with thin-film aluminum foil, is described by the equation:

$$U > \frac{hb}{l}\sqrt{\frac{\sigma_y}{\sigma E \ln(l/d)}}, \tag{4.27}$$

where: σ_y = allowable stress corresponding to elastic deformation of given material, σ = stress, and b = thickness of aluminum foil used in experiments.

For $b = 2.5$ μm, $h = 2$ μm, $l = 20$ μm (on the assumption that the σ_y of Al at elevated temperature equals 4.13 MPa), a polarization voltage sufficient to match bonded surfaces should be higher than 100 V.

The alignment of a surface bonded material anode can also be caused by glass flow at an elevated temperature in the presence of strong electrostatic pressure. All bondable glasses must be treated as solid-state fluids, whose viscosity seriously depends on temperature. For example, the viscosity of Borofloat 33 glass equals 10^{13} dPa·s at 560°C and becomes $10^{7.6}$ dPa at 815°C. A strong "clamping" force, reaching tens of thousands of MPa (see Table 4.10) can non-elastically deform glass at bonding temperature.

Viscous glass flows round the dust molecule and drowns it in glass volume. A minimal time τ_N, which is sufficient for the dust molecule to be absorbed, is expressed by the following equation:

$$\tau_N = \frac{3\eta H^2 P \pi l^2}{EU^2}, \tag{4.28}$$

where: η = viscosity of glass at the temperature of bonding, P = density of dust per surface unit, and H = height (diameter) of dust.

Dust molecules with dimensions equaling 1 μm, distributed uniformly with a density of 10^6 m^{-2}, are absorbed at a temperature of 450°C for $U = 300$ V after approximately 2400 seconds. So, single hard molecules of dust, which are the main cause of formation of extrinsic voids in the fusion bonding of silicon, are tolerated significantly better in silicon to glass anodic bonding. In practice, the "intake" by glass of dust and hard point impurities with considerable height (up to 2 μm) is rarely observed (Fig. 4.58). In spite of the elevated temperature, glass can preserve significant elasticity. Exceeding the proof stress is conducive to local cracks. Sharp ends of hard dust molecules cause point stresses in glass, which decreas its insulating power and result in local punctures connected with the formation of micro arcs and arc burning. Therefore the process of bonding should be prepared and performed in a dustless atmosphere, in local clean-rooms, similar to the Si–Si fusion bonding.

4.4.2.4. Chemical reactions, models of bonding

The interactions of physical forces, which lead to the alignment of silicon and glass surfaces, are accompanied by reactions that form permanent chemical bonding between sealed materials. It is commonly accepted that permanent bonding between silicon and glass is formed by strong siloxane Si–O–Si bonds, but the course of the process of the formation of bonds has not yet been fully explained. Two partially contradictory models exist: the first introduced by Baumann, Mach and Minzel [166], and the second presented by Schmidt, Nitzche, Lange, Grigull and Kreissing [157]. For ease of use, in this book, the first model will be named Baumann's, the second Schmidt's.

In Baumann's model the formation of Si–O–Si bonds is assumed, proceeding as an effect of multi-step processes.

Fig. 4.58. Hard impurity; the visible interferential rings that surround the impurity indicate the non-bonded area of silicon and glass.

The solid-state electrolysis of Na_2O under the influence of an electric field in hot glass generates sodium ions and free oxygen:

$$Na_2O \rightarrow 2Na^+ + O^-. \tag{4.29}$$

Simultaneously the dissociation of molecular water adsorbed on hydrophilic surfaces of silicon and glass occurs:

$$H_2O \rightarrow H^+ + OH^- . \tag{4.30}$$

Hydrogen coming from the decomposition of water diffuses to glass, where it takes part in the electrolysis of Na_2O according to the formula:

$$Na_2O + H^+ \rightarrow 2Na^+ + OH^-. \tag{4.31}$$

Oxygen coming from the decomposition of Na_2O and the OH^- groups from the decomposition of water migrate toward the anode where they react with silicon, which results in the formation of SiO_2 and ≡SiOH (≡ indicates that the hydrated silicon surface atom is bonded by three bonds to silicon atoms, its neighbors). Sodium ions are neutralized near the cathode. Sodium consumes the atmospheric oxygen or oxidizes at the cost of non-bonded oxygen taken from the glass matrix forming Na_2O in the region near the cathode. Next, Na_2O forms, with water from air, sodium hydroxide NaOH, which attacks the cathode and glass in the region near the cathode:

$$Na^+ + 4\bar{e} \rightarrow Na \tag{4.32}$$

$$2Na_2O + 2H_2O \rightarrow 2NaOH. \tag{4.33}$$

At the same time, on the bonded surfaces closely "clamped" by electrostatic pressure, the process of dehydration of partly oxidized surfaces proceeds:

$$\equiv Si{-}OH + OH{-}Si\equiv \;\rightarrow\; \equiv Si{-}O{-}Si\equiv + H_2O \tag{4.34}$$

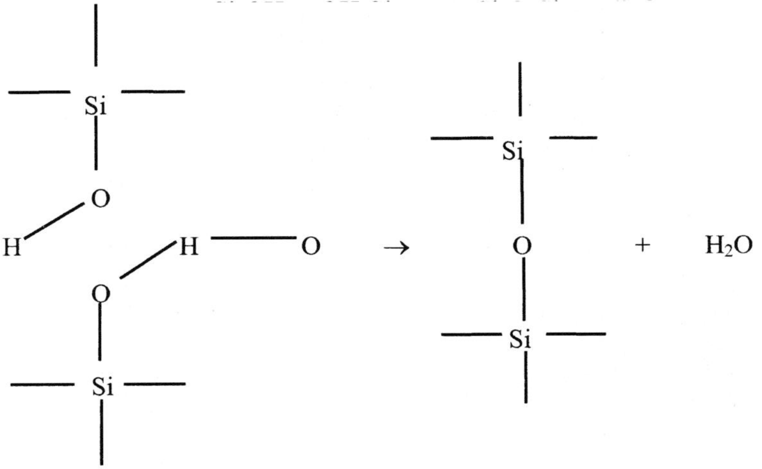

Baumann assumed, in his model, electrolysis of Na_2O and the presence of non-bonded oxygen, coming from this process, in glass [166]. Research carried out by Schmidt in paper [157] revealed that non-bonded oxygen, which could be a product of the decomposition of Na_2O, does not appear in borosilicate glass during bonding. What is more, he demonstrated that sodium ions form ion association complexes with $(BO_4)^-$ or $(AlO_4)^-$. According to that theory, the escape of sodium to the cathode has to be compensated by the inflow of hydrogen, which replaces sodium. The presence of H^+, in a quantity corresponding to the amount of sodium ions removed from the depletion layer in glass ($Na^+ \sim 0.95 \times 10^{17}\ cm^{-2}$, $H^+ \sim 1.1 \times 10^{17}\ cm^{-2}$), was confirmed with use of the ERDA method (***E**lastic **R**ecoil **D**etection **A**nalysis*).

Based on these results, Schmidt and co-authors have proposed a new mechanism of bonding. In the first step molecular water adsorbed at the hydrophilic surface of bonded glass decomposes under the influence of temperature and electric field:

$$H_2O \rightarrow H^+ + OH^- \quad \text{lub}\ 2H_2O \rightarrow H_3O^+ + OH^-. \tag{4.35}$$

Hydroxyl groups drift towards silicon:

$$Si + 4OH^- \rightarrow Si(OH)_4 + 4\bar{e}, \tag{4.36}$$

and then dehydration of the Si–OH bond occurs, resulting in the formation of siloxane bonds between silicon and glass:

$$Si(OH)_4 \rightarrow SiO_2 + 2H_2O. \tag{4.37}$$

Released water once more enters the cycle of decomposition and chemical reaction with silicon, until it is consumed. Hydrogen, being a by-product of molecular water decomposition, drifts in the electric field, through hot glass, toward the cathode, and replaces sodium in a $(BO_4)^-Na^+$ association complex:

$$\begin{aligned} &(BO_4)^-Na^+ + H^+ \leftrightarrow (BO_4)^-H^+ + Na^+ \\ &(AlO_4)^-Na^+ + H^+ \rightarrow (AlO_4)^-H^+ + Na^+. \end{aligned} \tag{4.38}$$

Sodium drifts in the direction of the cathode, where it is neutralized as described in Baumann's model. According to the presented model, the process of anodic bonding is activated by an electric field. This is a type of anodic oxidation of silicon, where surface water is an essential component. Electrostatic pressure, caused by the escape of sodium ions and the formation of a depletion layer, is also important. The electrolysis of Na_2O in glass does not occur.

However, it seems that the possibility of dissociation of Na_2O, compatible with Baumann's model [166], and confirmed indirectly by the change in composition of glass in depletion layer, cannot be excluded. It has been described in many works [151–155, 158, 162]. It is possible that oxygen, coming from Na_2O, bonds with hydrogen and hence does not appear in the non-bonded form, which is consistent with the results of Schmidt's investigation [157] (Fig. 4.59) and eliminates the contradiction between the two models presented above.

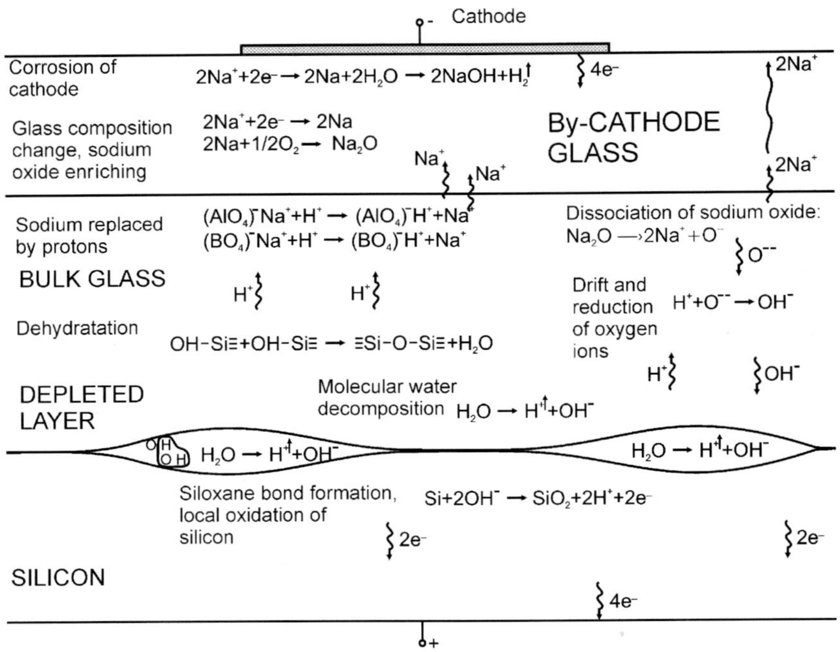

Fig. 4.59. Mechanism of bonding.

Anodic bonding is a three-step process. In the first stage, clean and dustless hydrophilic surfaces of silicon and glass, after a preliminary bonding and heating to a temperature of a few hundred degrees centigrade, form a weak bond (adhere to each other) due to the attraction of Van der Waals forces and long hydrogen bonds, similar to that observed for fusion bonding for two hydrophilic silicon substrates.

In the second step polarization is applied. Sodium drifts in the direction of the cathode, bonding current increases rapidly; a quasi-quartz depletion layer is formed in glass, near the glass boundary, from the silicon side. Bonded surfaces are electrostatically clamped. The distance between bonded surfaces approaches several parts of a nanometer. Molecular water existing at hydrophilic surfaces of silicon and glass decomposes into hydrogen and hydroxide OH^- groups. Hydrogen drifts through glass toward the cathode. The dissociation of Na_2O causes the generation of sodium ions, drifting under an electrical field toward the cathode, and oxygen ions, which react with hydrogen, forming additional OH^- groups. Those groups drift slowly toward the anode, where together with OH^- groups (a by-product of decomposition of surface molecular water) they oxidise silicon at surfaces of bonded wafers, forms ≡Si–OH–OH–SI≡ bonds, which replaces long hydrogen bonds existing at the very beginning of the sealing procedure. Hydrogen, being a product of the decomposition of water, replaces sodium in $(BO)_4^- Na^+$ and $(AlO_4)^- Na^+$ in glass. Dehydration of the surface occurs on the silicon–glass boundary. Weak

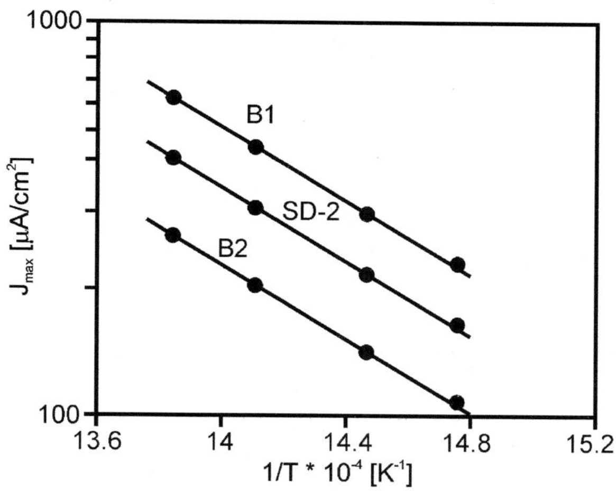

Fig. 4.68. Log $J_{max} = f(1/T)$, polarization voltage $U = 1000$ V, Borofloat 33 glass with thickness of 1 and 2 mm; the same curve obtained for 2 mm-thick SD2 glass is presented for comparison.

value is consistent with the data of Schmidt ($E_a = 0.97 \pm 0.03$ eV) and Carlson ($E_a = 0.95 \pm 0.03$ eV). Therefore it appears that the thermal activation of charge transport during anodic bonding of silicon to glass has been confirmed.

4.4.3. Technology of bonding

The anodic bonding of silicon to three types of borosilicate glasses, 7740 Corning, Borofloat 33 and SD2, has been investigated. All types of this process, which are useful in silicon micromechanics, have been examined. Glasses were selected in such a way that they should meet the requirements of as many micromechanical technologies as possible. The results obtained produced the possibility of determining the optimal method of anodic bonding, applied in the technology of sensors, actuators and other microsystem devices.

4.4.3.1. Silicon to glass bonding

4.4.3.1.1. Bonding in air, optimal parameters, current curves

In silicon micromechanics the anodic bonding of silicon wafers to glass plates is used most often, while the anodic bonding of various small constructional details made of these materials is applied more rarely. For these reasons bonding of complete wafers and plates has been researched first.

Bonding of all the examined glasses to silicon was satisfactory. Bonding current curves as a function of time showed a distinct maximum and a range of rapid decrease for optimal voltages and temperatures of bonding process (Table 4.11). The current attained its maximum value after 3–5 seconds, and then decreased by ten times after a few minutes. A current drop up to 0.1 I_{max} signified bonding of the whole surface, when the optimum parameters of bonding

Table 4.11. Optimal parameters of a good bonding of silicon wafers to glass, one-side polished Si wafers, bonding in air

Type of glass	Thickness [mm]	At temperature [°C]	Optimum U_p [V]	Time of complete bonding [min]
SD2	1	450	≥ 600	5
	3	450	≥ 1500	15
Corning 7740	1	350	≥ 1000	5
		450	≥ 500 (400)	5–10
Borofloat 33	2	350	≥ 1000	5

were applied. However, in many cases bonding of the whole surface was obtained even when current equaled 0.5 I_{max}. As the temperature and voltage were decreasing, the current curves were becoming more and more flat (Figs 4.69 and 4.70).

Research on the influence of conditions of anodic bonding and type of glass on the course of current characteristics, as a function of time, reveals that the process parameters of Corning and Borofloat glasses are similar. SD2 glass is characterized by lower values of current density. Consequently, a good bonding of Borofloat 33 and Corning glass can be obtained for the same or similar parameters of the process (temperature and polarization voltage), while SD2 glass requires higher polarizations or higher temperatures.

The conclusions reported above have been confirmed in broadened research of the impact of the conditions of bonding on the strength of silicon to glass bonding.

4.4.3.1.2. Quality and strength of bonding

Research into the influence of temperature and polarization voltage on the strength of bonding of silicon to different types of glass was carried out utilizing samples with surface area equaling a few square centimeters and tip cathode $r \sim 1$ mm made of stainless steel 1H18N9T. The strength of bonding was estimated by means of the method of destructive tests. The force with which square samples with area of 1 mm^2 were pulled out of the silicon–glass wafer was determined. About 500 μm-deep grooves were cut from the silicon side using a diamond circular saw. Tearing mandrel $\phi = 1$ mm in diameter was glued to silicon. So the prepared wafer was fixed by glueing to base with diameter bigger than the diameter of test sample (Fig. 4.71).

Results are presented in Table 4.12. Research showed that the strength of bonding equaling 20 MPa was obtained under the conditions represented by letter A, and that delaminating occurred during cutting of the silicon–glass bi-layer. Bonding B, with strength varying from 20 to 30 MPa, was uncertain. The current characteristic of the process was flat, without a clear current drop. The final current equaled about 0.5 I_{max}, non-bonded areas were distinguishable.

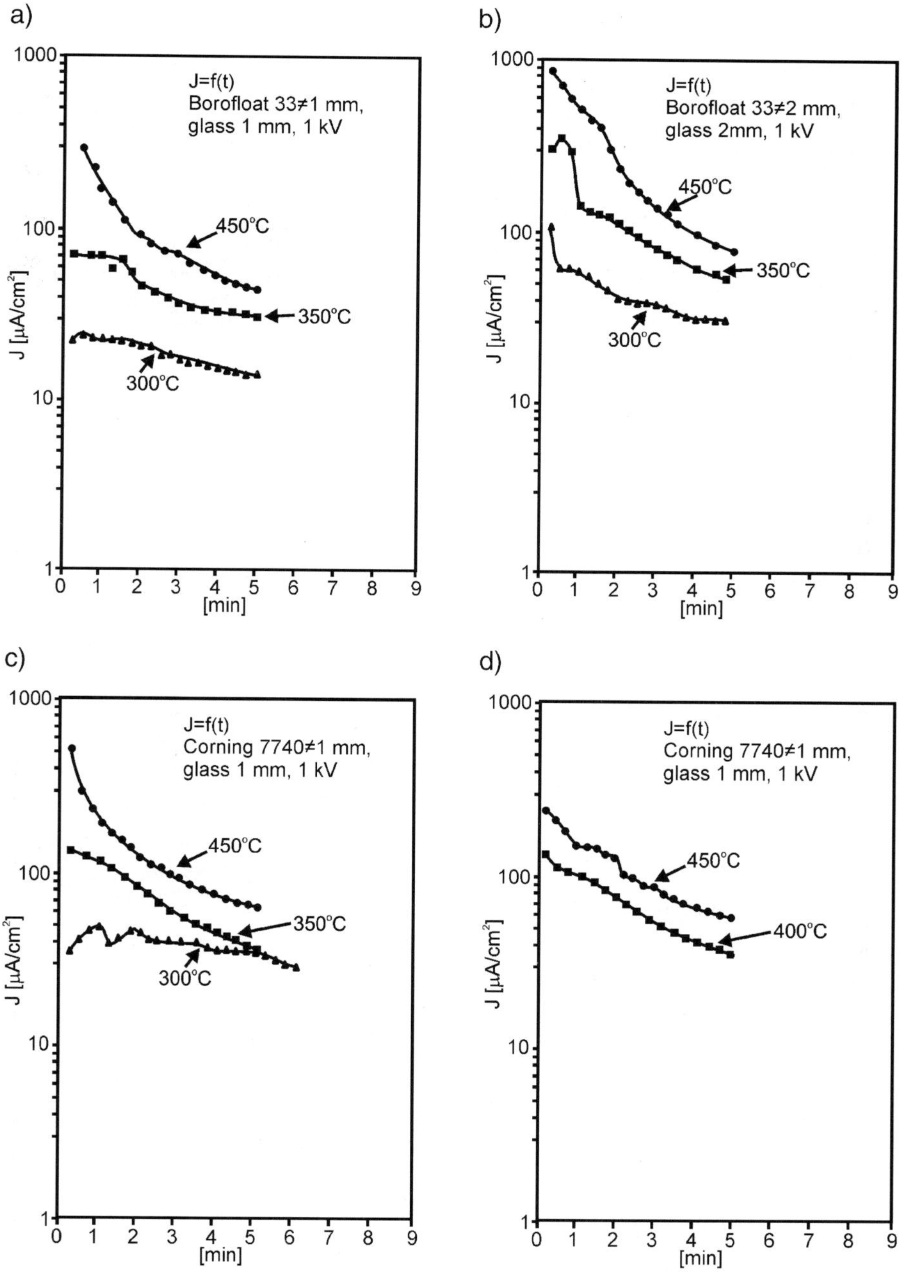

Fig. 4.69. Current characteristics of silicon to glass bonding, influence of temperature for different glasses, constant polarization $U_p = 1$ kV: a) Borofloat 33 $\neq$ 1 glass, b) Borofloat 33 $\neq$ 2 glass, c) Corning 7740 $\neq$ 1 glass, d) SD2 $\neq$ 1 glass.

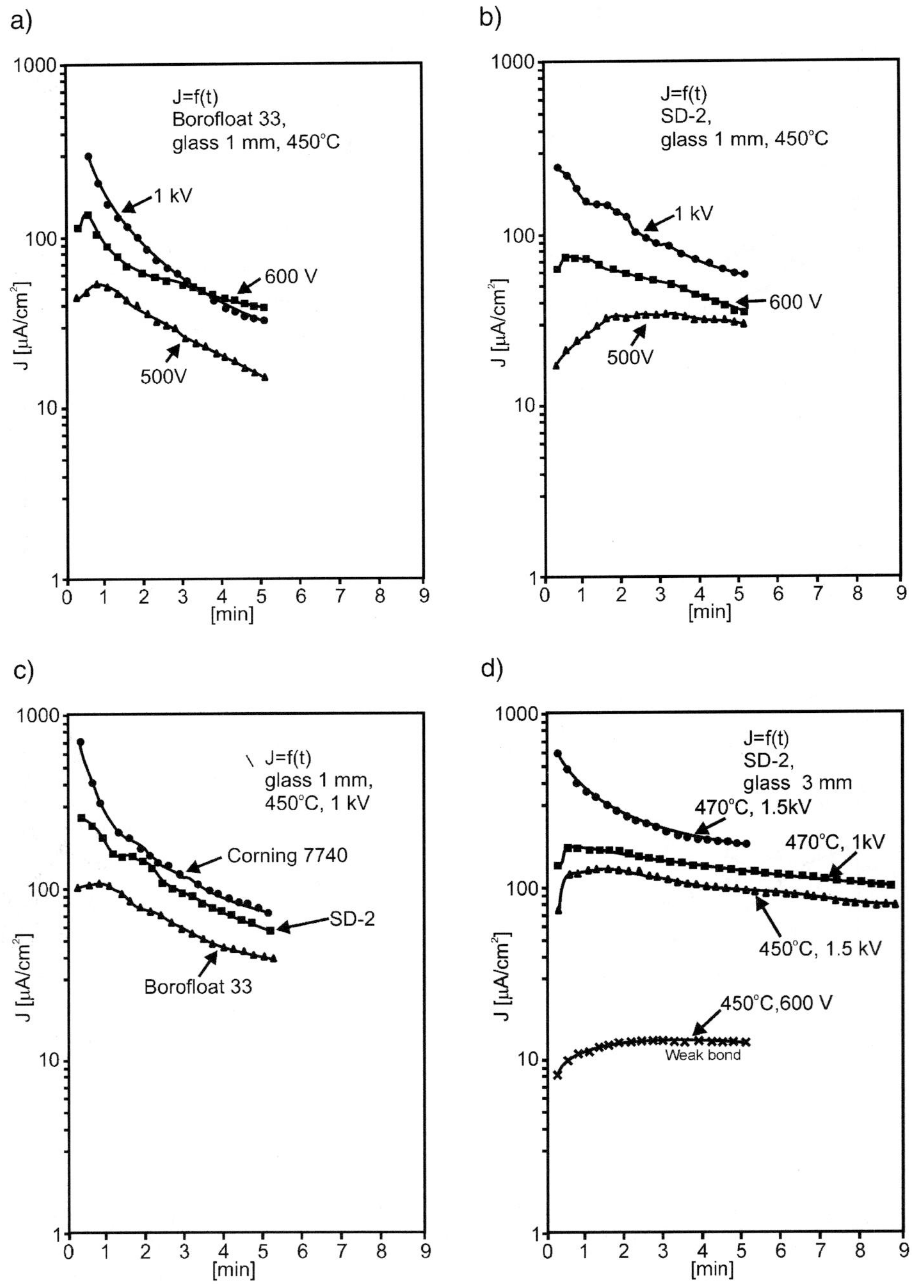

Fig. 4.70. Current characteristics of silicon to glass bonding, influence of polarization voltage for different glasses, and influence of glass type: a) Borofloat 33 ≠ 1 glass, b) SD2 ≠ 1 glass, c) SD2 ≠ 3 glass, d) Corning 7740, Borofloat 33 and SD2 ≠ 1 glasses. Careful selection of temperature ensured a good bonding.

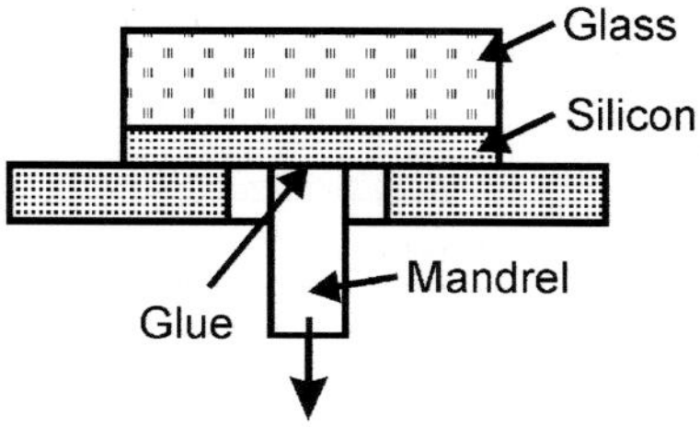

Fig. 4.71. The destructive pull tests method.

Table 4.12. Strength of silicon to glass bonding, method of tearing tests

Type of glass	Thickness [mm]	Temperature [°C]	Voltage [V] 500	700	900	1000	1200	1500
Borofloat 33	1	250	A	A	A	A	A	A
		300	A	A/B	A/B	B	B	B
		350	A	B	B/C	B/C	C	C
		400	B	B/C	C	C	C/D	D
		450	B	C	D	D	D	D
	2	300	A	A	A	A	A	A
		350	A	A	A/B	A/B	B	B
		400	A	A	B	B	B/C	C
		450	A/B	B	B/C	C	C/D	D
SD2	1	300	A	A	A	A	A	A
		350	A	A	A/B	A/B	B	B
		400	A	A/B	B	B	C	C
		450	B	B	C	C	C/D	D
	3	450	—	A	A/B	A/B	B	B/C
Corning 7740	1	250	A	A	A	A	A	A
		300	A	A/B	A/B	A/B	B	B
		350	A/B	B	B/C	C	C	C
		400	B	B/C	C	C	C/D	D
		450	C	C/D	C/D	D	D	D

Bonding C, with strength from 30 to 40 MPa, was satisfactory; small non-bonded areas were observed.

Very good bonding (marked D) was characterized by high strength (above 40 MPa) and rapid drop of bonding current in time. The final current met a condition attributed to the optimal conditions of bonding and equaled 0.1 I_{max}. The conditions of C, C/D and D bonding, under which a strong bonding was obtained, are marked with a heavy line.

Results concerning 2 mm-thick Borofloat 33 glass, which were obtained using the destructive test method and which contain precise numerical values that describe the strength of silicon–glass bonding, are presented in Table 4.13 [169].

Table 4.13. Strength of bonding (MPa) in relation to the parameters of bonding (point electrode, time: 10 min)

Temperature [°C]	Voltage [V]						
	500	700	900	1100	1300	1500	1700
300	—	—	—	—	—	2	5
350	—	10	15	15	17	25	26
400	15	16	23	29	30	33	40
450	20	23	30	38	40	44	46

Microscopic research on samples which had a bonding strength exceeding 30 MPa, showed the destruction of glass fragments and maintenance of practically intact silicon to glass bonding. Because of the fact that the glass bonding energy (γ) (equaling approximately 3 J/m^2) is lower than that of silicon, this proves a very high bonding energy exceeding 3 J/m^2. Go and Cho [167] reported that under similar conditions of bonding (temperature $T = 450°C$, voltage $U = 800$ V and time $t = 25$ minutes) the bonding energy exceeded 6 J/m^2. This value seems to be definitely too high, which may result from the utilized method of estimating γ that uses tests of built-in tip and is not very precise.

The strength of anode bonding of all the C, C/D and D tests equaled from 30 to 46 MPa. These values are over three times higher than the values reported in paper [164]. Data given in Table 4.12 are similar (at least as far as the strongest bonding is concerned) to the results presented in papers [170] and [171], which were obtained for silicon bonded to silicon through a thin layer of sputtered Pyrex glass. Data are also similar to that reported in paper [172]. The discrepancy between the data is mainly a result of the different test methods applied. It can also be an effect of the particularly scrupulous preparation of glass and silicon wafers in our own studies.

Experiments pointed out that a good or very good bonding of silicon to glass, with strength of bonding above 30 MPa, is marked by a high maximum current and a sudden drop in current characteristics. Flat current characteristics can be a good indicator of weak bonding, below 20 MPa, if the time of bonding was sufficiently long.

In further studies it was assumed that a good bonding of silicon to glass has to ensure a bonding strength above 30 MPa over the whole surface of a bonded wafer (plates). This is rather difficult to achieve, especially for point cathodes, because the strength of bonding of silicon to glass can be a function of the distance between the researched region and this electrode [107]. This phenomenon was investigated by means of the destructive method for 1 mm-thick Borofloat glass, optimally bonded to silicon. Sections with dimensions equalling 1 mm × 1 mm, situated at different distances from the tip cathode, were sampled. Samples prepared immediately after the process of bonding were used. Research revealed that the strength of bonding differed significantly for various distances, from the category of an excellent bonding (>40 MPa) near the cathode, to

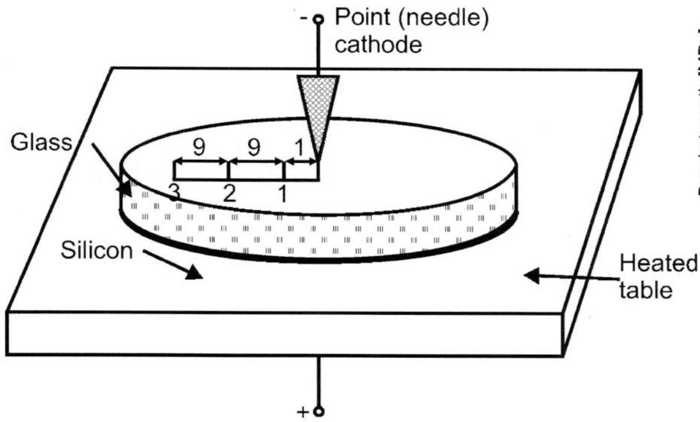

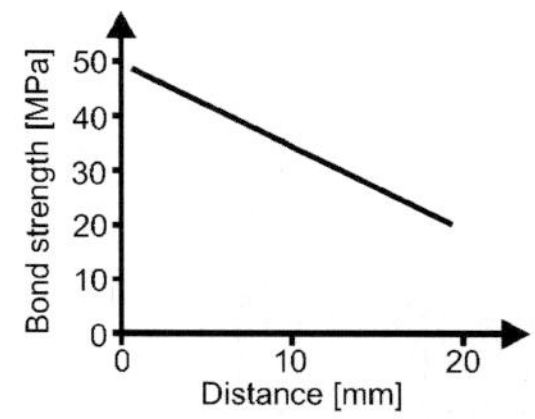

Point	1	2	3
Distance [mm]	1	10	19
Strength [MPa]	49	37	21

Fig. 4.72. Strength of bonding as a function of the distance from point cathode: 3″ silicon–Borofloat 33 wafer, anodic bonding $t = 450°C$, $U = 1500$ V, $t = 10$ min

weak bonding (~21 MPa) near the edge of the wafer (Fig. 4.72). The studies showed that bonding of big silicon and glass plates (3″ and more) requires a multi-tip or flat cathode.

The use of destructive tests is rather onerous; they cannot be applied directly on the workstand, therefore in our own studies a non-destructive method of research of the strength of silicon to glass bonding (the method of so-called built-in-tests) has been developed. In this method the cavities with known depth H_{Test} are etched on the surface of silicon. Most often the round patterns (diameter ϕ_{Test} from 25 μm to 2.5 μm) and the sets of rectangle-shaped patterns (width L_{Test} from 50 μm to 0.9 μm) are applied. Electrostatic force, which occurs during anodic bonding, causes local deflections of glass. If the force is sufficiently big, deflected glass touches the bottom of the cavity etched in silicon, which leads to the bonding of silicon to glass. It is easy to distinguish the cavities in which bonding has occurred, because the permanent bonding of silicon to glass causes a change in surface color: non-bonded areas remain silvery, while bonded regions become light blue. The influence of the dimensions of test cavities on the results of built-in-tests has been investigated in our own studies, as well as the range of the applicability of these tests for H_{Test} equaling from 820 to 850 nm, using the standard polarization U_{bond} (Fig. 4.73).

In our own studies the method of built-in-tests has been used for the auxiliary, fast estimation of the strength of silicon to glass bonding in the procedures of bonding, which are employed in the technology of many micromechanical sensors. The applied test pattern with depth $H_{Test} = 340$ nm (image in Fig. 4.74a) corresponded to a very good bonding with a strength exceeding 30 MPa.

Initial conditions of bonding

In order to start the process of silicon to glass electrostatic bonding, a minimum polarization voltage has to be applied to initiate the bonding wave. This voltage depends on the temperature of process – at the given temperature, if voltage is

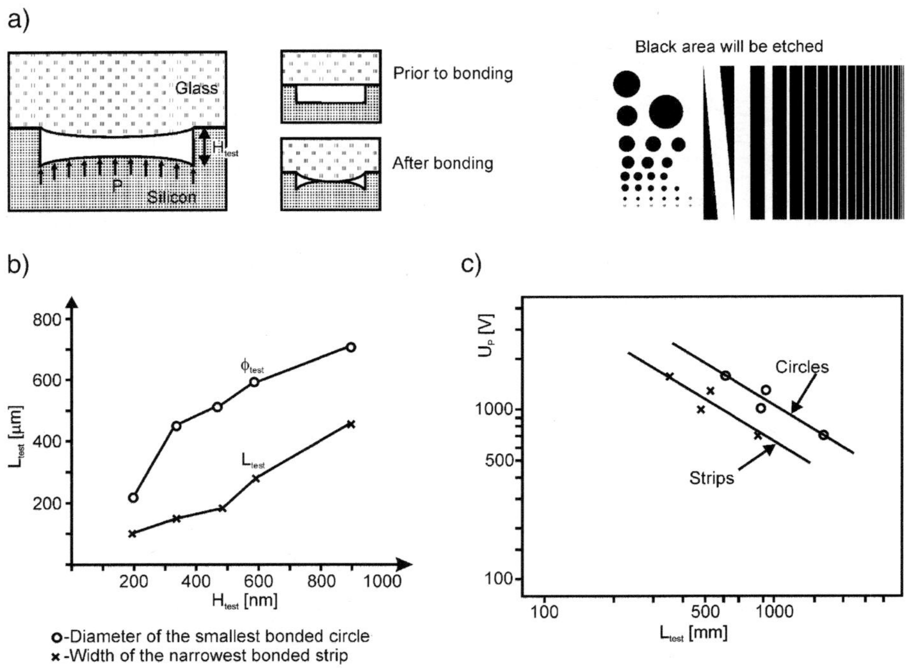

Fig. 4.73. Built-in-tests: a) principle of tests, b) results of measurements of the smallest, round (circles) and rectangular-shaped (strips) bonded structures after anodic bonding in relation to the depth of etched structures; ϕ_{Test}, $L_{\text{Test}} = f(H_{\text{Test}})$, ○ = diameter of the smallest bonded circle, × = width of the smallest bonded strip; conditions of bonding: $T = 450°\text{C}$, $U = 1$ kV, $t = 10'$, c) measurement results of the dependence of bonded surfaces voltage (U_{bond}) on the smallest dimensions of the diameters of circles and width of strips bonded during the process; $U_{\text{bond}} = f(L_{\text{Test}}, \phi_{\text{Test}})$, H_{Test} 820–880 nm, ○ = circles, × = strips.

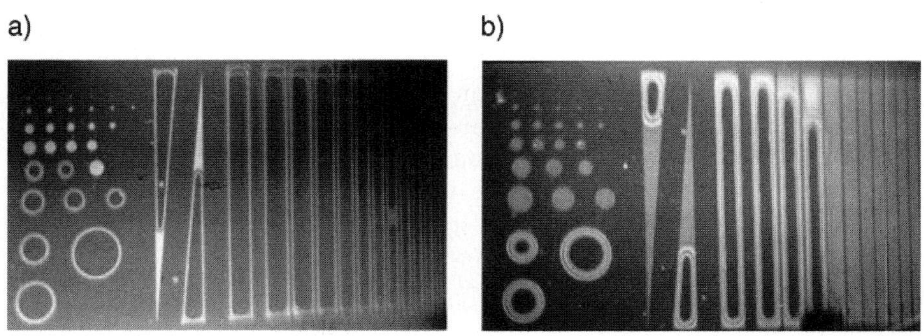

Fig. 4.74. Image of test structures after Si–Borofloat 33 ($\neq 1$ mm) anodic bonding ($T = 450°\text{C}$, $U = 1$ kV, $t = 10$ min): a) depth of etched pattern: $H_{\text{Test}} = 340$ nm, b) depth of etched pattern: $H_{\text{Test}} = 460$ nm.

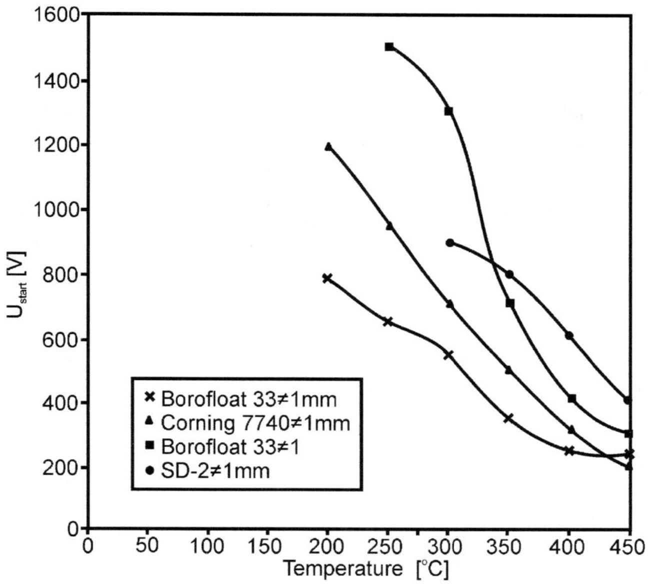

Fig. 4.75. Minimum bonding "start up" polarization voltage, which ensures the formation of a bonding wave, as a function of temperature: point electrode, bonding in air.

below the minimum value, bonding of materials will not occur. The curves presented in Fig. 4.75 determine the limit of the possibility of starting the bonding for different glasses. For the values on the left side of the curves anodic bonding does not occur, although bonding current increases slowly as a function of time after polarization is applied. When the polarization voltage reaches the values situated on the right side of the curves, bonding is initiated. The presented data show that Borofloat 33 glass bonds to silicon particularly well at reduced temperatures. For 2 mm-thick wafers the limiting polarization voltage, which causes the start of bonding, equals 2.95 kV at 300°C. If voltage exceeds 3 kV the breakdown of glass occurs*. Bonding cannot be obtained if the polarization is lower than 1000 V, at temperatures of 250–300°C, which are often recommended as the optimal range for bonding for the sake of the minimization of stresses built in the bi-layer [164, 94]. As far as this particular case is concerned,

* Baier, Schmidt, Straube and Horst in "Anodic bonding of low temperature using Li-doped glass": The 1997 Join. Int. Meeting of the Electrochemical Society and the International Society of Electrochemistry, Paris 1997, Proceed. Book PV-97-2, p. 2448 presented bonding of silicon to lithium glass (LiO_2–B_2O_3–SiO_2) with FeO and TiO_2 additives. Such glass at the temperature of 200°C is characterized by ionic conductivity higher than Pyrex-type glass. The authors report a successful bonding of silicon to glass directly and through SiO_2, SiON and Si_3N_4 at temperatures of approximately 200°C, with polarization of 2000 V. The strength parameters have not been given. It appears that the temperature of successful bonding of lithium glass is the lowest known. Lithium glasses are not used in the technology of silicon–glass microsystems.

the literature references are misleading, which may be a result of inaccurate measurements of the temperature of bonded wafers.

4.4.3.1.3. Bonding in vacuum, residual atmosphere

A successful silicon to glass anodic bonding carried out in vacuum ($p < 0.1$ Pa) does not differ significantly from bonding in air. However, the process in vacuum is longer, requires much time of the post-process annealing, and minimum polarization U_p should be about two times higher. For example, the optimal set of parameters of vacuum bonding in vacuum, taken from our own research, can be as follows:

- for 1 mm-thick Corning 7740 and Borofloat 33 glass: $t = 450°C$, $U \geq 1000$ V, 40 min.
- for 3 mm-thick SD2 glass: $t = 500°C$, $U \geq 1500$ V, 3 h.

The vacuum bonding of flat, unprocessed wafers does not pose any technical problems as opposed to micromachined wafers containing deeply etched patterns; for instance, reference cavities of absolute pressure sensors, whose gaseous atmosphere needs to be pumped out in order to obtain a very low referential pressure ($p < 10^{-3}$ Pa). Micromachined wafers require a special procedure of bonding. Activated (hydrophilic surfaces) wafers can be first contacted in air in dustless laminar chambers, then carried to the vacuum chamber and placed onto a heating table with the glass uppermost. Next, the vacuum chamber of the apparatus is closed and the upper glass wafer is lifted about 2–3 mm above the lower silicon wafer to allow the contents of closed spaces to be pumped out. After the pumping is finished, the upper glass wafer is brought into contact with the silicon wafer, the sandwich is heated to equalize their temperature, and then the polarization voltage is applied. Another method includes positioning of the silicon wafer onto the heating table inside the vacuum chamber followed by placing the glass wafer at mechanically operated spacers, lifting the glass wafer above the silicon one. Next the chamber is pumped to the desired vacuum level, spacers are removed, and wafers are contacted and bonded. Both of the above-described methods of anodic bonding apply distancing bonded wafers 2–3 mm while the gaseous atmosphere is pumped out. This is very important, because if distancing of wafers is not ensured, the gaseous atmosphere will not be pumped out from cavities situated in the central part of pre-bonded wafers. This was shown by Mack and others [173]. They bonded 3″ silicon wafers with a pattern of the etched cavities 1 mm × 1 mm × 0.3 mm arranged in a 2.4 mm × 2.4 mm grid (designed for application in absolute pressure sensors) to smooth Corning 7740 glass, without the use of wafer distancing. After 48 hours of pumping of the vacuum chamber at a 10^{-3} Pa level, real pressure in the centrally situated cavities equaled 10 kPa. Further continuation of pumping did not change this situation. This result showed experimentally that the impedance of a gas flow between pre-bonded flat, polished wafers, typically used in microsystem technology, is so large that any type of real-time pumping can ensure a high vacuum inside micromachined cavities. Wafers must be kept at a distance while pumping occurs. The cost of the dustless precise equipment for manipulation of wafers in vacuum greatly increases the price of the technical set-up necessary for the vacuum anodic bonding process.

The other important problem is control of bonding wave propagation and the end-point detection of bonding in vacuum. Typically, vacuum anodic bonding machines have no visual access for direct observation of bonding wave propagation. The technologist must rely on indirect factors of the course of the sealing process: current–time characteristic observation, control of sealing time, temperature and polarization voltage. This is sufficient for unprocessed wafers or shallow patterns, but cannot be used for deeply etched substrates. Bonding wave propagation can stop for several minutes at the edge of the deeply etched cavity (pattern). Stopping of propagation influences the current–time characteristic, usually decreasing the current significantly. This can be misunderstood as an end-point sign of bonding; but, after a sufficient period of time, the bonding wave usually overcomes the residual obstacle and "jumps", surrounding the deeply etched cavity. Current increases, and bonding proceeds further. Thus the author's advice is to make the testing process before the important one, and the direct observation of bonded sandwich, if possible. Current–time characteristics cannot be used as an ultimate bonding indicator.

In hermetic cavities formed by vacuum bonding of patterned silicon and glass wafers degradation of the vacuum occurs. This degradation is caused by several independent processes: the emission of gas from a bulk glass subjected to solid-state electrolysis under a strong electric field; gassing involved in the chemical reactions that proceed on the silicon–glass boundary during bonding [174]; diffusion of an external atmosphere through the glass cap, and desorption of gases from the surface of cavities possessing high surface to volume ratio [175]. The above-mentioned processes can together lead to the degradation of vacuum in orders of magnitude. Rogers observed the post-bonding degradation of vacuum inside the absolute capacitive pressure sensors (internal volume about $5 \cdot 10^{-11}$ m^3) from approximately 10^{-4} Pa to 100 kPa [145]. Mack and co-authors [173] have researched the composition and pressure of gases in the 0.09 mm^3 hermetic cavities formed in 380 μm thick Czochralski silicon wafers anodically bonded in vacuum to Tempax glass (Schott 8330), at 400°C, for polarization changed from 500 to 1000 V. The maximal current was limited, in order to make I_{max} independent of the conditions of process. Research revealed that the main gas product of anodic bonding is oxygen (partial pressure 85–95%) (in good agreement with anticipations of Carlson [152] and Baumann [166]), and mixture of CO_2, CH_4 with traces of H_2 and N_2 [173] (Fig. 4.76). Cross-wafer distribution of pressure level inside the etched spaces (3″ wafer, dimensions of cavity: 1.5 μm × 1.5 μm, net: 3.7 mm × 3.7 mm) does not depend on polarization voltage or on charge drifting through a bonded circuit [147]. It is supposed that oxygen generation occurs only at the very beginning of the bonding process, when the depletion layer is formed. The pressure level is not dependent on the ratio of area of referential space to the surface of frame that surrounds it, in contrast to direct fusion bonding (DBF). This leads to the

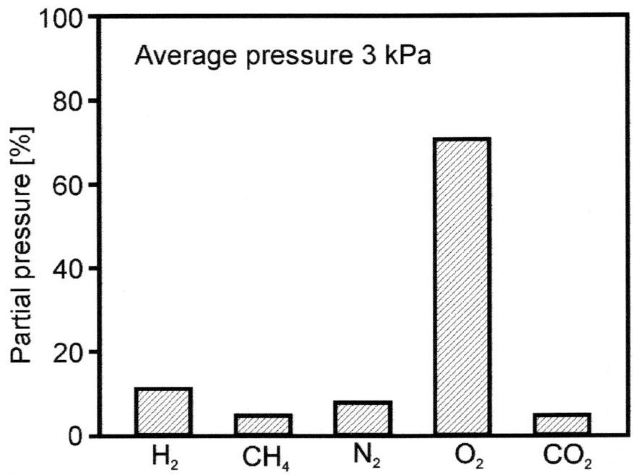

Fig. 4.76. Composition of the atmosphere inside test hermetically sealed by anodic bonding cavities. Partial pressure is expressed as the percentage of total pressure inside the referential space [147].

conclusion that the source of gassing is coming directly from the anodic bonding chemistry, and is not caused by the emission of gases from a bulk silicon (Fig. 4.77).

Permeation of gases from the environmental atmosphere through the glass was investigated in paper [174]. It was shown that the partial pressure of He in hermetic cavities increased from about 10^{-14} Pa to 10^{-4} Pa within 30 days for Pyrex-type glass and within 100 days for sodium glass.

The above-presented data led to the conclusion that the vacuum level in anodically bonded hermetic cavities is unstable, vacuum tends to worsen in

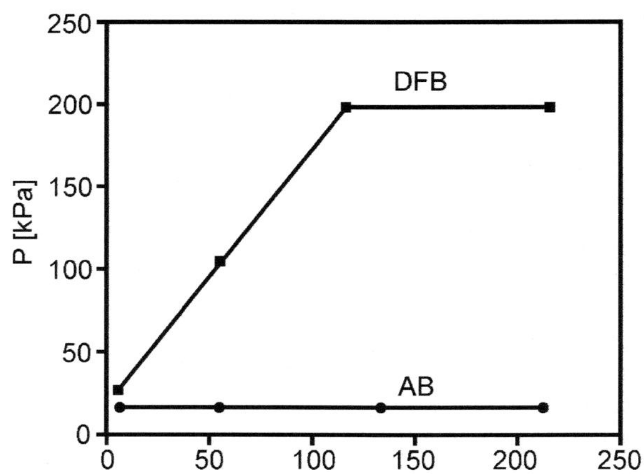

Fig. 4.77. Average pressure P inside the referential spaces as a function of area S of surrounding frame [173].

time. Two of the most important reasons are internal degassing and diffusion of gases from the external atmosphere (mainly He). Maintenance of vacuum in such spaces at low level requires the use of a non-evaporable getter (NEG) [175]. The getter is placed inside hermetic cavities and activated during anodic bonding. Recently, the SAES company (Milano, Italy) has announced printable getters for stabilization of vacuum (0.1–10 Pa) inside hermetic cavities of MEMS packages named PaGaWafer*. The getter material is patterned onto flat or micromachined silicon or glass wafers, it consists of zirconium alloy and ensures the best sorption of water, oxygen, mono-oxides and dioxides of carbon, nitrogen and hydrogen. The getter is activated in vacuum or a noble atmosphere at 300°C for 15 min. The SAES company advises using their solution for anodic bonding as well as for glass frit bonding and eutectic sealing in vacuum. The shelf-life of PaGaWafer getter is 5 years in nitrogen, it survives 500 cycles −40°C +150°C. The getter can be treated with caustic cleaning procedures SC1 or SC2 without losing its sorption properties.

Long-term stabilization of the vacuum or gaseous atmosphere inside hermetic cavities plays a crucial role in the fabrication of MEMS-type vibrating sensors, the moving parts of which are extremely sensitive to damping caused by residual, unwanted gases. Degassing and long-term drift of a composition of gaseous atmosphere inside a vacuum-bonded cesium atomic clock MEMS cell is the main reason for unstable work of this device [176]. The application of PaGaWafer getters seems to be the only acceptable solution.

4.4.3.1.4. Post-bonding wafers sandwich deformations

Silicon and glass wafers sandwich bonded at high temperature deform after cooling down to the ambient temperature. The deformation – most often bowing (bending), waving and warping – depends on process parameters (Fig. 4.78), type and thickness of glass and pre- and post-bonding treatment. Deformations are caused by several factors. The most important is bowing of bonded sandwich

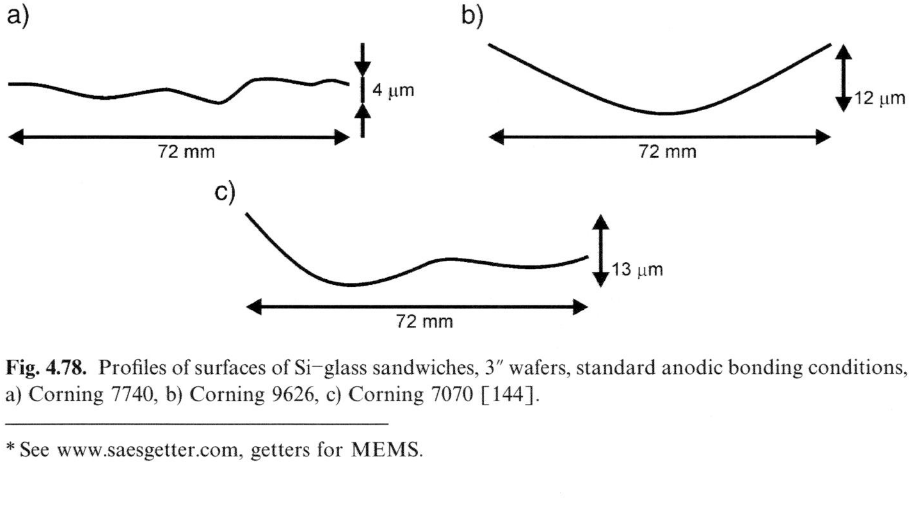

Fig. 4.78. Profiles of surfaces of Si–glass sandwiches, 3″ wafers, standard anodic bonding conditions, a) Corning 7740, b) Corning 9626, c) Corning 7070 [144].

* See www.saesgetter.com, getters for MEMS.

involved in mismatch of thermal expansion of bulk glass, depletion layers and silicon. Depletion layer and bulk glass have different compositions (and composition influences mechanical parameters). Many authors have shown that the depleted layer is more quartz-like than borosilicate glass [152, 154, 162]. The change in composition results in a significant change in the coefficient of linear thermal expansion. Mismatch of the coefficients of linear thermal expansion of depletion layer α_{gZ} and glass α_g can lead to the formation of concavities ($\alpha_{gZ} > \alpha_g$) or convexities ($\alpha_{gZ} < \alpha_g$) of wafers. The impact of the depletion layer on bowing of wafers was investigated by Shoji and Esashi [176]. They researched the influence of the bonding time of 1 cm^2 samples of silicon and varying types of Corning borosilicate glasses bonded at 400°C and 1000 V, on sandwich geometry. Samples bonded longer, with thicker depletion layer, were more deformed (Table 4.14).

In the described tests, the thickness of the depletion layer that corresponded to the changed parameters of bonding was determined by measuring the etch rate of glass in a buffered aqueous solution of HF with composition: H_2O:HF:NH_4 (1:1:10) [175, 177] (the etch rate of the depletion layer is higher than noted for bulk glass [151] (Fig. 4.79). It was found that the thickness of the depletion layer and the bowing of sample wafers were proportional to the total charge transported inside the silicon–glass sandwich (Fig. 4.80).

The bowing of bonded sandwich depends – beyond the total transported charge – on the type of bonded glass. Compressive forces evolve in the depletion layer formed in Corning 7740 glass and cause the concavity of glass ($\alpha_{gZ} < \alpha_g$). In SD2 glass tensile forces appear that lead to the convexity of wafers ($\alpha_{gZ} > \alpha_g$) (Fig. 4.81). In the worst case, bowing can reach 30 μm for a 3 mm-thick 3″ silicon–glass sandwich bonded typically at 400°C.

The local increase of temperature, usually ascribed to the local reheating of wafers (Joule heat) that appears as the bonding wave propagates, equals about 50°C [166]. This insignificantly influences the dynamics of bonding, but – according to Rogers [144] – can lead to the warping of wafers. The warping of wafers (airscrew) is observed for Corning 7070 glass below 420°C; it occurs also in 7740 glass and other similar glasses. Bowing of wafers can be minimized by shortening bonding procedures [164, 177] or by applying a reverse polarization (minus at Si, plus at glass) right after the bonding [136]. The "opposite"

Table 4.14. Research on the influence of bonding time on bowing of wafers made of Pyrex-type glass (experimental conditions) [176]

Sample	Time of bonding [min]	Total charge Q [mC]	Measured bowing [μm]
A	30	400	6.3
B	10	220	3.8
C	2.5	125	2.5
D	1.0	65	2.3
E	0.5	15	1.3

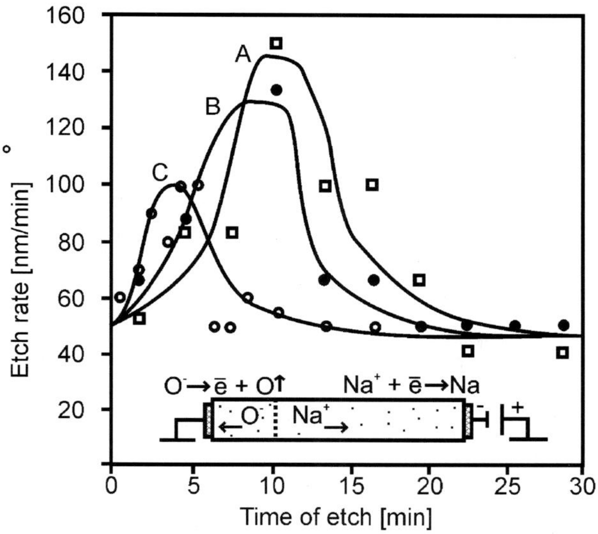

Fig. 4.79. Etch rate of the depletion layer as a function of bonding time; A, B, C designations according to Table 4.14 [177].

polarization cannot be applied for too long, because this is accompanied by negative side-effects. When a superfluous "reversed" sodium ion stream reaches silicon, chemical reactions proceed at the silicon and glass inter-layer, especially reactions of silicon and sodium and/or silicon and by-products of solid-state electrolyses of the bulk glass. Silicon and glass suffer corrosion, gaseous products are liberated. Residual defects, voids and delaminations are formed (Fig. 4.82).

Bowing of wafers can be minimized by ensuring of a uniform distribution ($\pm 5°C$) of temperature and electric field across bonded wafers. Planar cathodes (or ringed) should be utilized [155]. Flat – after anodic bonding – sandwich can be obtained by means of the introduction of an intentional mismatch of silicon and glass coefficients of expansion or by increasing the temperature above the matching point of linear coefficients of expansion of glass and silicon. The application of temperature above 520°C is possible but Corning 7740 and 7070, as well as Borofloat 33 glasses, should not be cured for longer times at this temperature, exceeding their transformation point. Bonding of 3″ diameter, thin ($d < 2$ mm) Corning 7740 glass to standard silicon wafer, at the temperature of 450°C, provides a deflection smaller than ± 2 μm [144]. Bowing of wafers is also an effect of the non-uniform distribution of temperature across the sandwich of bonded wafers. According to Shoji, double-side reheating (front- and back-side heaters) of a silicon–glass assembly reduces by many times the deflection of wafers [175] (Fig. 4.83). What is more, good bonding can be obtained for lower U_p polarization. Small deflection is obtained when thick glasses ($d > 3$ mm) are utilized. However, the higher bonding temperature required during the bonding of such wafers, leads to a large growth of internal

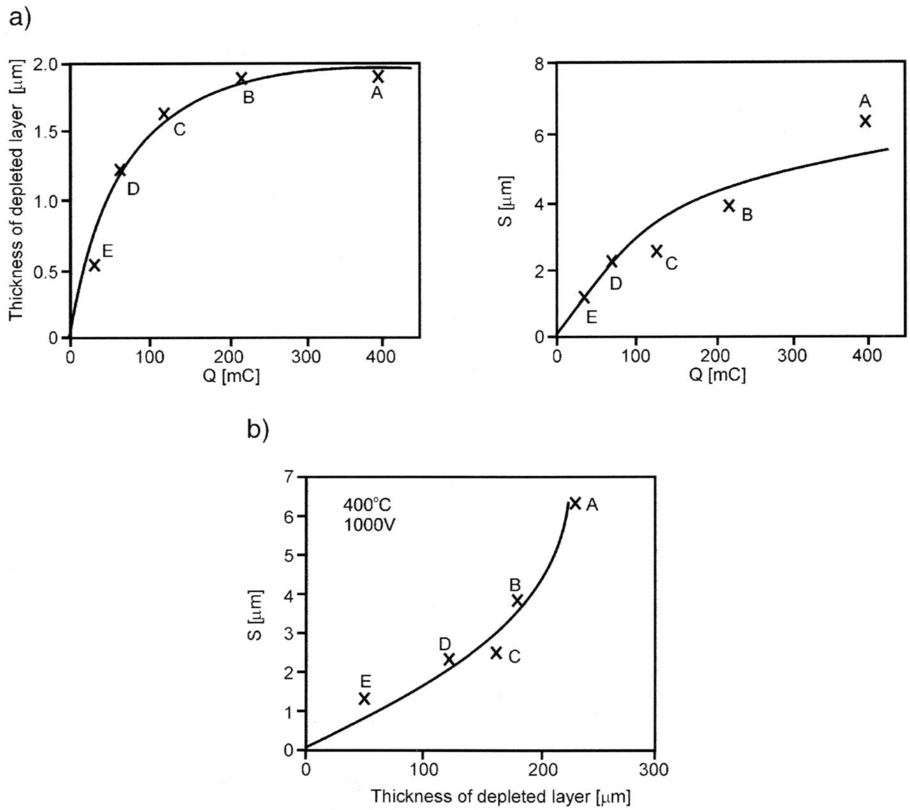

Fig. 4.80. Bowing S of wafers, area of 1 cm^2, $U = 1000$ V, $t = 400$°C: a) thickness of depletion layer and bowing as a function of total charge Q, b) bowing as a function of depletion layer thickness. A-E designations according to Table 4.14 [177].

stresses built in the silicon–glass sandwich, after the assembly is cooled down to the environmental temperature. These compressive stresses can transform into tensile stresses (Fig. 4.84).

Experimental results in paper [164], led to the conclusion that the stress-free anodic bonding of Pyrex-type glasses (Corning 7740, TEMPAX Schott 8380, Borofloat 33) requires low temperatures, approximately 300°C. The commencement of bonding at such a temperature is difficult (compare Table 4.11 and Fig. 4.75), and bonding quality is poor. The process needs a high polarizing voltage, but a high polarization voltage causes electrical breakdowns: through the glass layer or ionized air.

4.4.3.1.5. Dispersion of parameters of glass

The coefficient of thermal expansion of monocrystalline silicon α_{Si} does not depend on the producer or delivery batch. The statistical dispersion of thermal expansion coefficient α_g can occur in glass. This is a very important obstacle in

The phase segregation of SiO_2 and B_2O_3 which occurs in glass volume below the critical temperature T_g [37] can be the other factor involving higher bulk hardness of the bonded glass.

In our own studies a considerable increase in glass hardness has been observed in the layers near the glass–silicon boundary. Hardness of glass at a distance to 100 μm from the silicon–glass boundary increased from about 0.4 to 1.3 GPa according to the Knoop scale*. The growth of glass hardness after bonding near the anode was easily perceptible during the separation of silicon–glass wafers into small pieces using the method of cutting with a high-speed diamond saw (Fig. 4.88). Cutting of glass after bonding was difficult, and the diamond saw-blade disks wore very quickly. Deep cutting of multilayers with various hardness values led to the ovalization of the front of the circular saw and its seizing in grooves. Especially high wear of saw-blade disks (in our own studies) was noticed during the cutting of SD2 glasses. It was observed that, in order to cut bonded wafers of silicon and glass, it is necessary to select diamond saw-blade disks wider than those usually used. Moreover, to cut SD2 glass one should utilize a rotational speed about three times higher in comparison with the cutting of 7740 and Borofloat 33 glasses of similar thickness.

The cost of the separation of bonded silicon–glass wafers forms a very important cost factor in the fabrication of silicon/glass microsensors. For instance, the cutting of a bonded 3″ silicon and glass wafer with total thickness 2.38 mm (2 mm of glass, 0.38 mm of silicon) into squares with dimensions 2.5 mm × 2.5 mm, which is a standard module used in the technology of micromechanical pressure sensors, causes the complete destruction of a cutting device, the price of which equals a few hundred euros.

Cutting using a circular saw can be replaced with diamond scratching and breaking of bonded silicon–glass plates. A well-bonded sandwich can be separated like monolithic plates and does not separate on the bonding boundary.

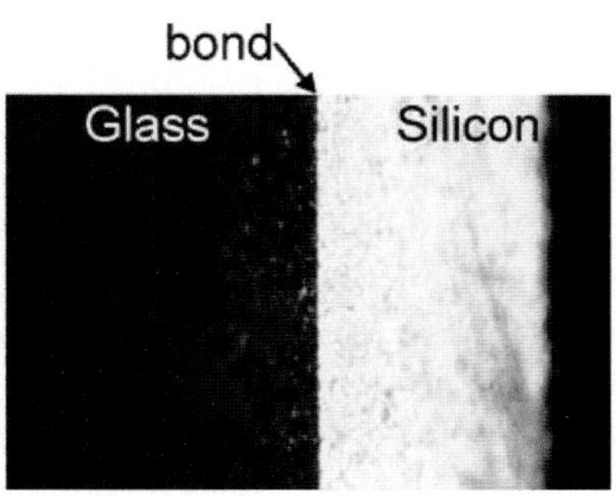

Fig. 4.88. Cross-section of silicon–glass sandwich, SEM 1000 ×. High-speed circular saw with diamond saw-blade disk DQ3 was applied.

---

* Holland says that the Vickers' method is not very accurate, and is not useful in the determination of glass hardness. The influence of the high pressure at the diamond end of a measuring instrument can result in local microstructural changes in glass, and therefore in incorrect results [37].

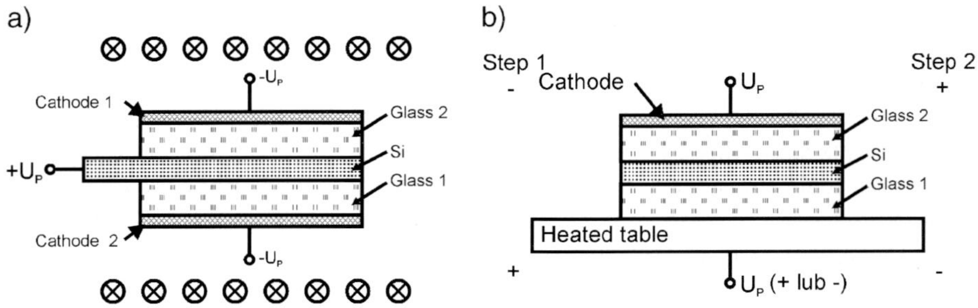

Fig. 4.89. Principle of multi-layer bonding: a) three-electrode assembly, b) two-electrode assembly; simultaneous bonding.

This method cannot be applied for separation of stress-sensitive sensors and micosystems.

4.4.3.2. Special techniques

4.4.3.2.1. Multi-layer bonding

In some applications bonding of a stack of many wafers of silicon and glass is required. It is possible to pile silicon–glass–silicon–glass, etc. wafers. Such bonding is a sum of two types of multi-layer bonding: silicon–glass–silicon or glass–silicon–glass. Multi-layer bonding is applied first of all in the μTAS to form the plates of chemical reactors, to bond interconnections and to fabricate the mechanisms that control the flow of liquids or gases – valves, pumps etc.

Multi-layer bonding of silicon and glass plates can be carried out in the glass–silicon–glass assembly in a three- or two-electrode configuration [178] (Fig. 4.89).

Three-electrode configuration

Simultaneous bonding of two glass wafers to silicon is performed in a three-electrode assembly. A positive potential is applied to silicon, and a negative potential is supplied to glass through two flat cathodes (Fig. 4.89a). The described process can be treated as a composition of two independent processes of silicon–glass bonding. The applied parameters (temperature, voltage, time) are similar to the parameters of a single process. During the bonding two depletion layers evolve at the same time in glass, transport of Na^+ ions toward the cathodes occurs along with the bonding of materials. Maintenance of a constant temperature in a cross-section of the bonded sandwich introduces technical difficulty, as it requires heating of both top and bottom glass wafers. The sealing obtained does not differ in quality from the known results of silicon–glass procedures of bonding.

Two-electrode configuration

In the two-electrode method the simultaneous polarization of silicon and two glass wafers involves two different processes: for example, when the top glass wafer is polarized negatively according to the bottom glass wafer (Fig. 4.89b), silicon is anodically bonded to the top wafer, while at the same time the bottom glass is subjected to so-called cathode bonding, first described by Wallis in 1968 [131]. After the polarity changes, the reverse of sodium ion transport occurs in the top glass. Sodium ions, which enrich the top glass surface area from the side of silicon, and destroy the depleted layer, are subject to multiple chemical reactions changing the color of the surface of silicon (brown spots, green points) and producing spots of the micro-electro-erosion of glass.

At the same time the bottom glass, polarized negatively, is anodically bonded to silicon. However, because of the fact that, during the first step of the process sodium drifted toward the surface of the bottom glass from the side of silicon, sealing of the bottom wafer is more difficult; many defects and voids are formed on the bonded interface (Fig. 4.90).

Step-by-step

The stacks of glass–silicon–glass plates etc. can be bonded by step-by-step, two- or multi-step anodic bonding. Let us assume the bonding of two glass wafers to a silicon substrate (Fig. 4.91). First bonding of the first glass wafer to silicon is carried out, according to the procedures discussed above in this book. Next, the bonded sandwich is placed "upside down" on a heating table, the second glass wafer is put on silicon, and the anodic bonding procedure is done, sealing the second glass wafer to the opposite side of the silicon substrate. In the second process of bonding a higher polarization voltage has to be applied (increased by approximately 50% in comparison with bonding of the first glass), and the process should last as short a time as possible. A flat cathode should be preferred. Longer processes than absolutely needed result in generation of a number of small-area voids, appearing at the bonded interface of the first glass

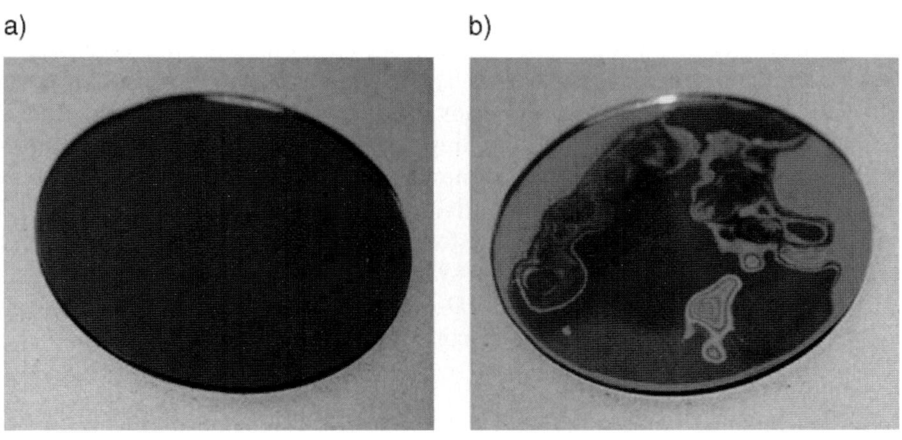

Fig. 4.90. 3″ silicon and Borofloat 33 ≠ 1 mm wafers after bonding: a) anodic bonding – excellent quality of bonding can be observed, b) cathode bonding – numerous defects and bubbles of non-bonded areas are distinguishable.

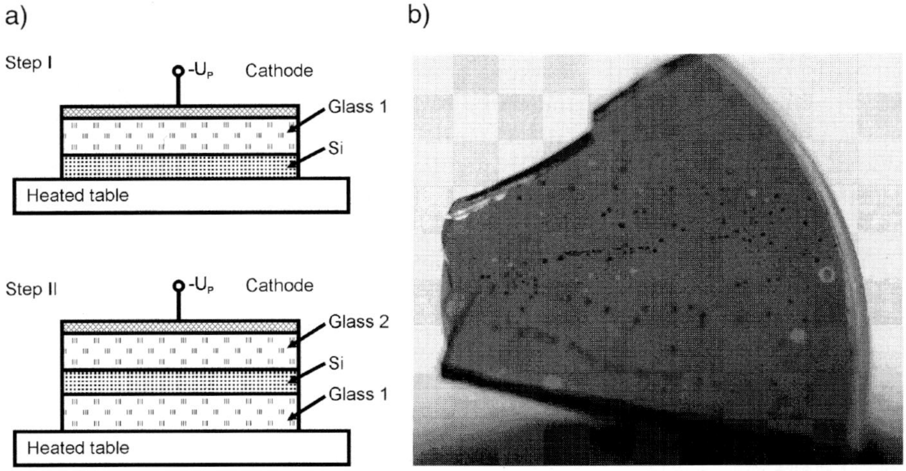

Fig. 4.91. Glass–silicon–glass bonding, two-step process: a) principle of process, b) defects of surface on the boundary of silicon and the first of two bonded Borofloat 33 ≠ 1 substrates.

and silicon. Sometimes small brownish corrosion points may be observed, especially for large-area samples (Fig. 4.91b). Such defects did not evolve in small samples.

4.4.3.2.2. Selective and lateral bonding, small details

Bonding of glass to silicon wafers with movable micromechanical structures requires the use of special techniques. The electrostatic force between silicon and glass, pulling delicate structures, is usually sufficient to contact the movable silicon detail to the glass surface, causing undesirable bonding of elements at the point of contact of materials, which immobilizes the detail (Fig. 4.92a). The above-mentioned effect is a real nuisance in the technology of mass accelerometers, pressure sensors and microvalves. It makes simple assembly of other types of silicon/glass micromachines with movable silicon or metal parts impossible.

There are a few methods of avoiding this problem: deposition of a non-bondable thin-film layer under the moving part; introduction of a thin-film metallic/semiconductor contact negatively polarized; roughening (dulling) of surface or employing cathodes in the form of conductive intermediate frames controlling the bonding current (Fig. 4.92b).

Experiments have shown that a CVD-deposited densified, thick ($d > 300$ nm) layer of Si_3N_4, effectively protects silicon and glass from bonding (Table 4.16).

Lateral bonding

In lateral bonding the silicon detail polarized positively is positioned onto the front surface of the heated glass wafer. A planar, thin-film metallic cathode is localized near the bonded detail, at the same side of the glass wafer. Bonding is based on the lateral flow of sodium ions in the glass (Fig. 4.93). Such a

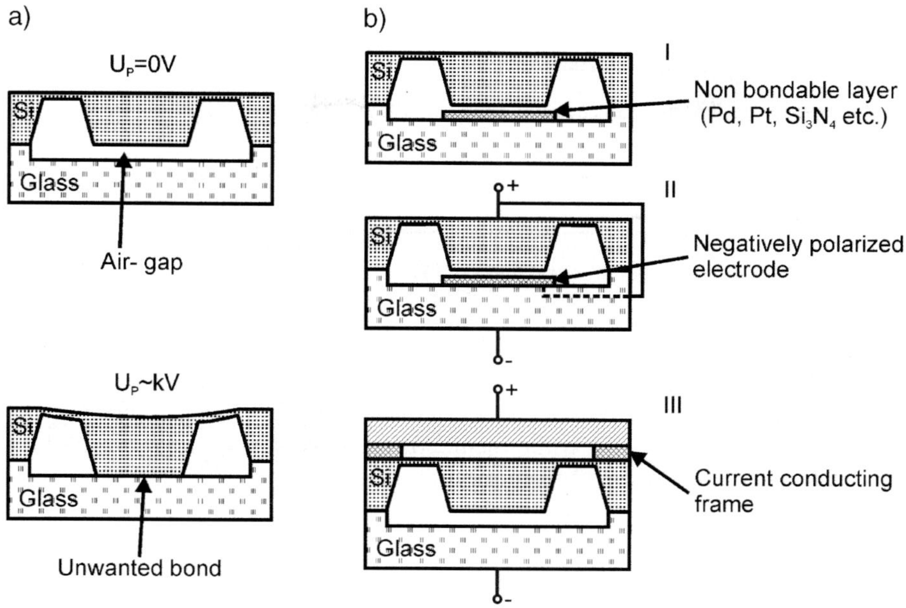

Fig. 4.92. Bonding of corrugated membranes: a) deflection of membrane caused by the electrostatic attraction force, b) three methods of bonding: with a layer that blocks the bonding (I), with an additional electrode (II), with a frame that controls current propagation (III).

Table 4.16. Selective bonding of silicon to glass

	Temperature [°C]					
U [V]	385	400	415	425	450	Type of surface
1500	–	–	–	–	–	Dull silicon
1700	–	–	–	–	–	Dull silicon
2000	–	–	–	–	–	Dull silicon
1500	–	–	–	–	–	PCVD nitride densified
2000	–	–	–	–	+/–	PCVD nitride densified
1500	–	–	–	+	+	PCVD nitride non-densified
2000	+/–	+	+	+	+	PCVD nitride non-densified

+, successful bonding; –, lack of bonding; +/–, uncertain or weak bonding.

configuration of electrodes, along with the use of thin glass wafer ($d < 0.5$ mm), results in successful bonding of silicon details to a silicon–glass–silicon sandwich, and is particularly recommended in the assembly of integrated micro-fluidic systems.

Small details

Bonding of small details needs special instrumentation – grips, positioners – and properly selected cathodes. Breakdowns connected with the flow of charges

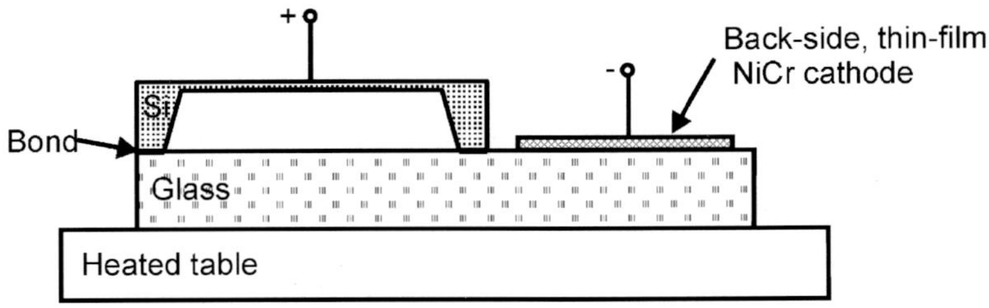

Fig. 4.93. Lay-out of electrodes in lateral bonding [179].

on the glass surface can pose some difficulties while bonding very small details. Bonding of high-aspect ratio glass details is easy and their insulating properties sufficient. This method of bonding of small details has been applied in the construction of silicon–glass components for gas microchromatographs and for microfluidic systems. A gas/fluid connection fabricated by means of this method is presented in Fig. 4.94.

4.4.3.3. Bonding of silicon to glass through thin layers

Anodic bonding through thin layers was described as early as 1969 by the discoverers of anodic bonding – Wallis and Pommerantz [134]. Two different types of this method are used in microsystem technology:

- the classical process of bonding of silicon to borosilicate glass through a thin-film dielectric (SiO_2, Si_3N_4, Al_2O_3, SiC) or metallic (Al, Ti, Ta) layer covering the silicon substrate [180–186]
- the process of bonding of silicon to silicon through a glass layer deposited on one of the sealed materials [78, 94, 171–174, 189–194].

4.4.3.3.1. Bonding through SiO_2, SiO_x

Bonding of silicon covered with thin dielectric layers to glass is applied widely in microsystem technology. The dielectric layer can be made of any material bondable to glass. This is deposited or grown onto the bonded silicon wafers. The sealing process is characterized by dynamics of the process somewhat different from the bonding of brand silicon. First, because thick dielectric layers covering silicon reduce electrostatic pressure; second, because dielectric layers introduce significant changes in bonding chemistry.

Electrostatic pressure reduction is described by equation (4.48) [141]:

$$p = p_{\mathrm{d}} \frac{\varepsilon_{\mathrm{g}} \cdot d}{\varepsilon_{\mathrm{g}} \cdot d + \varepsilon d_{\mathrm{ox}}}, \tag{4.48}$$

where: p_{d} = electrostatic pressure formed during the bonding of silicon to glass, ε_{g} = permitivity of glass, d = distance between bonded wafers (thickness of glass), ε = permitivity of dielectric, d_{ox} = thickness of the thin dielectric layer covering the anode.

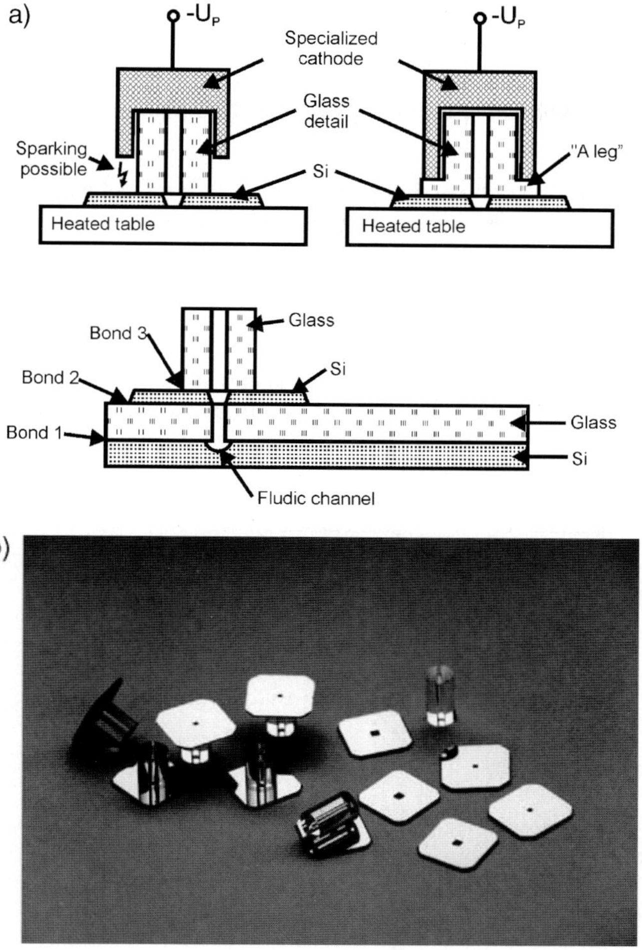

Fig. 4.94. Glass connection for gas/fluidic microsystems: a) two configurations of cathodes and a cross-sectional view of the assembled multi-layer fluidic device, b) parts of connections before and after bonding of glass tube and silicon orifice.

The influence of different dielectric layers on electrostatic pressure was investigated by Plaza and co-authors [180] using built-in tests [181]. They reported that thermal oxide with thickness from 0.025 μm to 0.4 μm does not significantly affect electrostatic pressure during bonding of test areas at 400°C and polarization equaling 1000 V. The thickest layer bondable to glass established by them was 0.6 μm.

Cozma and Puers reported a smaller value of thickness of the bondable SiO_2 layer, about 0.4 μm [164], Baire and co-authors approx. 0.5 μm [186]. The bondability of silicon wafers covered with silicon dioxide thin layers has been verified – in comparison to the reported literature data – experimentally for 1 mm-thick Borofloat 33 glass. Tests were done for:

- thermal high-temperature silicon dioxide, grown in steam at 1050°C,
- silicon dioxide deposited by the PECVD method,
- silicon dioxide (quartz) magnetron sputtered in argon from a quartz target,
- SOG glaze, sol-gel, spin-on deposited onto a silicon wafer.

Wafers were washed and activated (RCA1 procedure), dried and bonded at 450°C for polarization above 500 V. The polarization level was selected in order to obtain a good bonding. Current–time characteristics of the bonding process for wafers covered with thermal oxide with thickness from 0.05 to 0.4 μm are presented in Figs 4.95 and 4.96. Wafers formed spontaneous bonds very easily, the wave of bonding propagated quickly, and sealing was complete. These results were in good agreement with other authors' reports [141, 164, 180].

Tests made for very thick thermal oxide layers ($d_{SiO_2} > 1$ μm) have documented that sealing is easy [182] if a special procedure is applied. Oxidized wafers

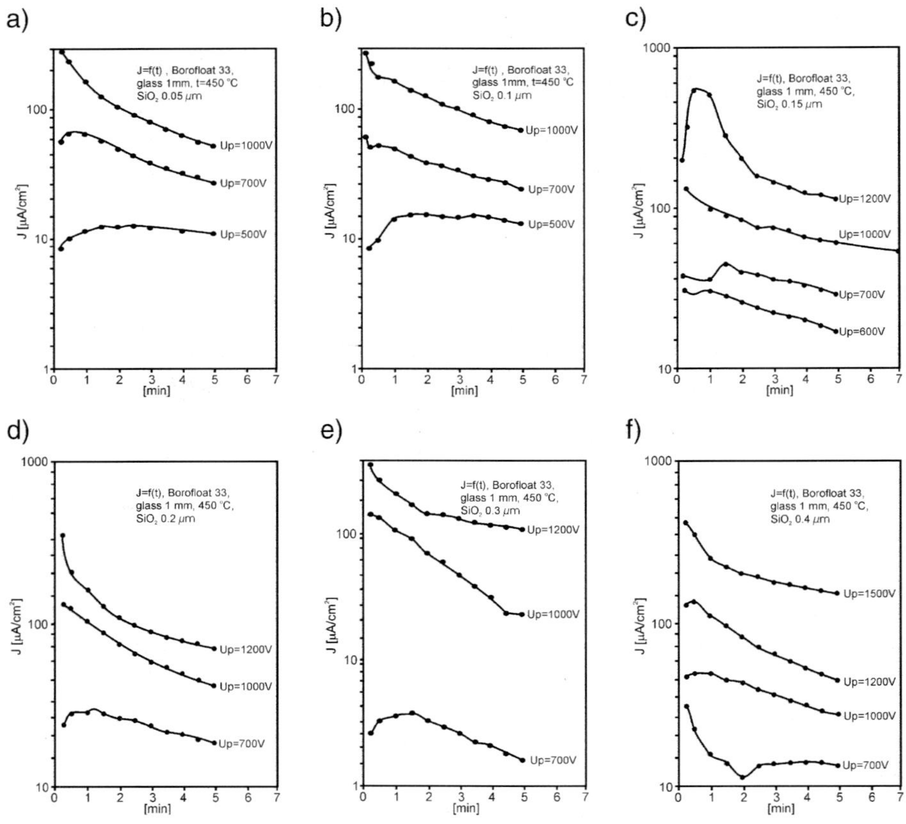

Fig. 4.95. Bonding current density as a function of time of bonding at 450°C of Borofloat 33 glass, 1 mm thick, to silicon covered with thermal oxide with thickness equalling: a) $d_{SiO_2} = 0.05$ μm, b) $d_{SiO_2} = 0.1$ μm, c) $d_{SiO_2} = 0.15$ μm, d) $d_{SiO_2} = 0.2$ μm, e) $d_{SiO_2} = 0.3$ μm, f) $d_{SiO_2} = 0.4$ μm. Polarization voltage U_p is a parameter.

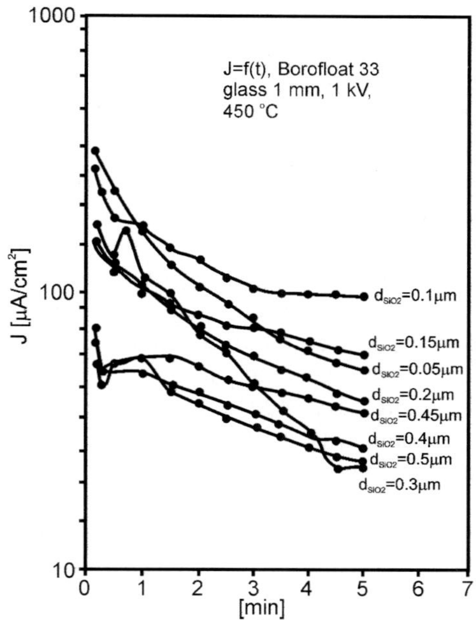

Fig. 4.96. Bonding current density as a function of time of bonding at 450°C of Borofloat 33 glass, 1 mm-thick, to silicon covered with thermal oxide with different thickness for constant polarization voltage U_p.

undergo washing and activation RCA1 procedure, and are next stored in DI water for at least 30 minutes. Glass is prepared in a standard procedure, boiled in 30% H_2O_2 and stored in DI water. Wafers are contacted in DI water or dried in a N_2 dust-less filtered jet, and positioned at the bonding table, glass on the top.

Initially the bonding is reversible, even for the polarization and temperatures that ensure a good bonding through thinner (0.3–0.6 μm) SiO_2 layers. Although the bonding wave propagates very quickly and the wafers adhere to each other, after the polarization is switched off many non-bonded areas appear, keep enlarging and merge, which can lead to the separation of wafers. Irreversible bonding is obtained using the flat metal cathode with intermediate glass electrode with clamp (about 0.2 kg for a 3″ wafer) and – first of all – by means of the prolonged formation of bonding. The most permanent bonding was obtained in the heating process lasting for over 1 hour at 405°C and with full polarization applied. With use of the above-mentioned recommendations, the uniform and strong Si/SiO_2-glass bonds through SiO_2 thermal oxide layers with a thickness reaching 2 μm were obtained, without bubbles or delamination of layers [183, 184].

During the process of silicon to glass bonding through SiO_2 layers, which are so thick, the bonding current decreases very slowly over time. The shape of current–time characteristics is almost independent of polarization voltage. The minimum value of current, even after annealing of bonded sandwich for many hours, does not decrease below 0.3–0.2 I_{max} (Fig. 4.97a), which can testify to the intense charge exchange on the interface of SiO_2 and glass that can be connected with the improvement of stoichiometry of oxide and "consumption" of the large amount of molecular water trapped inside a thick layer of the thermal oxide. The shape of current–time characteristics of bonding through thinner layers is dependent to a large degree on the polarization voltage (Fig. 4.97b).

Current characteristics of bonding of silicon wafers covered with plasma oxide and quartz sputtered in Ar to 2 mm-thick Borofloat 33 glass plate (washing and activation as described for thermal oxide) are presented in Fig. 4.98. Bonding of wafers was easy to perform and repeatable, the strength of bonding of well-connected wafers equaled at least 30 MPa.

Bonding of silicon covered with a thin layer (100–200 nm) of spin-on, P115 A-type glaze made by the company FILMTRONIC INC (USA) to Borofloat 33 glass was carried out at 450°C. The sol of the glaze was spin-on deposited for 30 seconds at a speed of 400 turns per minute, then baked for 2 minutes at a temperature of 75°C, and cured at 140°C for 2 minutes and finally for 4 minutes at 220°C.

Once the glaze gel was formed the wafer was activated by boiling in 30% H_2O_2, dried in an N_2 stream and bonded to a glass substrate. Very good sealing,

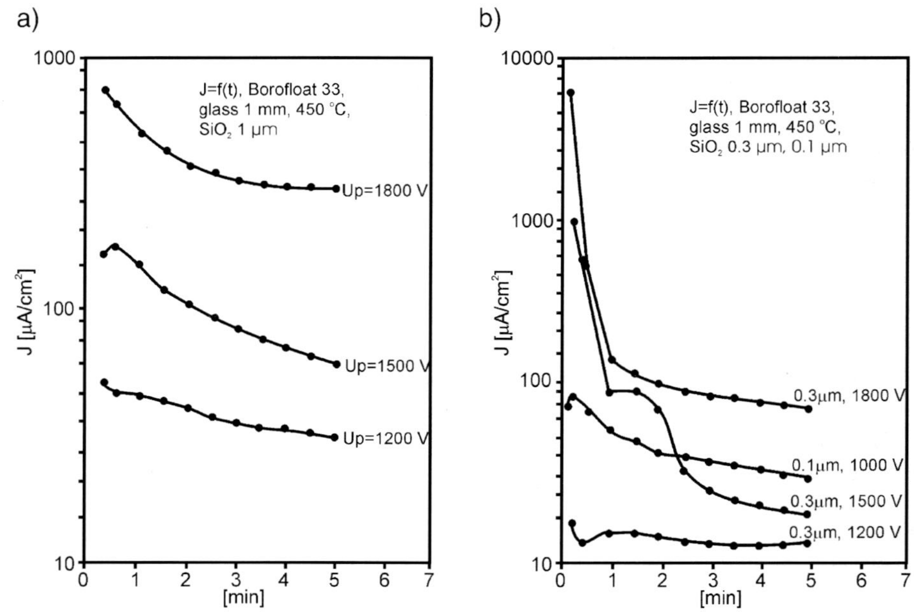

Fig. 4.97. Bonding current density as a function of time of bonding at 450°C of Borofloat 33 glass, 1 mm-thick, to silicon covered with a) 1 μm-thick thermal oxide, b) thermal oxide with thickness equaling 0.3 μm and 0.1 μm. Polarization voltage U_p is a parameter.

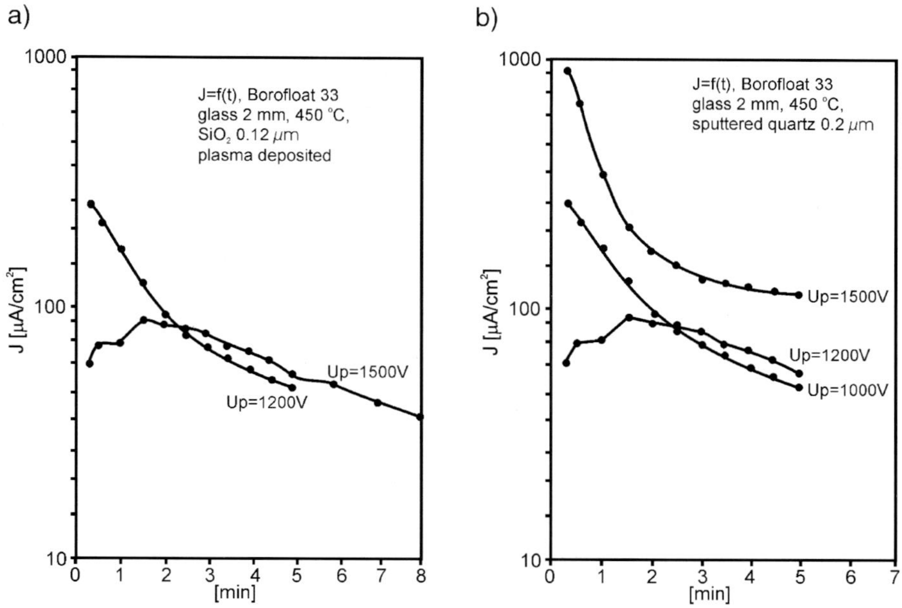

Fig. 4.98. Current density of silicon to glass bonding through silicon dioxide layers as a function of time, $t = 450°C$, 2 mm-thick Borofloat 33 glass, a) plasma oxide with thickness equaling 0.12 μm, b) 0.2 μm-thick sputtered oxide.

which did not separate under the razor blade test, was obtained accurately at 450°C for 1300 V. This process needed very accurate temperature setting; a small decrease in temperature to 400°C made wafers almost non-bondable, many voids were visible, and samples were easily separated under the razor blade test. A small voltage drop (to 1000 V) also worsened the quality of bonding even at a temperature of 450°C.

The optimum conditions of silicon to glass bonding through oxide layers obtained by means of different methods are listed in Table 4.17. The strength of sealing was analyzed using destructive pulling tests and built-in tests (Fig. 4.99).

4.4.3.3.2. Bonding through SiO_2/Si, SiO_2/SiC layers

Anodic bonding of silicon substrates covered with a double-thin-film-layer of thermally grown silicon dioxide layer and silicon carbide layer, or thermally grown oxide and a thin layer of silicon doped with a few selected impurities to Borofloat 33 and Corning 7740 1 and 2 mm thick wafers, was studied in order to extend the technological usefulness of the anodic bonding method in microsystem technology [185].

First, 1 μm thermal wet silicon oxide was grown at 1150°C in steam at 380 μm-thick, 3 Ωcm, n-type (100) oriented silicon wafers. Layers of silicon carbide or doped silicon or pure silicon were deposited onto the hot ($t = 300°C$)

Table 4.17. Optimum conditions of bonding of Si/SiO_x–Borofloat 33 glass, $d = 2$ mm

Type of oxide	Thermal oxidation	Magnetron sputtering	Plasma oxide	SoG	Pure silicon
Thickness [μm]	0.3	0.2	0.12	0.2	—
Optimal parameters [°C, V, min]	450, 1200, heating for 1 h with applied voltage	450, 1000, heating for 30′ with applied voltage	450, 1000, heating for 30′ with applied voltage	450, 1300, heating for 10′ without voltage	450, 1100, heating for 20′ with applied voltage
Strength of bonding [MPa]	38	32	38	—	38

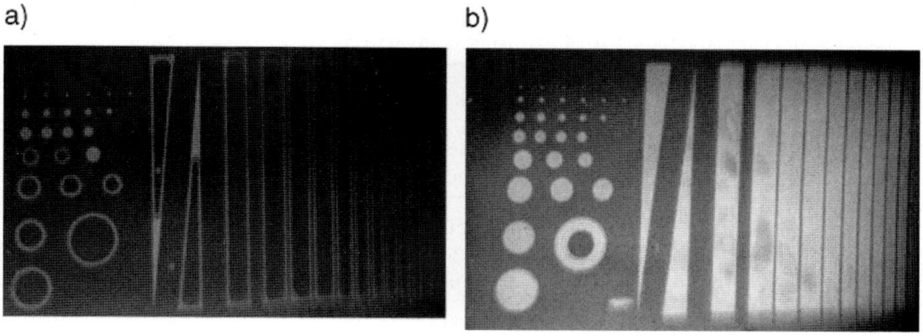

Fig. 4.99. Built-in tests, $H_{test} = 500$ nm: a) bonding of Si–Borofloat 33 ($\neq 1$ mm) without interlayer, $t = 450°C$, $U_p = 1000$ V, b) bonding of Si–SiO_2–Borofloat 33 ($\neq 1$ mm) through 1 μm-thick thermal oxide layer. The sealed area is dark.

oxidized substrates by the magnetron sputtering method from the mosaic target. Parameters of the deposition process were tested and optimized from the point of view of the properties of layers applied in anodic bonding. For comparison, silicon layers deposited by means of the CVD technique on 0.2 μm-thick thermal silicon oxide were also researched.

The strength of bonding was estimated by inserting a razor blade between bonded wafers and/or by breaking the substrates. If bonding was good, the glass did not separate from the silicon (in both tests), which indicated the strength of bonding equaling approximately 30 MPa.

Results of the performed test processes of anodic bonding of glass to chosen multi-layers are listed in Table 4.18. The list contains results achieved for wafers bonded immediately after removal from the vacuum device, as well as for wafers subjected to the partial or complete cleaning procedure. In order to improve the quality of surface samples an additional process of curing of the SiO_2/SiC multi-layer in N_2 at 600°C for 1 hour was also applied. Layers of doped silicon were prepared for bonding according to following procedures:

- without preparation, layers bonded immediately after they were taken out of the vacuum device,
- boiling in H_2O_2, washing in DI water,
- washing in hot trichloroethylene, acetone, Piranha, boiling in 30% H_2O_2,
- washing as above, curing at 600°C for 30 min, boiling in boiling 30% H_2O_2.

Glass wafers were always prepared by the same procedure: degreasing, washing in a 2% chromic acid cleaning mixture at 80°C, washing in DI water, boiling in 30% H_2O_2.

Si(V) layers bonded weakly to glass after a longer time of bonding using a flat electrode with weight (unless the effect of yellowing of glass occurred). A positive result of bonding of pure silicon layers was obtained when the complete procedure of cleaning and heating was applied. Palladium-doped layers did not bond to glass, in contrast to the silicon layers deposited using the CVD method

Table 4.18. Anodic bonding of $Si/SiO_2/Si(C, V, Pd, B)$ – glass, SiC, Si(B, V, Pd), positive result + indicates an achieved bonding strength higher than 30 MPa

Layer	Thickness [μm]	Temperature of substrates during the deposition [°C]	Parameters of bonding	Preparations	Strength tests: Razor blade	Strength tests: Breaking
Si(B)	~3	~300	$T = 450°C$	a	–	–
			$U = 2–2.1$ kV	b	+	+
			$t = 5$ min			
Si(V)	~3	~300	$T = 450°C$	a	–	–
			$U = 1.8–2.3$ kV	b	+/–	–
			$t = 5–15$ min	c	+/–	–
SiC(B)	~3	~300	$T = 450°C$	b	+	+
			$U = 2–2.2$ kV	c	+	+
			$t = 5$ min	d	+	+
Si(Pd)	~3	~300	$T = 350–450°C$	b	–	–
			$U = 1.5–2$ kV	c	–	–
			$t = 5$ min			
Si	~3	~300	$T = 450°C$	b	—	—
			$U = 1.5–2.2$ kV			
			$t = 5–20$ min			
Si	~1	~30	$T = 450°C$	c	+	–
			$U = 1.2–1.8$ kV	d	+	–
			$t = 5$ min			
Si_{LPCVD}	0.5	~800	$T = 450°C$	c	+	+/–
			$U = 1.3–1.8$ kV	d	+	+/–
			$t = 5$ min			

with glass plate, which did bond, but during the strength tests exhibited a weak adherence to thermal oxide, and bonding retreated.

The best results, which meet the requirements of microsystem technology, were obtained for silicon layers doped with carbon and boron-doped SiC layers. Current characteristic of bonding had a typical course with high I_{max} and a rapid drop in time (Fig. 4.100).

Complete bonding of a $Si–SiO_2–SiC(B)$ multilayer to glass occurs only when the surface of the layer is uniform and clean. Dust and small inequalities of surface cause large local voids. Bigger inequalities with dimensions of a few micrometers lead to the local cracking of glass.

4.4.3.3.3. Bonding through Al or Al_xO_y layers

Al layer

Monolithic aluminum and glass can be bonded, but after cooling to the ambient temperature the glass sample will be destroyed as an effect of the large thermal mismatch of these two materials ($\alpha_{Al}/\alpha_{Si} = 8$). However, stresses that destroy monolithic samples are transferred well through a thin layer of aluminum, which

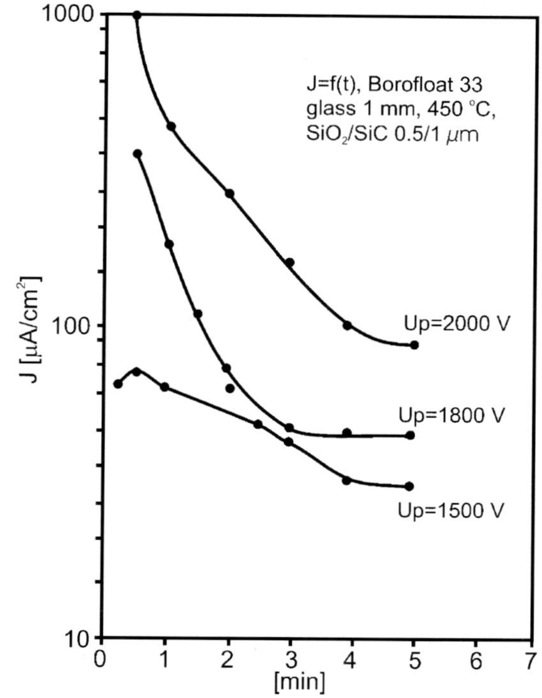

Fig. 4.100. Current density of bonding of silicon to glass through SiO_2/SiC interlayers, $d_{SiO_2} =$ 0.5 μm, $d_{SiC} = 1$ μm, Borofloat 33 glass $\neq$ 1 μm, $t = 450°C$.

makes anodic bonding of silicon covered with a thin-film aluminum layer possible. Nese and Hanneborg [170] report that the process of bonding proceeds with more "difficulty" in comparison to standard silicon–glass bonding. The bonding current tends to be delayed in relation to the moment at which polarization voltage is applied, relaxation can even reach 60 seconds. The final strength of sealing is limited mainly by the adherence of the aluminum layer to silicon. The strength of bonding does not exceed 16 MPa; most often it equals 6–10 MPa. A significant increase of bonding strength (of 30–40%) is observed after the bonded sandwich is stored in water for 300 hours.

In the author's experiments a thin-film layer of aluminum was evaporated in vacuum or magnetron sputtered in argon under a pressure of 1 Pa. Prior to deposition, cleaning glow discharge was carried out under a pressure of 10^{-2} Pa and with a significant discharge power.

Good bonding was achieved for metallic glossy, smooth, 500 nm-thick evaporated layers. Layers which were sputtered did not bond to glass. Bonding proceeded at a temperature of 450°C using 1 mm-thick Borofloat 33 glass. A distinct maximum and a rapid drop were observed in the current characteristics of bonding (Fig. 4.101). The majority of bonded wafers did not fail during the razor blade test. Accidental delaminating was observed after the destructive

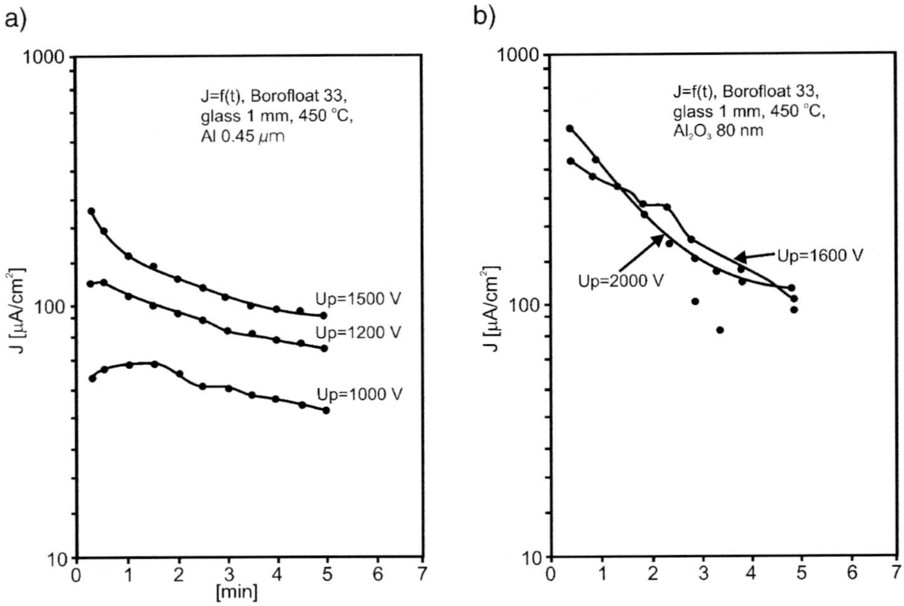

Fig. 4.101. Current density of bonding of silicon to Borofloat 33 glass 1 mm-thick, $t = 450°C$: a) through an Al layer with thickness ~0.4 μm, b) through an Al_xO_y layer with thickness ~80 nm.

breaking tests. Curing of silicon substrates covered with aluminum layers before bonding, at a temperature similar to the critical eutectic temperature of Al–Si (about 577°C), increased the adherence of the aluminum layer to the silicon substrate, but thermally treated samples did not bond to glass.

Al_2O_3 layer

The result of bonding depends considerably on the fabrication method of the Al_xO_y layer. Pure sapphire-colored stoichiometric layers obtained by magnetron-sputtering in oxygen of the aluminum target did not bond to glass no matter what process parameters were applied. Brown layers, with excess aluminum, were bondable to glass under very high polarization – 2–3 kV, when the color of the Al_xO_y layer changed after bonding from brown-metallic to pure sapphire. The current characteristics of Si/Al_xO_y–glass bonding through 80 nm-thick Al_xO_y layers are presented in Fig. 4.101b. Bonded samples passed the razor blade tests.

4.4.3.3.4. Bonding through a Si_3N_4 layer

Bonding of thin silicon nitride layers deposited on silicon substrate to monolithic glass is a difficult process [180, 186]. A thin Si_3N_4 layer does not significantly affect the pulling force caused by electrostatic attraction of wafers [180], but under the typical bonding conditions (450°C–550°C, 1000 V), voids evolve on the Si_3N_4–glass boundary. Curing does not significantly improve the strength of bonding: bonded wafers separate quite easily, even after annealing of bonded samples for many hours in air at a temperature of 450°C [186]. High-temperature annealing of silicon substrates covered with a thin silicon nitride layer at a temperature of 1050°C, in steam, for 1 hour [186] before sealing, significantly impoves the strength of bonding. During annealing, a thin, 12 nm-thick, oxynitride $Si_xN_yO_z$ layer is formed on the surface of Si_3N_4, which ensures very good bonding. This process is similar to the method developed for Si/Si_3N_4–Si fusion bonding [60]. The bonding procedure needs an elevated polarization voltage, and flat cathodes should be applied.

4.4.3.3.5. Silicon to silicon bonding through thin glass layers

Two silicon substrates or details can be anodically bonded through a thin glass layer. A thin layer of sputtered borosilicate glass [78, 94, 171–174, 186–193] or borate glass, or sodium spin-on glass [194] is deposited onto a surface of one of the bonded wafers or details. The glass layer thickness varies from under a micrometer to several micrometers; typical parameters of the process, as well as its dynamics, differ significantly from the standard silicon–glass wafer bonding. Polarization is significantly lower, typically $U_p = 30–300$ V, temperature $t = 25–450$°C, and the bonded sandwich needs to be annealed (Table 4.19).

The course of the bonding process is a consequence of the fact that in the silicon–thin-film glass–silicon stack, silicon plays the role of the planar cathode, attacked by sodium ions. As mentioned above, the glass layer is deposited onto one of the wafers (the upper one, according to Fig. 4.102). This wafer is polarized negatively. After the polarization is applied the sodium ions drift in the direction of the cathode. The current characteristic of Si–Si anodic bonding through a thin borosilicate layer does not differ from the known curves of Si–monolithic glass bonding [171, 175] (Fig. 4.103).

Enrichment of the region near the cathode by sodium and the formation of Na_2O (change in composition) can occur, as well as corrosion of the silicon cathode by sodium and its chemical reaction products. If the upper silicon substrate is covered with a thin layer of thermal oxides, sodium ions migrate to SiO_2. This phenomenon can destabilize parameters of electronic systems made in silicon, that forms the cathode. The thin film layer of Si_3N_4 deposited on silicon "under" SiO_2 blocks the drift of Na^+ ions and eliminates this undesirable effect. Chemical reactions responsible for forming the siloxane bonds are

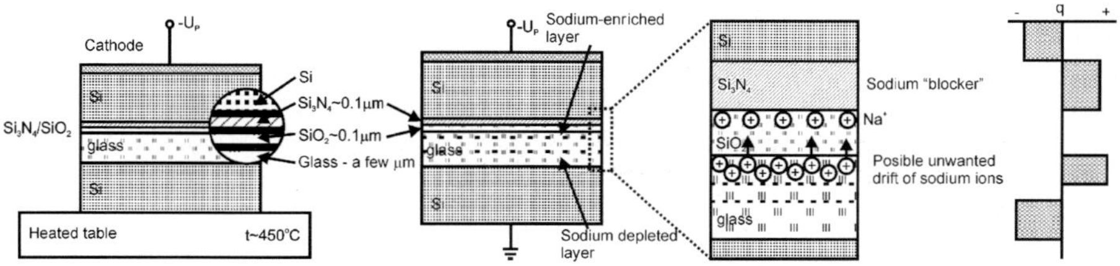

Fig. 4.102. Si–Si anodic bonding through a thin glass layer: layout of wafers and distribution of charges.

Table 4.19. Parameters of silicon to silicon-anodic bonding through a thin layer of borosilicate glass

Thickness of glass d_g μm	Polarization V	Temperature °C	Type of glass	Curing	Other applied layers	Comments	Refs.	Year
4	50	450–550	7740	500–800°C O_2, N_2, H_2O	—	The best curing: annealing in steam before bonding, bonding designed for pressure sensors	188	1972
4	50	350–550	7740	650°C, H_2O		For pressure sensors, high stability	94	1985
2–5	70–300	160	Unknown	—	Al 0.05–1 μm on the base of Ti	Probably a typical laboratory glass	190	1987
0.5–4	30–60	25	Iwaki 7570	—	—	Bonding at room temperature, fusible glass	191	1990
0.5–4	60	25	Iwaki 7570/ Corning 7059	—	Al, ITO	Bonding silicon-thin layer of 7570 glass – thin layer of Al or ITO – monolithic glass 7059	191	1990
3	50–200	400	7740	—	SiO_2 0.1 μm Si_3N_4 0.1 μm	For pressure sensors, bonding force 2.2–3 MPa	173	1991
2.5	50–200	400	7740	—	SiO_2 0.3 μm Si_3N_4, Si, Al	Bonding force 5–25 MPa, start of bonding for $U = 20$ V	171/ 192	1992/ 93
0.5–4	8–10(20)	450	Schott 8330	550°C, nitrogen, 3.5 h	—	Applied for the fabrication of very shallow (< 100 nm) and deep (> 100 μm) cavities covered by membrane	7	1994

Continued

Table 4.19. *Continued*

Thickness of glass d_g µm	Polarization V	Temperature °C	Type of glass	Curing	Other applied layers	Comments	Refs.	Year
0.1–0.4	1–4	450	Schott 8330	550°C, nitrogen	—	Very low bonding voltage	189	1994
7	200	200	Schott 8329	450/30′ 500/10′ air	—	Slow increase of voltage reduces stresses from 30–40 MPa to < 10 MPa	172	1995
0.4–2.5	35–60	135–160	7740	—		Applied for the fabrication of vacuum microtriode (FETRODA), bonding force 1–8 MPa	187/ 193	1997
5	15–100	200–400	Schott 8329	350°C/15 air		Bonding force 25 MPa for $t > 300$°C, glass deposited using electron beam	188	1998
5	300	200	Borofloat 33	—	—	> 30 MPa, cool slowly research	own	1999

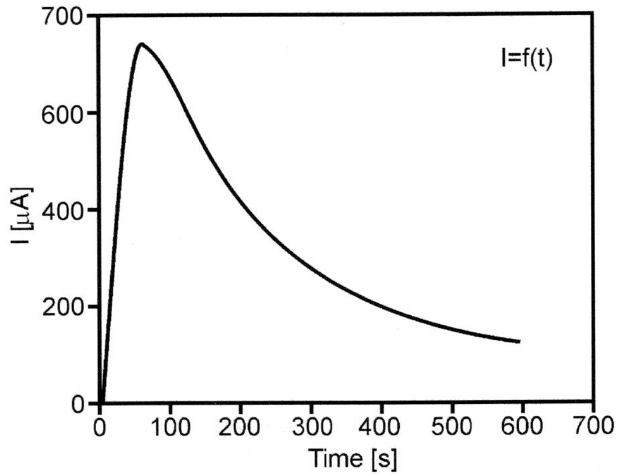

Fig. 4.103. Bonding current as a function of time, Si–Si bonding through Corning 7740 glass with thickness $d_g = 1$ μm, conditions of the process: 400°C, 200 V, 10 minutes without heating [171].

the same as for bonding of monolithic glass. However, in a very thin glass layer the thickness of the evolving depletion layer can be comparable to the thickness of the glass layer. Anthony researched the lower limit of glass thickness sufficient for successful bonding [141]. He reported that layers thinner than depletion layer do bond, and he suggested that layers with thickness similar to intrinsic oxide could be bonded anodically.

The method of obtaining the thin layer exerts a big influence on the parameters of Si–Si anodic bonding through thin layers of sputtered glass. There are two applied methods, usually ion (magnetron) sputtering of the glass target [7, 94, 171, 187, 188, 190, 191, 193] or electron beam evaporation of the material [172, 189]. Glass thermally matching in a monolithic form to silicon is used, most often Corning 7740, but sputtering of glass changes the properties of this material, introducing the misfit of temperature coefficients of expansion of silicon and sputtered Pyrex-like glass, reaching 50% [171]. What is more, sputtered glass layers suffer from built-in stresses, tensile or compressive. Type of stresses depend on the method of layer deposition and conditions of temperature post-deposition treatment. Additionally, built-in stresses can appear in the sputtered glass while bonded, as well as stresses being introduced by mismatch of SiO_2 and Si_3N_4 layers that cover the bonded wafers. For this reason thermal matching of monolithic glass, which functions as the target, is less important. Moreover, the use of glass not matched thermally is allowed, if the glass has other parameters favorable from the point of view of bonding quality.

Esashi utilized fusible Iwaki 7570 glass [191]. This type of glass has both the temperature of stress-releasing annealing (342°) and a melting point (440°C) lower than the Corning 7740 glass, for which these values equal respectively 590°C and 821°C [191]. Iwaki 7570 glass ensures higher resistance of the thin layer to breakdowns; its resistivity is 10^{17} Ωcm, about one hundred times higher than the resistivity of Corning 7740 glass. During bonding, a bigger charge accumulates in a layer of this glass, which increases the electrostatic force, because the relative permittivity of Iwaki 7570 glass is over three times higher than the relative permittivity of Corning 7740 glass that equals $\varepsilon_g = 4.6$. In the original work of Esashi [191], Iwaki 7570 glass was deposited onto a silicon wafer by means of magnetron sputtering in an Ar atmosphere with O_2 (30%) under a pressure of 0.6 Pa. Easy Si–Si bonding was obtained at the ambient temperature (20°C), with a small polarization applied (about 20 V) and using pressure pulling the bonded wafers (160 kPa), supporting the sealing.

According to Shoji, a stress-free, 1 μm thick, layer of 7740 glasses may be obtained by magnetron sputtering of a monolithic target in an atmosphere of 90% Ar + 10% O_2, under a pressure of 3 Pa [176]. Correct stoichiometry is attained by curing the deposited layer at 565°C in steam. Berenschott and co-authors used Schott 8380 glass, which has properties similar to Corning 7740 glass [7]. Layers are sputtered in argon or in argon + 10–25% O_2, under a pressure of 1 Pa. The applied power density equaled 2.7 W/cm^2. In thin glass layers, obtained using the method described above, stresses appeared, which were significantly reduced from 128 MPa to 5 MPa (layers sputtered in Ar, deposition rate 286 nm/h) after curing in nitrogen at 550°C for 3.5 h. Brooks and co-workers recommend high-temperature stress-releasing annealing of deposited layers, before bonding, in steam at 500–900°C [188].

Magnetron sputtered Borofloat 33 glass deposited onto one of the bonded silicon substrates, in 3 Pa Ar, covered with a thermal oxide layer 0.8 μm thick, has been applied in our own research. After deposition the glass layer was cured in steam at 450°C followed by annealing in nitrogen at 500°C. Wafers were activated under the RCA1 procedure; bonding was carried out in air using a planar glass–metal electrode with clamp. Very good bonding of wafers covered with a 5 μm-thick glass layer was obtained at 200°C for 300 V. Pulling tests revealed a crack in monolithic silicon, but the bond remained intact, which showed that the strength of bonding was over 30 MPa and corresponded to a bonding energy γ equaling at least 2.5 J/m^2.

According to the literature data [7, 94, 171, 172, 187–191, 193] the conditions of bonding need to be selected for the specific assembly of thin glass and dielectric layers (Table 4.19). It is not possible to formulate a general technological procedure, similar to the process of bonding of silicon to monolithic glass. Generally, anodic bonding through a thin glass layer with thickness equaling only 0.1–0.4 μm can be carried for a very small polarization 1–4 V, at a relatively high temperature (450°C), bonding through a circa 1 μm thick layer of fusible glass at room temperature needs polarization equaling 30–60 V. Thicker glass layers, 4–7 μm, require similar polarization voltages (15–100 V) in the temperature range from 13°C to 400°C.

4.4.3.4. High-temperature anodic bonding

High-temperature silicon-to-silicon anodic bonding through a thin SiO_2 layer [13, 175, 141, 195] is a peculiar process. A depletion layer does not evolve in

SiO_2 during the bonding of wafers. Electrostatic pressure is low, but sufficient to initiate a bonding wave. The motion of movable charges, sufficient to start the bonding, occurs effectively over 800°C; polarization voltage does not exceed 30 V. Shoji reported that the process of bonding proceeds at a temperature of 850°C for 45 minutes [175]. Frey and co-workers carried out bonding of silicon wafers 3″ in diameter, covered with a 0.35 μm-thick thermal oxide layer. They bonded a pure wafer to an oxidized wafer and two oxidized wafers to each other [13]. The process was performed in nitrogen at a temperature of 1150°C.

Bonding of a Si–SiO_2/Si assembly proceeded only when a negative polarization voltage was applied to the oxidized substrate. Polarity was not relevant in Si/SiO_2–SiO_2/Si bonds. Bonding current as a function of time for polarization 20 V had a typical course (Fig. 4.104).

According to paper [13] the total charge transported during high-temperature anodic bonding corresponds to the charge of volumetric density of oxygen ions immediately after the oxidation of "fresh" wet thermal oxides ($10^{19}/cm^3$). This value is a few orders of magnitude bigger than the concentration of Na^+ attributed to pure thermal oxides. Such transport of charges shows that high-temperature Si–Si anodic bonding through thermal oxides is very similar to the process of silicon-to-silicon fusion bonding (DFB).

Frye and co-workers reported a complete, void-free, bonding of an oxidized silicon wafer to silicon substrate [13] (Si/SiO_2–Si) at 1100°C in the presence of an electrostatic field, and then annealed at 1200°C without polarization, for 10 min. Excellent quality sealing of two oxidized wafers (Si/SiO_2–SiO_2/Si), was obtained at 1100–1200°C for 20 V.

Washing of wafers in NH_4OH before bonding, as well as carrying out the bonding in wet (steam added) nitrogen, help the formation of a good high-temperature anodic bond. High-temperature anodic bonding does not lead to structural defects in SiO_2. Therefore, thermal oxide can be used after the process of bonding as a good insulating layer. There are no stresses in silicon; the process of bonding does not cause any impurities. Si–Si anodic bonding through a thermal oxide layer, developed by Anthony [150], is recommended as an

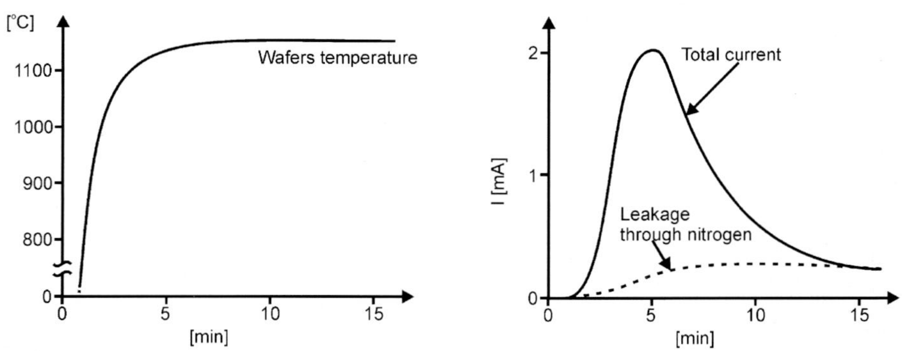

Fig. 4.104. Current versus time and temperature as functions of time curves. Si–Si high temperature anodic bonding through a 3.5 μm thick layer of thermal oxide. After paper [13].

industrial method of fabrication of SOI substrates, alternative to the DFB method.

4.4.3.5. *Glass–FeNiCo alloy bonding*

Alloys of iron, nickel and cobalt (Fe–Ni–Co), called Kovar, are commonly used in microelectronics. Sealing of Kovar to glass, one of the oldest, classical methods of fabrication of vacuum valves and bulbs, is applied commonly in high-vacuum technology. In this method, metal details are "flooded" by glass melted in fire, resulting in strong sealing of metal and glass. Unfortunately, this simple "fire" method cannot be used in microsystem technology.

Wallis and co-workers discovered the anodic bonding of silicon to Kovar* in 1971 [196]. The basic criterion of glass selection to anodic bonding to Kovar is the compatibility of linear coefficients of thermal expansion of these materials in the range from at least the ambient temperature to the optimal bonding temperature. Moreover, it is necessary to take into consideration the compatibility of coefficients of silicon and glass α_{Si} and α_g to the coefficient of metal α_m. The triple compatibility, thermal matching of α_{Si}, α_g and α_m is obtained for Pyrex 7740 glass (and other derivative glasses) and for KV10 Fe Ni 30% Co 13% Kovar alloy of Sumimoto or similar alloys offered by different producers [124, 197] (Fig. 4.105).

In our own studies very good quality bonding of 1–2 mm-thick Corning 7740 and Borofloat 33 glass to Kovar Fe Ni 19% Co 17% was obtained [198]. The optimal parameters of bonding, consistent with the literature data [196, 197], are as follows: 360°C, 1200–1500 V for 10 minutes. SD2 glass did not bond to Kovar. Metal substrate or detail was polished to attain a roughness less than

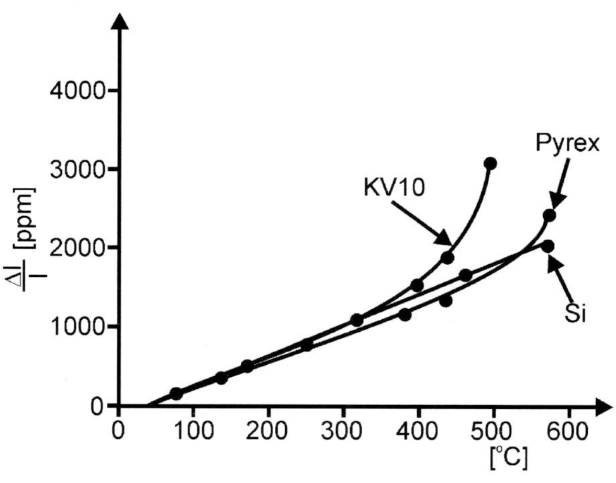

Fig. 4.105. Unit elongation Δl/l of Kovar KV10, Pyrex glass (Corning) and silicon as a function of temperature. After [197].

* Wallis was then an employee of P.R. Mallory & Co. Inc., Burlington, Mass, USA. That is why in early works anodic bonding was sporadically called Mallory's method or process.

1 μm (mirror surface), next it was degreased in detergents and etched in an aqueous solution of HF (1–5%). The glass was then degreased, etched in chromic acid, boiled in 30% H_2O_2 and stored in DI water. The surface of metal details was formed before bonding by annealing in air at 450°C for about 5 minutes, in order to form an oxide layer about 100 nm thick [196]. This process was carried out before the actual bonding of details, directly on the heated table of the anodic bonding set-up. After bonding, the color of the metal–glass interface changed from metallic bright gray as observed before sealing to metallic brown (in case of weaker sealing) or to metallic brown-brown (in case of good, strong sealing). The current characteristics of excellent glass–Kovar bonding are presented in Fig. 4.106a/b.

Destructive pulling tests and/or razor-blade tests led to the destruction of glass in volume, not influencing (destruction) the bond; this could be interpreted as a very high bonding energy γ, exceeding 3 J/m^2.

4.4.4. Application of anodic bonding

Within the space of the following years the application of anodic bonding has been increasing constantly, and for the present this method of bonding of

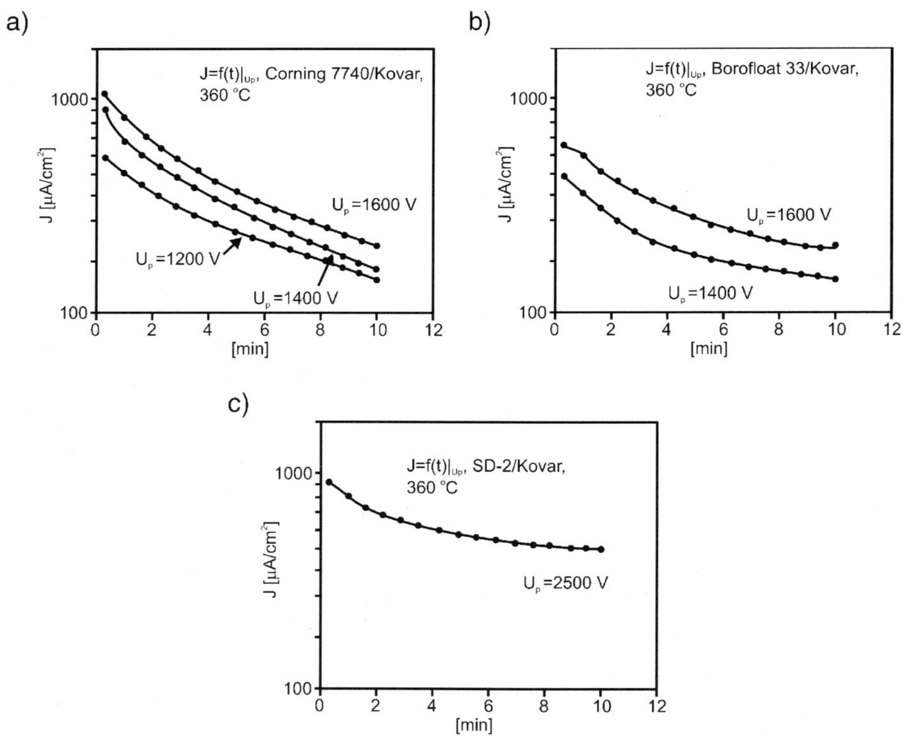

Fig. 4.106. Bonding current density as a function of time. Bonding of Corning 7740 (a), Borofloat 33 (b) and SD2 glasses (c) to Kovar Fe Ni 19% Co 17%.

materials is commonly used in microsystem technology. The market niche value of microsystems exceeds many billions of euro [199], and tends to grow by 20% a year or faster.

In this section, chosen examples of micromechanical sensors, actuators, microsystems and microinstruments made using anodic bonding are briefly described. The contents of the section are based mainly on the author's and his co-workers' experiences. This presentation is the author's view only, and cannot be treated as an encyclopedia of anodic bonding application in microsystem technology.

4.4.4.1. Sensors and actuators

4.4.4.1.1. Accelerometers/other sensors

Many different sensors, actuators and other microsystems are fabricated using the method of silicon-to-glass anodic bonding. For example, multi-layer anodic bonding has been applied in the production of a micromechanical capacitive accelerometer [200] (Fig. 4.107). Two silicon parts containing movable elements sensitive to acceleration were sealed by means of direct bonding DFB. Then, both sides of the sealed component were bonded to two covers in two successive, selective processes of anodic bonding to Pyrex-type glass.

References [201]–[208] describe accelerometers and angular position sensors, mainly capacitive, in the fabrication of which the multi-layer glass-silicon–glass bonding of micromachined silicon wafers to 0.25 mm/1 mm thick glass Corning 7740 wafers was used. Typical parameters of the process are: 400°C, 600 V. Glass–silicon–glass air-bag sensors with a resonant beam, thermally activated to vibration, are presented in paper [207]. At present this new solution of the Sensonor company (Norway) is being applied widely in motorization technology (over 30 million pieces a year) (Fig. 4.108).

Esashi has described different silicon–glass micromechanical sensors fabricated using varying types of anodic-bonding procedures [175]. Selective silicon to Pyrex glass anodic bonding through thin Pt/Ti layers has been applied in the fabrication of pressure and vacuum sensors. Bi-layer selective SOG anodic bonding has been utilized in capacitive pressure sensor manufacturing, and an accelerometer has been produced using multi-layer Pyrex–silicon–Pyrex bonding [208].

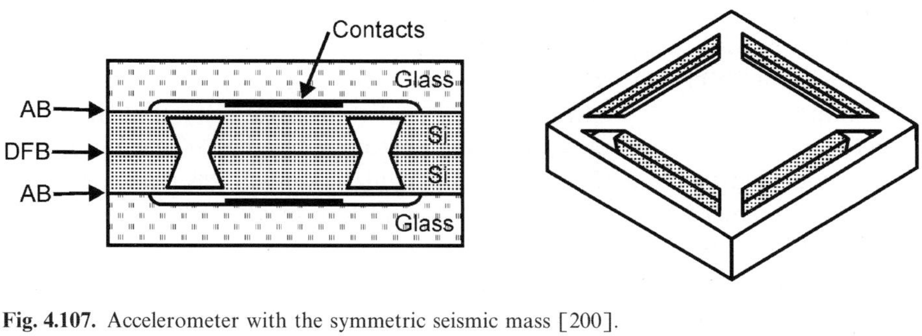

Fig. 4.107. Accelerometer with the symmetric seismic mass [200].

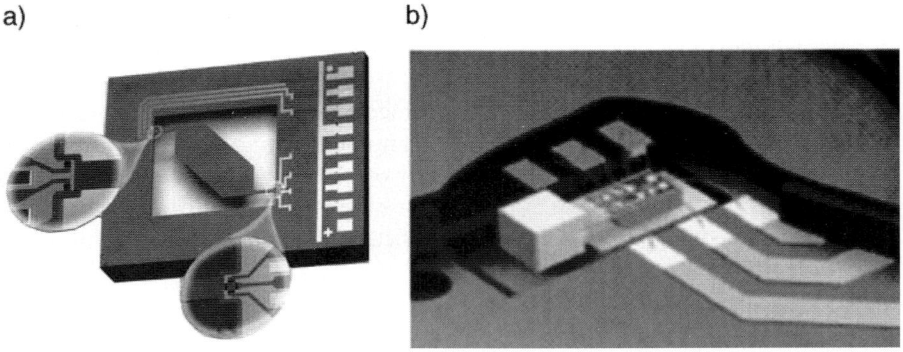

Fig. 4.108. Accelerometers made by the Sensonor company: a) structure of the sensor, b) assembly method. From the catalog of the Sensonor company [207].

In the detector designed for ice detection in aviation [209], a silicon micromechanical structure with membrane, electrostatically activated to vibration, is bonded to the base made of Pyrex glass. A movable membrane of the sensor stiffens under the influence of icing. Stiffening is detected (using the capacitive method) as a change of resonance vibration frequency of the micromechanical structure.

An infrared sensor with a vibratory silicon element, made inside the reference cavity closed in reference space filled with getter was discussed in paper [210]. The vibratory element, heated by infrared radiation absorbed by getter, changes its resonance vibration frequency. The structure of the vibration meter with mainframe is anodically bonded to Pyrex glass. The sensitivity of sensor equals 500 ppm/μW, quality factor $Q = 20\,000$, and vibration frequency 100 kHz. Anodic bonding of silicon to glass and silicon to silicon through thin glass layer has been applied in the fabrication process of a micromechanical microphone [211] (Fig. 4.110). A thin sound-sensitive membrane was double-side etched in a silicon wafer and, after suitable assembly, it functioned as a movable sheet of an air capacitor. A stationary, immovable sheet of this capacitor was formed by a perforated, thin glass plate covered with a thin metal layer.

Silicon masks, utilized in the deep etching of glass in alkaline solutions, bonded to borosilicate glass, were presented in paper [212]. Reference [213] describes sensors of CO_2 content in blood, in the construction of which glass–silicon multi-layers bonded through a thermal oxide layer were applied. Fusion bonding of silicon through a Si_3N_4 layer and double-bonding of silicon to Pyrex glass were used in the fabrication of a pneumatic-optical gas detector that is discussed in paper [214]. In this sensor gas trapped inside a reference chamber absorbs the energy of infrared radiation, which causes an increase in its temperature and pressure. The micromechanical structure of the sensor which is detected by means of a capacitive method (Fig. 4.111).

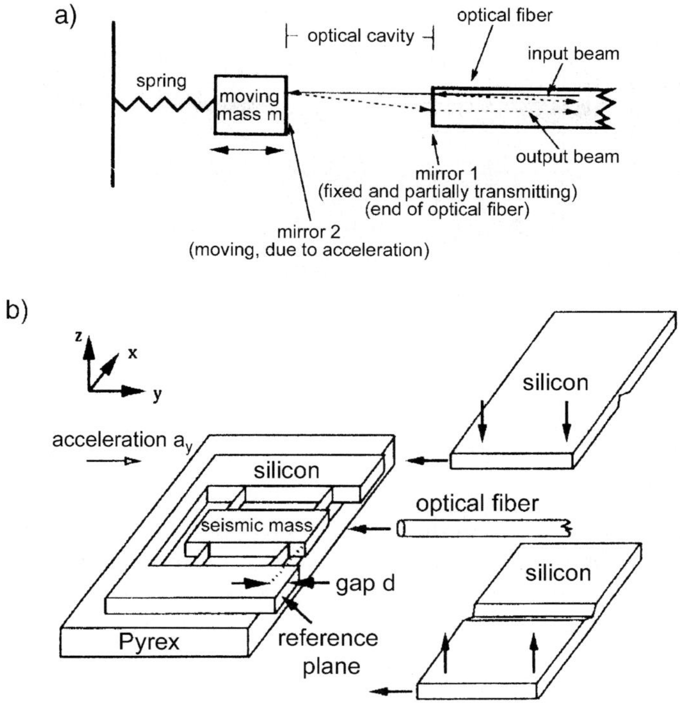

Fig. 4.109. Accelerometer made of three layers of glass and silicon [208]: a) principle of the sensor, b) schematic view of optomechanical sensor.

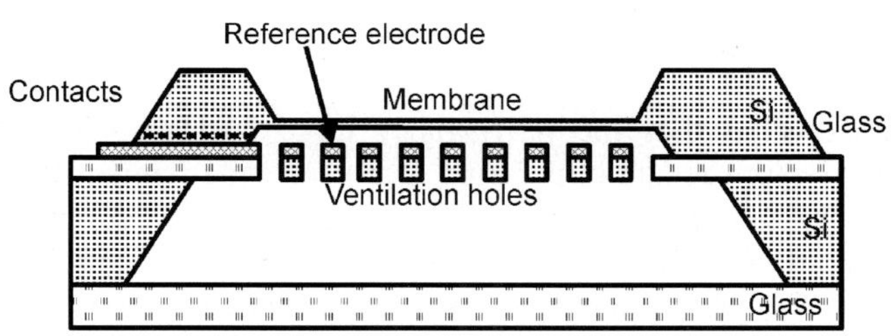

Fig. 4.110. Silicon-glass microphone. After paper [211].

4.4.4.1.2. Piezoresistive pressure sensors on glass

Anodic bonding is most often used in the fabrication of silicon pressure sensors [215, 216]. The reason for this lies in the fact that micromechanical structures of silicon piezoresistive pressure sensors with a thin membrane cannot be directly attached to any type of package; metallic, ceramic or plastic, because direct hard-epoxy glueing involves generation of very strong mechanical effects, being an effect of mismatch of the coefficients of thermal expansion of silicon and the material of the package. The mismatch introduces serious out-of-control deflection of membrane and unwanted stress-related incorrect signals (Fig. 4.112).

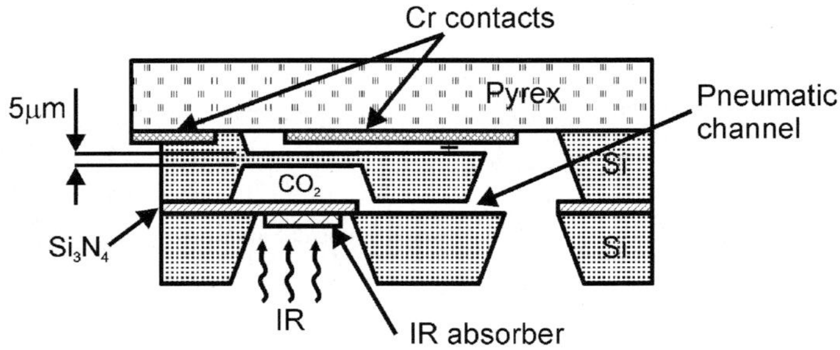

Fig. 4.111. Gas sensor. After paper [214].

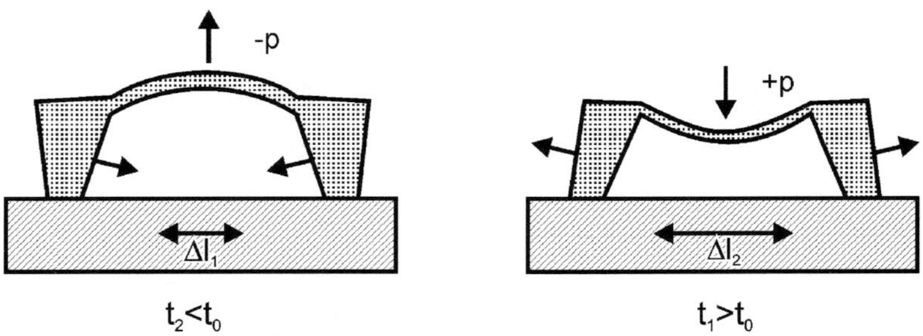

Fig. 4.112. Mismatch of thermal expansion of silicon and a package influence a flat silicon membrane; t_o is a temperature corresponding to a flat membrane, the thermal expansion coefficient of material of a package is assumed to be larger than the coefficient of silicon.

The incorrect signals can be comparable to, or dominate over, the non-falsified pressure-induced output metrological signals expected for the given type of sensor.

It can be expected that the particularly large false signals will occur in piezoresistive pressure sensors operating in a very wide temperature range, for example in sensors applied in motorization technology. For instance, MAP sensors (**M**anifold **A**bsolute **P**ressure), which need to be reliable for at least 100 000 km or 10 years according to automobile industry requirements [217–220], operate at temperatures from −40°C to +120°C, and sporadically temperature can exceed + 150°C.

Cheap, non-accurate pressure sensors, applied in toys and simple pressure transducers (tire pressure control, on-off pressure detectors) can be packaged using soft deformable glue, by "flowing" glueing of sensor structures [221] or may be hard-epoxy glued at one of the edges of sensor structures (Fig. 4.113). Both methods ensure the independent displacement of structure and package involved by thermal expansion, decrease the deflection of membrane caused by thermal incompatibility of silicon and package, but cannot be used for packaging of metrological sensors.

Metrological sensors are always packaged indirectly; they are situated on high "columns" of Pyrex-like glass (Fig. 4.114). The micromachined, processed

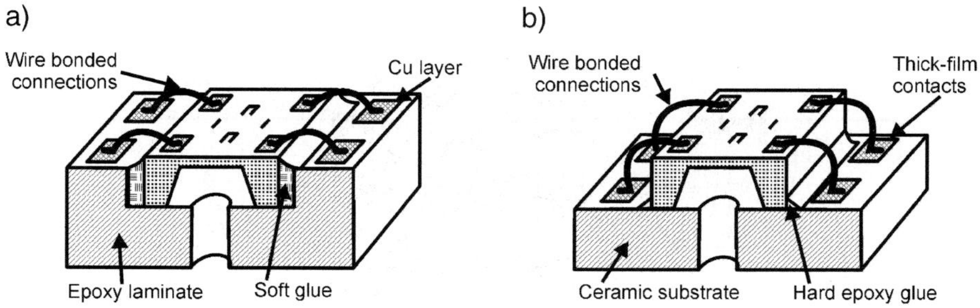

Fig. 4.113. Substitute methods of assembling of pressure sensor structures: a) using flexible gel glue, b) edge assembling.

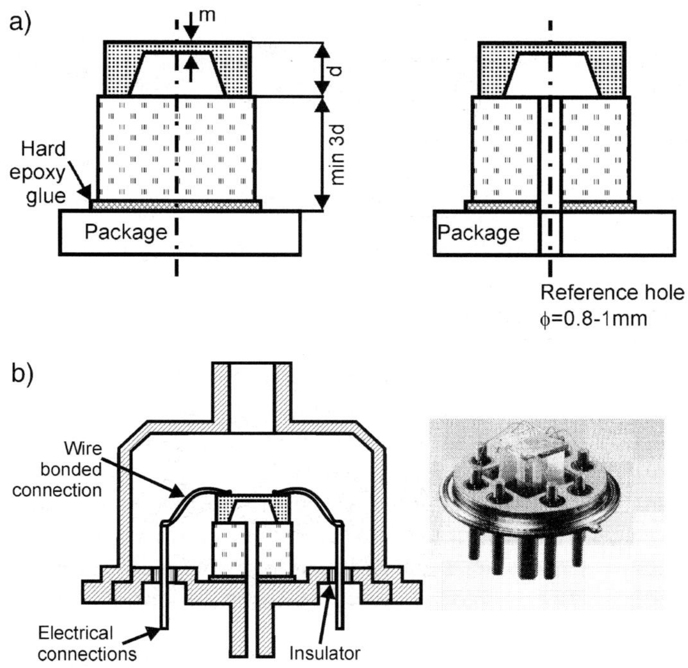

Fig. 4.114. Assembly of piezoresistive pressure sensors: a) structure on glass – half section, absolute sensor on the left, relative sensor on the right, b) structure on glass in the TO5 package.

silicon wafer, containing some hundreds of the sensor structures, is anodically bonded to the Pyrex-like glass substrate, and divided into individual silicon-chips. Chips are hard-epoxy glued to a package. The glass support mechanically insulates the silicon structure with a thin membrane from the package in such a way that stresses generated at the interface of glass and package are suppressed inside the glass and do not affect the membrane. Usually, thermally induced stresses are negligible for glass supports higher than 1 mm (Fig. 4.115).

Although the above-mentioned schematic description of the packaging of

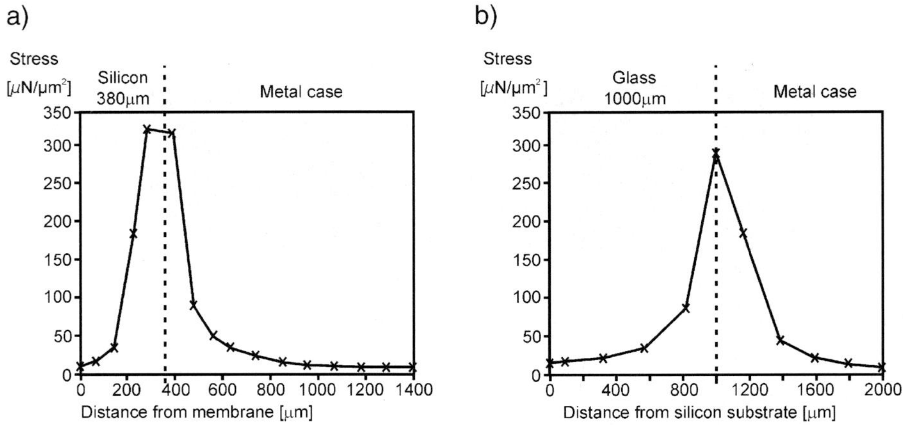

Fig. 4.115. Von Misses' stresses – result of modeling: a) as a function of distance from the membrane surface, for a silicon wafer attached directly to a metal package, b) as a function of distance from the surface of silicon–glass contact, for a glass plate attached directly to a package, there is visible decay of stresses at a large distance from the front.

pressure sensors seems to be very simple, the algorithm of preparatory procedures and anodic bonding of silicon wafers to glass, as well as of the controlling operations connected with the division of bonded sandwich into specific structures, is quite complicated. The pressure sensor structures are first subjected to microelectronic procedures forming piezoresistors, thin-film metallic connections and wire-bonding fields, micromachined in order to form membranes and tested (Fig. 4.116). Next, silicon and glass wafers are bonded, sewn into chips, controlled visually, tested, and packaged to the final package. Silicon wafers with structures of the relative pressure sensors are bonded to the machined Pyrex–glass wafer with an array of drilled holes; wafers with structures of the absolute pressure sensors are bonded in vacuum to an unprocessed, flat glass wafer (Fig. 4.117). Both types of pressure sensor structures "on a glass" are shown in Fig. 4.118.

Low-pressure sensors with over-load protection are most often designed in a form of the square structure with a bossed membrane, selectively bonded anodically to a thick glass. The boss is recessed a few micrometers from the surface of the glass (Fig. 4.119). Pressure overloading the sensor puts the boss in contact with glass, protecting the membrane against mechanical destruction.

Dies of very accurate, low-range pressure sensors have to be bonded selectively to a very thick glass, one-by-one. After anodic bonding, electrical wire connections are wire-bonded, and an encapsulating cap is added. Such a pressure sensor on a thick glass support, packaged to the TO8 package, is shown in Fig. 4.120. The sensor was designed to operate in the range of 20 kPa. The overload factor equals 100. The 2 mm × 2 mm bossed membrane is distanced 5 μm from the surface of a glass support (Fig. 4.120). Structures were attached

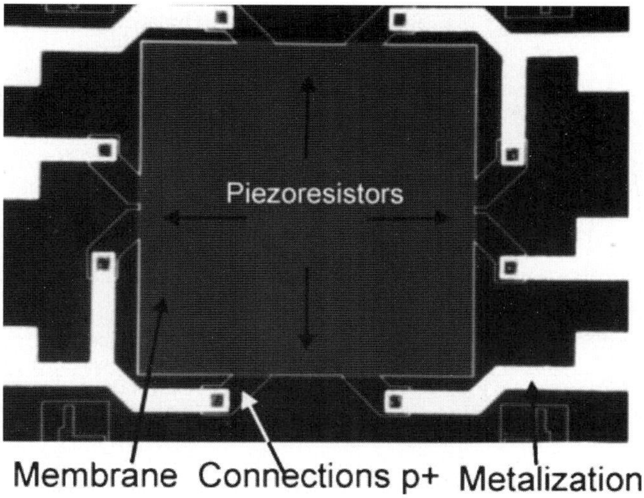

Fig. 4.116. A bird's eye view of the pressure sensor for 100 kPa [222]. Obtained by courtesy of M.Sc. eng. Jan Koszur and M.Sc. eng. Paweł Kowalski from the Institute of Electron Technology in Warsaw, Poland.

to a $\phi = 5$ mm round support made of Borofloat 33 glass, 6 mm-thick by selective bonding using the conductive frame method.

Anodic bonding of silicon to glass has been applied in a new, vertical assembly of pressure sensor structures. The essence of the principle of this solution is the vertical (to the surface of a package) location of the elongated structure that keeps the membrane of the sensor away from the stress-inducing zone, at a sufficiently large distance from the package [223, 224] (Fig. 4.121). Absolute sensors are bonded in vacuum to a flat Pyrex-like glass wafer, relative sensors are bonded to the micromachined glass wafer with etched referential channels. Electric connections between the chip and the contacts at the package pad (laminate, thick-film substrate) can be made by means of the semi-flip-chip method (Fig. 4.122).

The best compensation of temperature-induced stresses is attained for thin glasses (contrary to the classical solution) [225]. A sufficient reduction of thermal stresses is obtained for structures of typical membrane-type pressure sensors with membrane dimensions equaling 1000 μm × 1000 μm × 20 μm, situated at a distance of 4.2 mm from the package (high aspect ratio chip with height equaling 5 mm).

4.4.4.1.3. Capacitive sensors

Glass–silicon capacitive sensors (pressure sensor, sensor of moment of force, microphone, connected pressure and temperature sensors, intelligent pressure sensor) have been described in many works [226–230]. An example of a simple, micromachined, anodic-bonded glass–silicon capacitive pressure sensor is shown in Fig. 4.123. In this sensor the thin silicon bossed membrane and thin-film

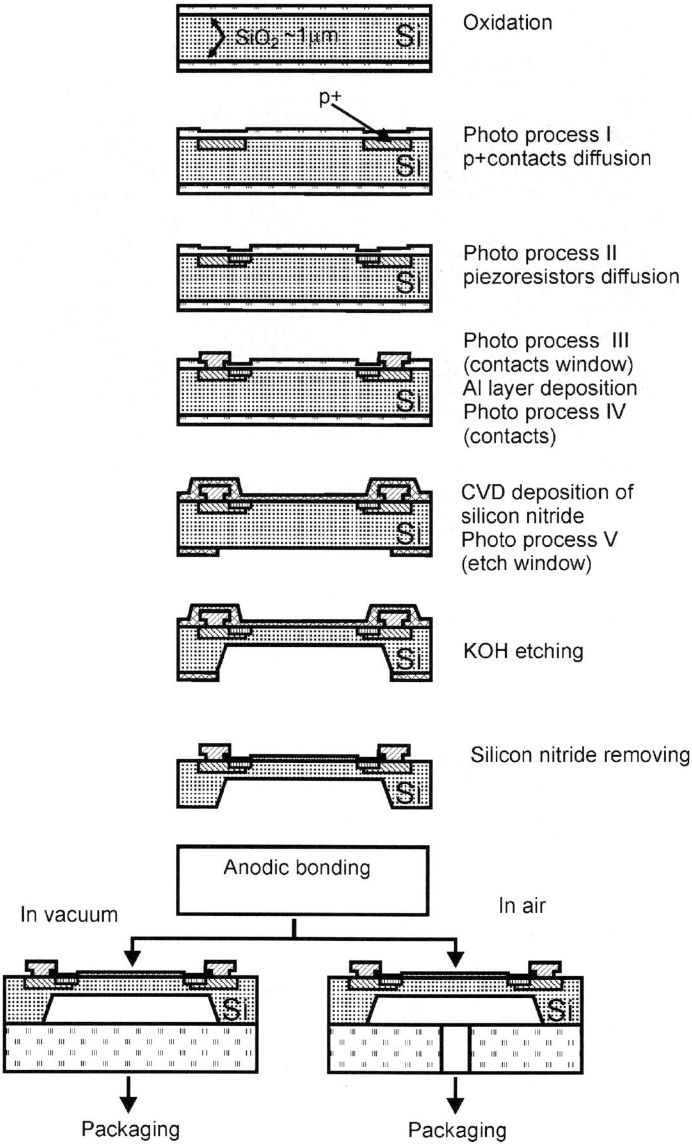

Fig. 4.117. Fabrication process of a pressure sensor die.

metal electrode deposited on a glass substrate form the air capacitor. The pressure of fluids (or force pressure) deflects the membrane and modulates the capacitor capacitance.

The silicon structure of the sensor had dimensions equaling 10 mm × 8 mm × 0.38 mm with a 5 mm × 5 mm bossed membrane EMSi wet-etched in (100) silicon substrate, in 5M KOH at 70°C. The face plane of the bossed

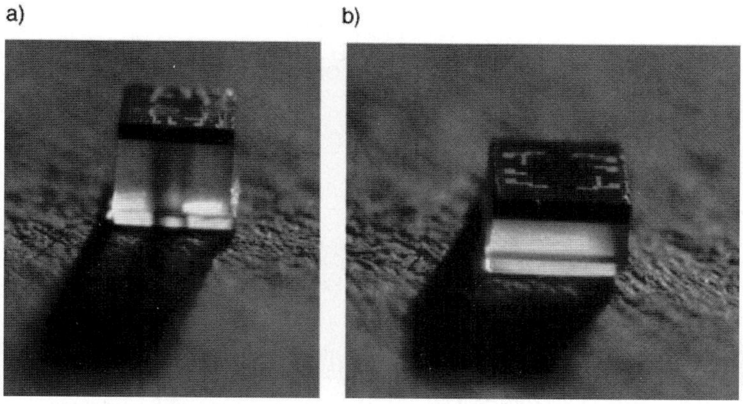

Fig. 4.118. Structures of 100 kPa silicon piezoresistive pressure sensors "on a glass": a) relative sensor – referential channel can be observed, b) absolute sensor.

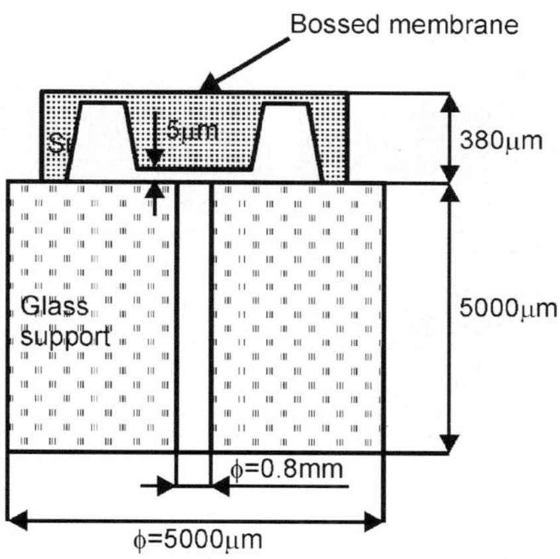

Fig. 4.119. Cross-section of low-pressure sensor structure on glass.

Fig. 4.120. 20 kPa silicon pressure sensor die on a glass column.

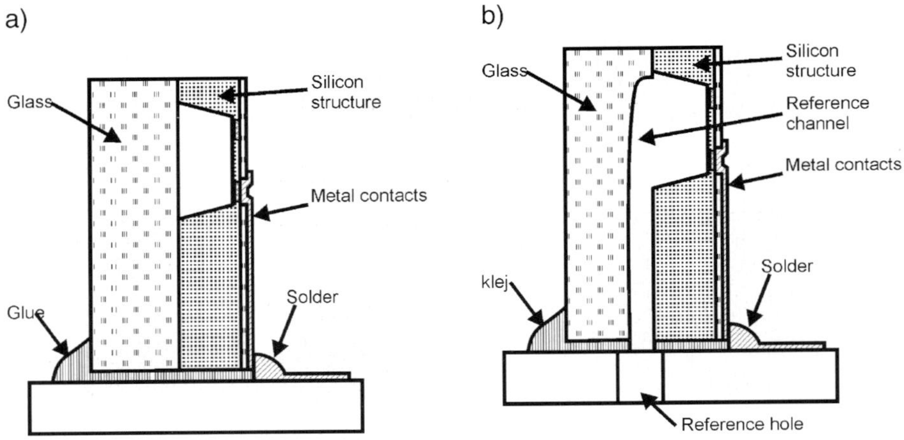

Fig. 4.121. Vertical assembly of sensors: a) absolute version, b) relative version.

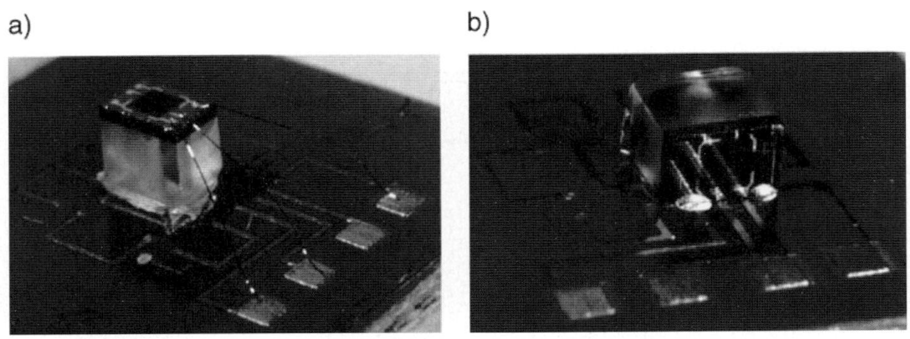

Fig. 4.122. Comparison of assembly methods: a) classical method with electric wire connections, b) vertical method – semi-flip-chip connections.

membrane was next recessed 15 μm in relation to the plane of the wafer surface by short selective isotropic etching of the substrate. A reference hole 0.8 mm in diameter was drilled in a glass substrate. A Cr–Ni–Au electrode was deposited and patterned photographically on the glass substrate. A Ti–W–Au contact was deposited on the front side of the silicon structure. A thin SiO_2 layer was then deposited on this formed surface by means of reactive magnetron sputtering. Next, the silicon structure was selectively bonded to the glass substrate through SiO_2 (420°C, 400 V, 10 minutes, frame cathode) [231, 232].

A view and the characteristics of the pressure sensor so obtained is presented in Fig. 4.124.

4.4.4.1.4. *Optoelectronic pressure sensor*

Multi-layer bonding of small glass details to silicon has been applied in the technology of an optoelectronic pressure sensor [233] (Fig. 4.125). The fabrication process consists of the front-side deep KOH etching of a (100) silicon

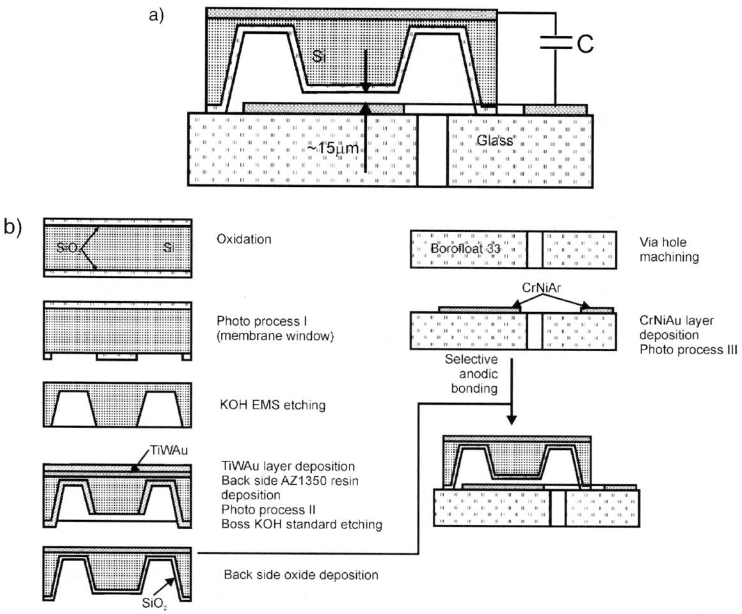

Fig. 4.123. Capacitive pressure sensor: a) cross-section of structure, b) fabrication method.

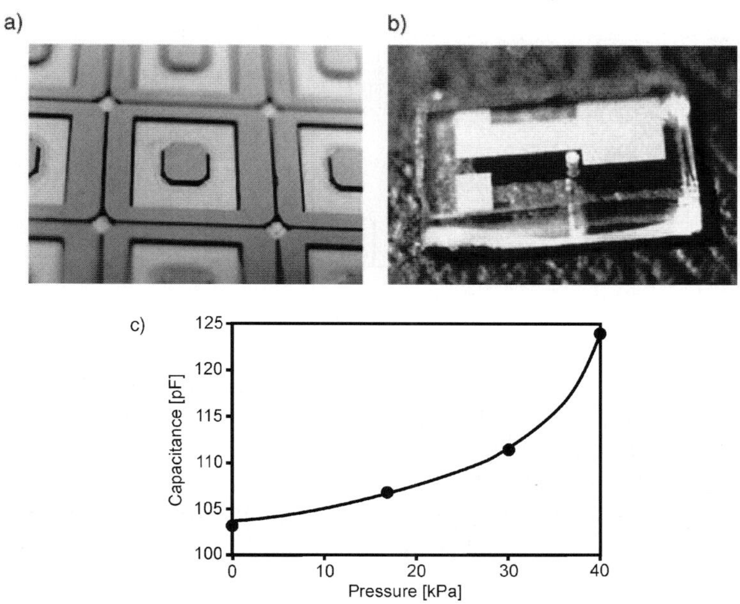

Fig. 4.124. Capacitive pressure sensor: a) bossed membranes at a silicon wafer, b) the glass substrate with electrodes and reference hole, c) output signal characteristics of the sensor.

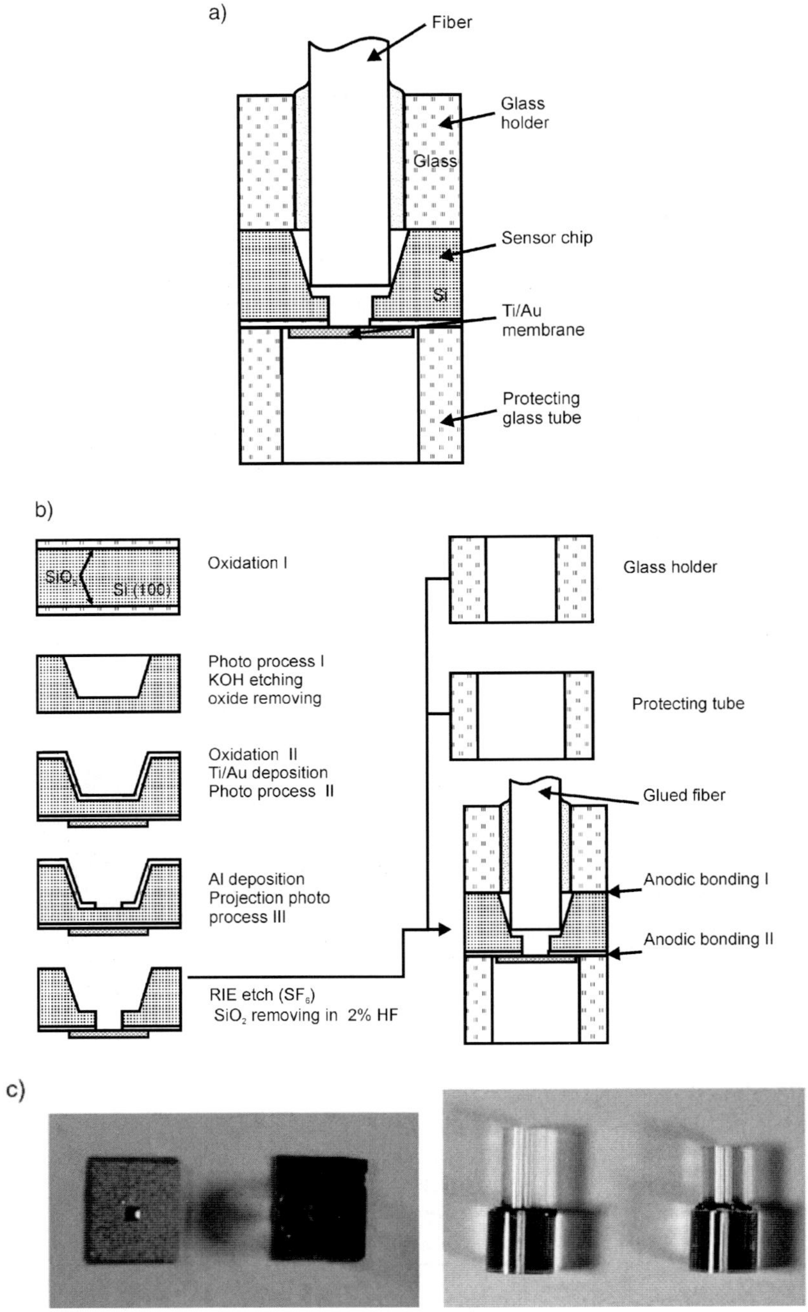

Fig. 4.125. Optoelectronic pressure sensor: a) construction scheme, b) fabrication diagram, c) details and assembled sensor without fiber [233].

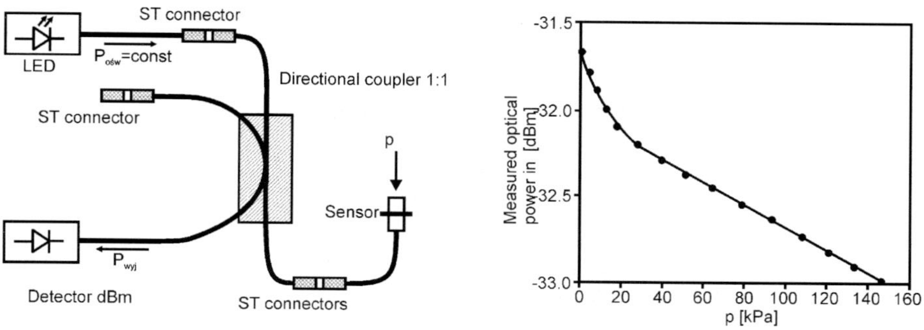

Fig. 4.126. Test set-up and sensor output signal characteristics [238].

wafer, forming a 20 μm-thick silicon membrane, followed by wet high-temperature oxidation of the wafer and back-side deposition and patterning of the thin-film Ti–Au layer. Next, a small etching window is front-side formed using the projection photolithography method, and a Ti–Au membrane is realized by a front-side PERIE etch in SF_6. The processed wafer is sewn into chips. The single chip is front-side anodically bonded to the first glass tube ID 0.8 mm and back-side bonded to the second glass tube ID 1 mm. An optical fiber is positioned in the first tube. In this sensor a very thin metal membrane Ti–Au bi-layer is deflected by the pressure of fluids, which modulates the reflection of light supplied to the membrane through the optical glass fiber (Fig. 4.126). Sensors of this type, characterized by a reasonably high sensitivity, are nonlinear but completely spark-safe.

4.4.4.1.5. Pneumatic micromachine

Microactuators and micromechanical engines constructed as silicon–glass structures have been discussed in many works, among others in refs [210] and [234–237]. Research into microengines and microactuators leads to the development of new electromechanical microdevices, whose future applications are often difficult to predict. Studies on hard disk drives and microgenerators of electrical energy are particularly important, especially because the latter are expected to replace the sources of power used at present in portable devices.

Construction of a Very Small Machine shown in Fig. 4.127 [238] based on a well-known, commonly applied, water consumption meter with optical readout (Fig. 4.128). In this measuring instrument, flowing water turns a turbine. The arms of the rotating turbine modulate a light stream coming from an LED. In the light detector impulses of electric signal are generated proportionally to the rotational speed of the turbine, and thus to the flow of water.

The miniaturized version of the water consumption meter consists of three elements: silicon base, aluminum turbine and glass cover. The 31 μm (±0.5–1 μm) deep liquid-gas channels, were front-side dry PERIE etched in SF_6 in (100) silicon substrate 1. The illumination window was back-side etched in KOH. The window was etched until a thin (8–10 μm ± 1 μm), semi-transparent

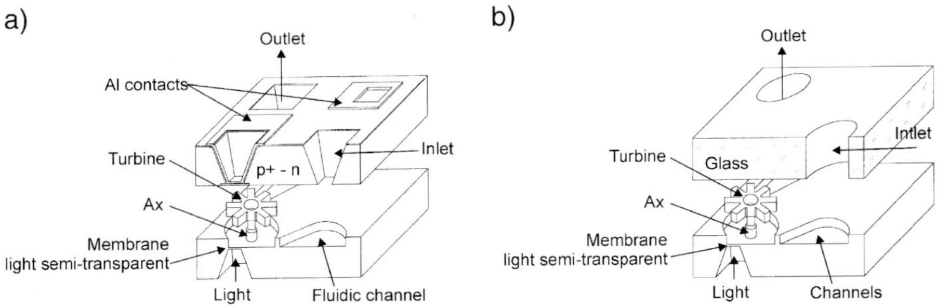

Fig. 4.127. Silicon micromachine: a) flowmeter, b) simplified version – pneumatic microengine.

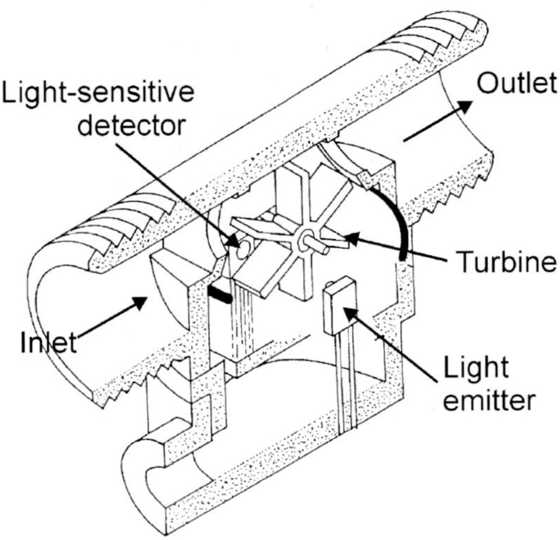

Fig. 4.128. Water consumption meter.

to light membrane, was formed. An aluminum microturbine was fabricated in a separate technological process (substrate 2). Two 0.8 mm holes were electro-eroded [239] or mechanically drilled in a 1 mm-thick glass SD-2 cover (Fig. 4.129).

Assembly of the microdevice (Fig. 4.130) was carried out manually under a microscope. The turbine was fixed on the axle using an electrified human hair. The glass cover was anodically bonded ($t = 350°C$, $U = 500$ V) to the silicon base. The assembled micromechanism was tested (Fig. 4.133c) at a pneumatic set-up equipped with an electronic gas flowmeter and a video-camera. The described miniaturized device worked effectively as a flowmeter and as a pneumatic microengine.*

* The described device is the smallest and the most complex working mechanical micromachine which has ever been produced in Poland.

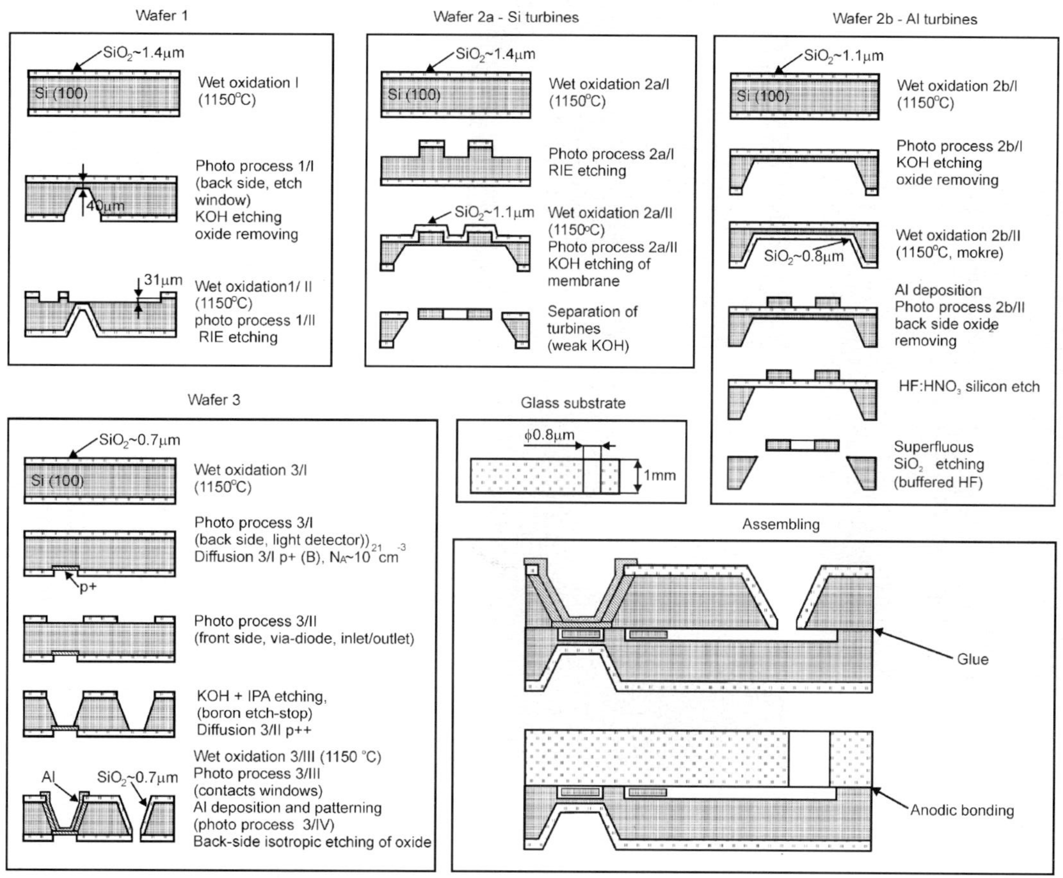

Fig. 4.129. Fabrication process of silicon micromachine – flow-chart.

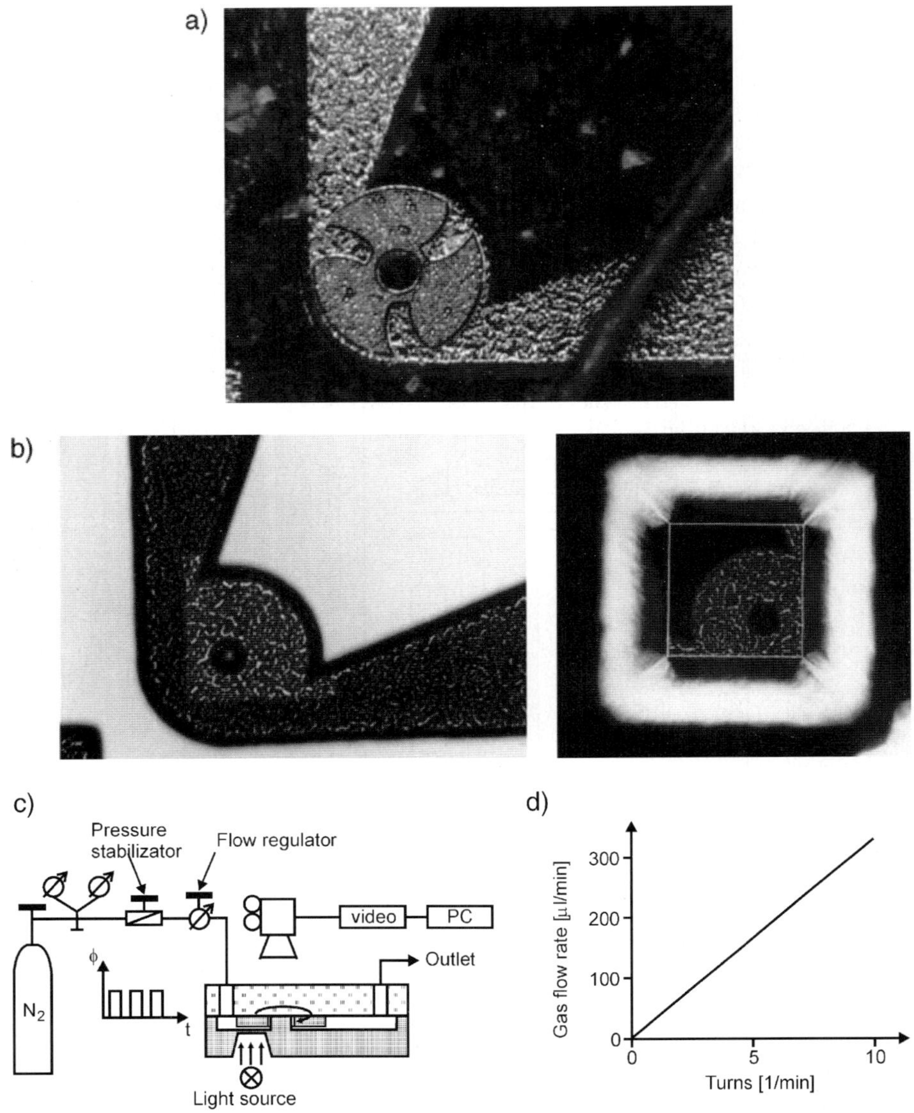

Fig. 4.130. The glass–silicon micromachine: a) front side view of the assembled device; turbine ϕ = 98 μm in diameter localized at the axis of rotation, gas channels and the human hair can be observed, b) front- and back-side view of the silicon base; illuminating window, axis and gas channels can be seen, c) testing system, d) characteristic of flowmeter, flow as a function of turbine rotation.

4.4.4.2. Gas and fluidic devices and microsystems

The concept of an integrated analytical microsystem called μTAS, made using microengineering methods, from silicon, glass and some other microelectronic-compatible materials, assembled mainly by means of silicon to glass anodic bonding, has been proposed at the beginning of the 1990s by Manz, Graber and Widmer [240] and Bergveld [241] (Fig. 4.131). From that time many types of silicon–glass or glass–glass multi-layer gas and liquid chemical microsystems have been developed; microreactors, gas and liquid chromatographs, biomedical and biochemical analyzers, components for combinatorial chemistry and many others [242–245].

A good example of a well-developed, multi-layer biomedical on-chip analyzer equipped with pH, pressure and temperature meters, capable of automatic sampling and purging of fluid samples, was described in paper [246]. This analyzer was constructed using Si–Si fusion bonding (applied in the fabrication of a thermally activated micropump) and silicon–glass–silicon–glass anodic bonding (which formed measuring channels and chambers) (Fig. 4.132).

The ongoing developments in chemical microsystems has created increased interest in research on gas–liquid maintenance on a micro- and nano-volume scale. Various silicon–glass valves, pumps and other fluidic devices (jet-printers) have been discussed repeatedly in the literature [247–253]. Such devices are constructed using silicon to glass anodic bonding in the multi-layer technique. Selective bonding and bonding through thin-film dielectric layers were also applied (Fig. 4.133a).

Other popular anodic-bonded glass–silicon devices are heads of ink-jet printers widely used for black-and-white or color printing (Fig. 4.133b) [253, 254]. A unique micro-syringe with steam actuator, used for intramuscular injections, produced utilizing typical anodic bonding of silicon to Pyrex glass, was presented in paper [251].

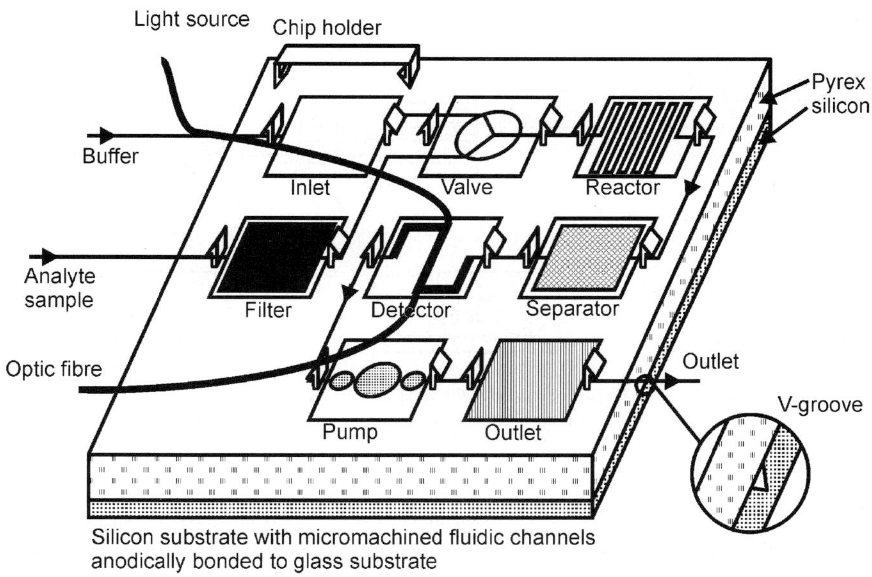

Fig. 4.131. Analytical silicon–glass microsystem – according to Bergveld's conception [241].

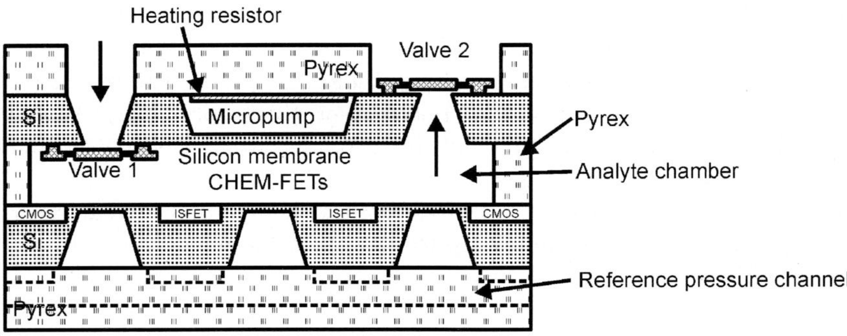

Fig. 4.132. Cross-section of "on-chip" analyzer [246].

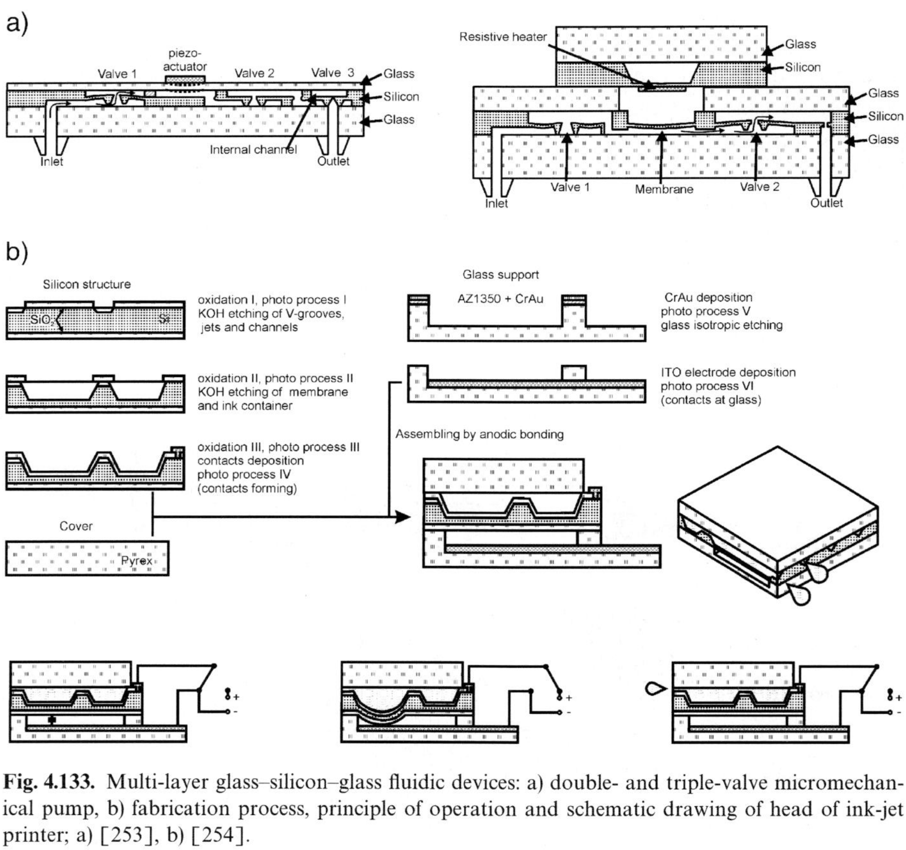

Fig. 4.133. Multi-layer glass–silicon–glass fluidic devices: a) double- and triple-valve micromechanical pump, b) fabrication process, principle of operation and schematic drawing of head of ink-jet printer; a) [253], b) [254].

Anodic bonding of silicon to Pyrex glass at 390°C, 400 V for 15 minutes has been applied in the manufacture of a chemical microsystem presented in paper [255]. The microsystem contains two pumps, flowmeters, micromixers and channels that distribute analytes. Anodic bonding of silicon to Pyrex glass was used in the first integrated gas chromatograph in its first laboratory version as early as 1978 [256]. In the device that is produced at present [257, 258], anodic bonding is used for assembly of silicon–glass multi-layer injection systems containing two valves, mixer chamber, system of distribution channels, as well as for fabrication of the integrated silicon–glass thermal conductivity detector TCD, similar, in part, to the technical solutions shown in papers [259] and [260]. Pumps, valves, injectors, TCD sensors and many other silicon–glass devices are described in the following section of this book.

4.4.4.2.1. Valve-less pump

Anodic bonding of silicon to glass through a thin layer of thermal oxide has been used in a fabrication procedure of the silicon–glass valve-less micromechanical pump, based on the solution of Olsson and co-authors [261, 262]. In this pump two ejectors and the pumping chamber are etched in a silicon wafer (Fig. 4.134).

An external piezoelectric or magneto-electric actuator moving periodically actuates a silicone membrane in the pumping chamber. Liquid is squeezed out of the pumping chamber and flows through the ejectors. Conductance of a liquid flow through the widening ejector is a little bit higher in comparison to the reverse flow. The pumping efficiency is given by:

$$\varepsilon = \frac{\sigma_+ - \sigma_-}{\sigma_+ + \sigma_-}, \tag{4.49}$$

where: σ_+ = hydrodynamic conductance in the direction of ejector widening, σ_- = hydrodynamic conductance in the opposite direction, the ejector contraction.

Coefficients σ_+, σ_- and ε are strictly dependant on the geometrical relations inside the pump, among others on the angle of ejectors, and the length and width of nozzles (Fig. 4.135) [263].

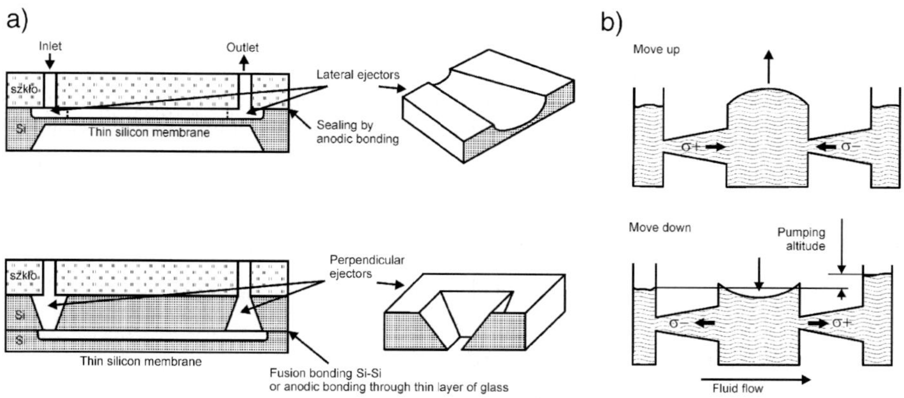

Fig. 4.134. Ejector pump: a) cross-section of structure, b) principle of work [261].

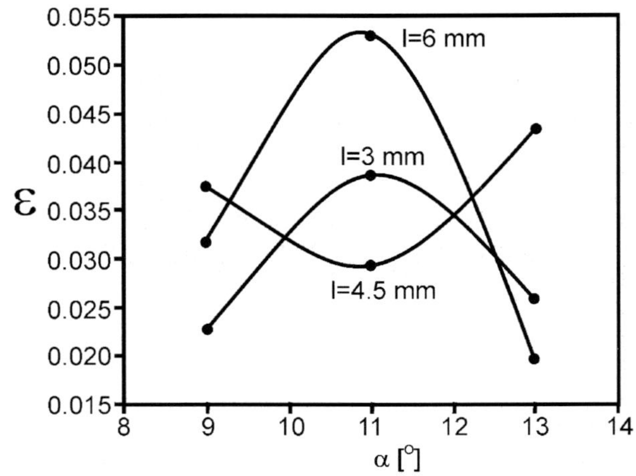

Fig. 4.135. Pumping efficiency ε as a function of angle of ejectors with different length l [263].

In the solution presented in Fig. 4.136, ejectors and bossed membrane were etched isotropically in aqueous solutions of $HF:HNO_3$ at ambient temperature and next wet-oxidized in steam at 1050°C to form a 0.8 μm-thick silicon oxide layer. Two 0.8 mm holes were mechanically drilled in a glass cover made of 1 mm-thick Borofloat 33 glass. Fluidic connections were made of Borofloat 33 glass tubes (height $h = 5$ mm, diameters: $ED = 4$ mm, $ID = 0.8$ mm) bonded ($t \sim 450$°C, $U_p \sim 1.5$–2 kV) to silicon orifices with via-holes etched in 30% KOH. A glass cover with the connections was anodically bonded ($t = 420$°C, $U_p = 650$ V) to silicon covered with a thick thermal oxide layer. A PZT or electromagnetic actuator was hard-epoxy glued to the bossed membrane in its center (Fig. 4.137).

The pump showed a linear decreasing characteristic of debit of water pumping, from 21.5 to 14.5 μl/min, for the activation frequency changed from 13 to 26 Hz (Fig. 4.138).

4.4.4.2.2. Microvalves

Silicon–glass–Kovar

A metal–glass–silicon double-seat, switching valve, made of silicon, Pyrex glass and Kovar alloy was shown in reference [197]. A Kovar block was bonded to a 1 mm-thick Pyrex glass substrate with drilled holes. Next, bonded parts were electrostatically bonded to a silicon structure, in which valve seats and movable sealing membranes were formed. Selective anodic bonding through thin Pt–Ti layers to 0.25 mm-thick Pyrex glass stiffened movable components. Finally, all parts were anodically bonded to the 1.2 mm-thick Pyrex glass base (300°C and 1400 V) (Fig. 4.139).

Silicon–glass valve with Teflon® membrane

A pneumatically operated silicon–glass valve consists of three plates; silicon body and two glass covers (Fig. 4.140) [264]. Valve-seat and fluidic channels

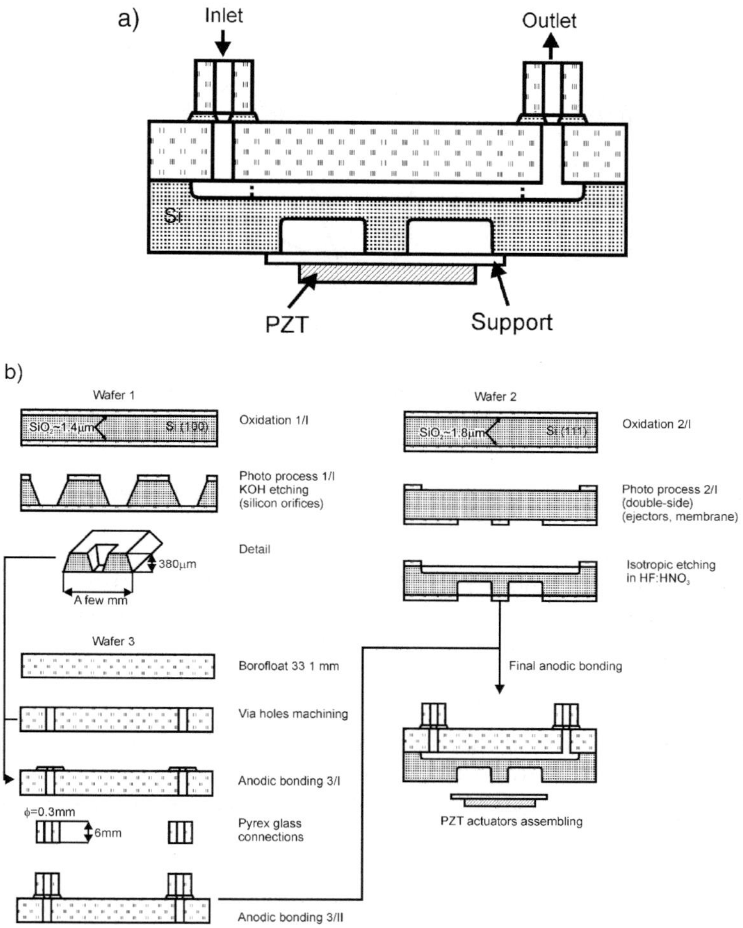

Fig. 4.136. Silicon–glass pump: a) cross-section, b) fabrication method.

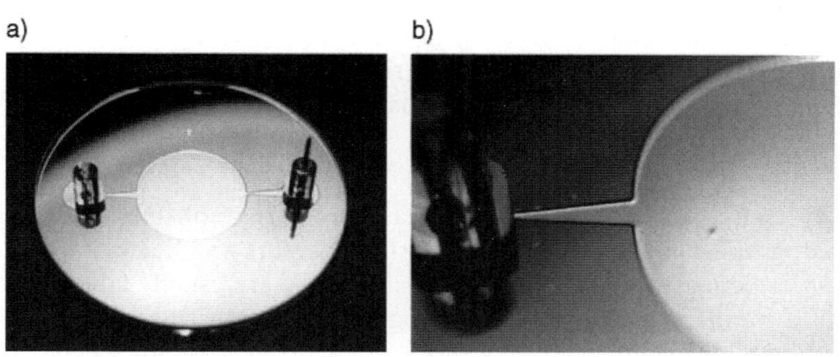

Fig. 4.137. Silicon–glass pump: a) general view, b) ejector.

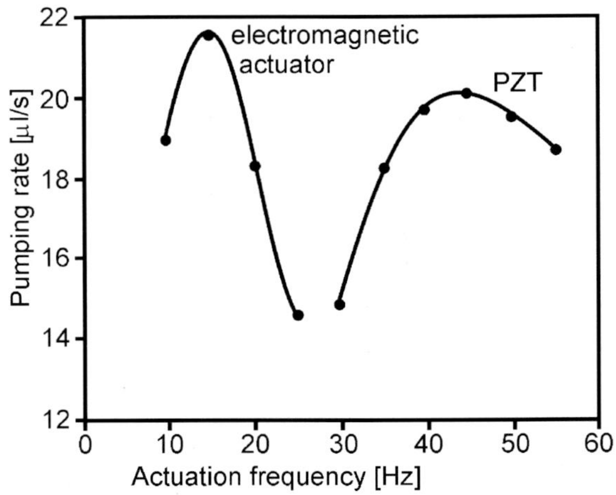

Fig. 4.138. Pumping efficiency as a function of frequency of membrane activation [263].

and via-hole conducting liquid are wet etched in KOH in a double-side polished (100) oriented silicon wafer. A shallow chamber is isotropically etched; a via-hole is mechanically drilled in the top glass cover. Two via-holes are drilled in the bottom glass cover. Borofloat 33 glass fluidic tube connections are bonded through a silicon orifice to both glass covers, next the bottom glass cover is bonded (450°C, 600 V) to the back-side surface of the silicon body. A flat gasket, that is a 30 μm-thick Teflon® membrane, is placed at the valve-seat and the top glass cover is bonded (350°C, 900 V) to the front-side of the silicon plate. The edges of the Teflon® are squeezed during anodic bonding, and thereby the valve is sealed. Components of the valve are listed in Table 4.21, while a view of details from different fabrication steps is presented in Fig. 4.141.

The working principle of the valve is as follows. When inlet pressure P_1 is lower than steering pressure P_s the Teflon® membrane clamps to the valve-seat and the valve is closed. Decrease of P_s below P_1, or applying the underpressure $-P_s$ results in lifting of the membrane, and the valve opens.

Selected characteristics of the microvalve are presented in Fig. 4.142. The gas flow cut-off is obtained for steering pressure P_s higher by 10–15% than inlet pressure P_i. The flow cut-off is fast and repeatable like the switching on of the valve. The dead volume of the valve equals 20 pL.

4.4.4.3. Components of integrated gas chromatograph

The key components of the simplest gas chromatograph (GC) are: injector, chromatographic column(s) and detector(s). Injectors and detectors – the most often used thermal conductivity detectors (TCD) – are usually fabricated in the form of a mechanical device. Chromatographic columns made in the form of spiraled glass or metallic capillary tubes, a few meter long, are easily available today.

Terry, Angell and Jerman developed the first gas chromatograph in the late

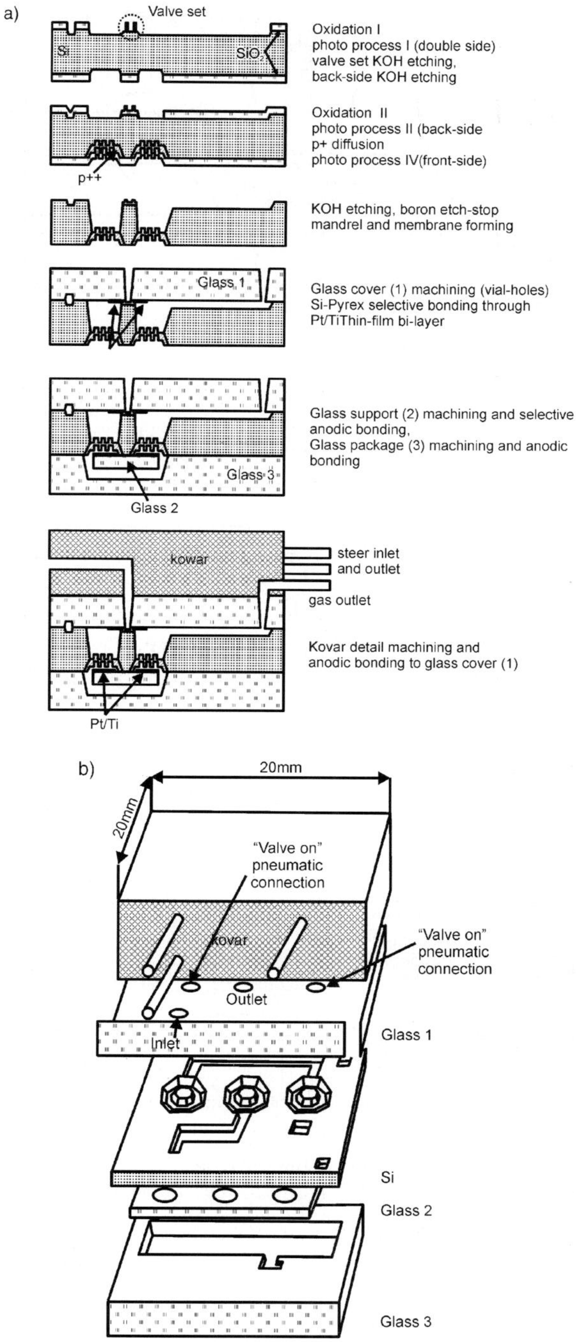

Fig. 4.139. Multi-layer bonding (Kovar–glass–silicon (Pt–Ti)-glass) in the construction of a microvalve: a) fabrication diagram, b) structure of microvalve [197].

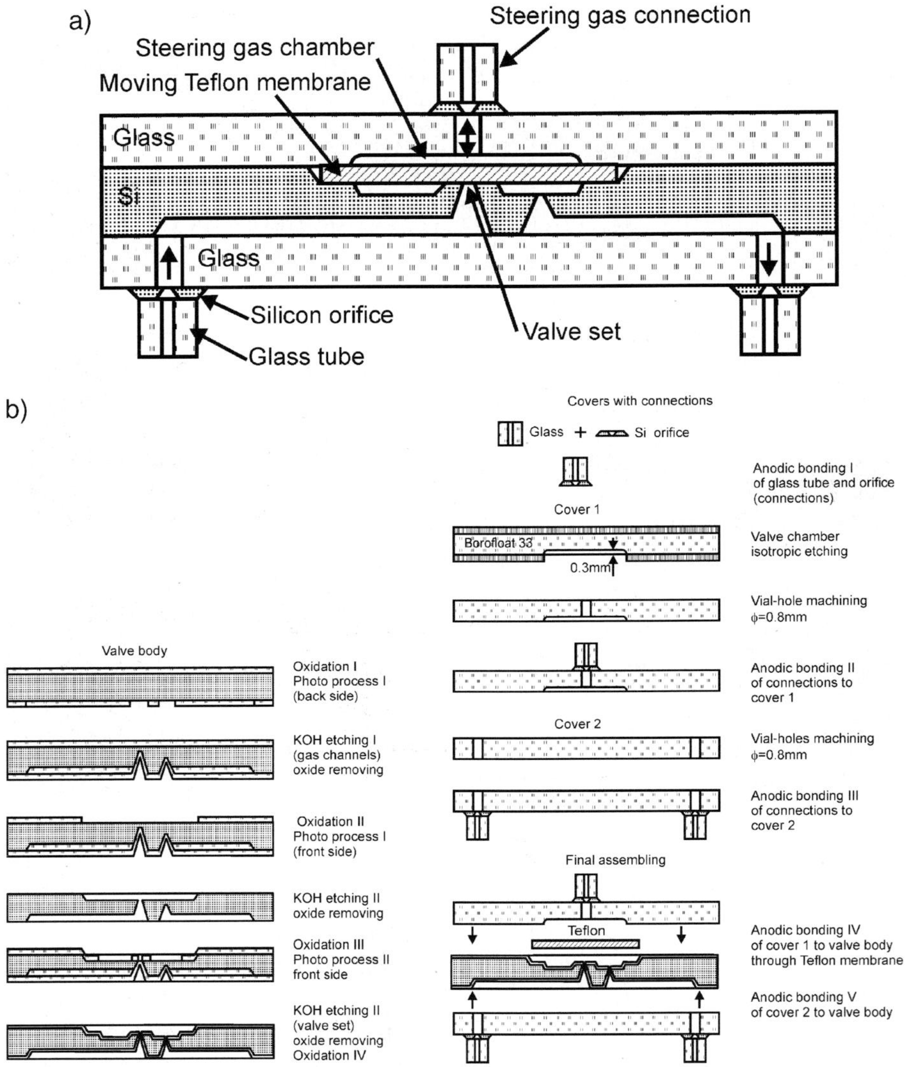

Fig. 4.140. Microvalve: a) cross-section view, b) fabrication scheme.

Table 4.21. Components of the valve: view and description

Name of component	Parts per 1 valve
Silicon orifice	3
Glass tube and silicon orifice	3
Silicon plate with a valve-seat	1
Top glass plate	1
Bottom glass plate	1
Teflon® membrane	1

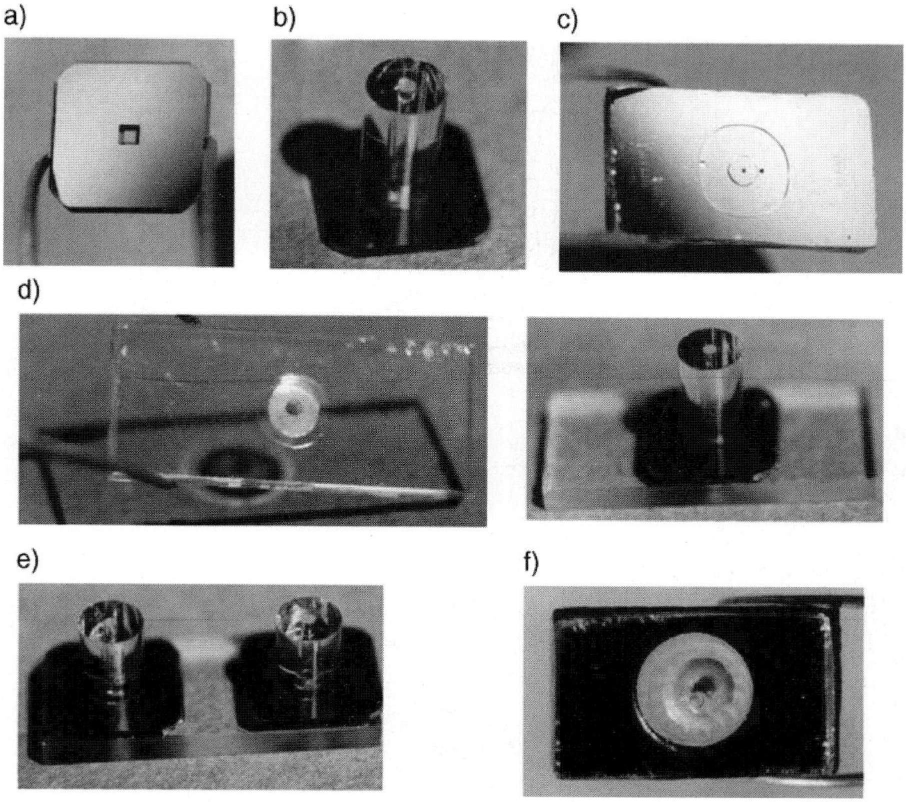

Fig. 4.141. Microvalve, constructional details: a) silicon orifice, b) gas connection, c) the body; valve-seat and gas ports can be seen, d) top glass plate before and after bonding of gas connection, e) bottom glass plate with bonded gas connections, f) Teflon® membrane in the valve-seat observed through the top glass plate (for the clarity of the picture the gas connection was removed).

1970s [256]. Today portable gas chromatographs, which are sometimes called "integrated", utilize at least two important, extremely miniaturized, components made using anodic bonding; microengineered silicon/glass injectors and thermal conductivity detectors (TCD). The injector injects small quantities of an analyte – a mixture of gases – into the stream of stand-by gas flowing inside the chromatographic column. Moving together with a stand-by gas, the mixture separates into particular fractions which are detected at the end of chromatographic column by the TCD.

The TCD detects gas mixture components by measuring heat transport between a micro-heater and micro-thermo resistor positioned inside the gas channel. Heat transport is strictly proportional to the amount of a component, and its time of retention indicates the type of component. All parts of a TCD are extremely miniaturized in order to minimize its dead volume and the response time.

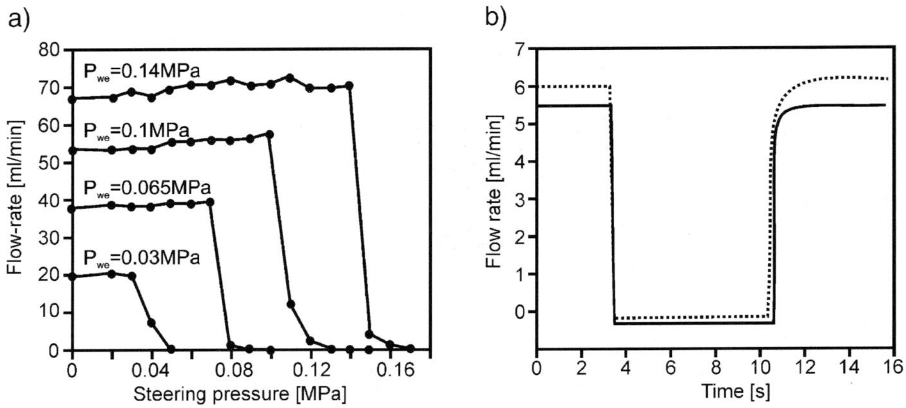

Fig. 4.142. Microvalve characteristics: a) cut-off characteristic, b) dynamic on-off characteristic; the dashed line represents the output signal of custom-made fast flowmeter.

4.4.4.3.1. Injector

The pneumatic scheme of the injector and co-working external gas system – which was described in paper [265] – is shown in Fig. 4.143. The injector chip is built of two symmetric dosing loops (DL1, DL2), one pneumatically operated

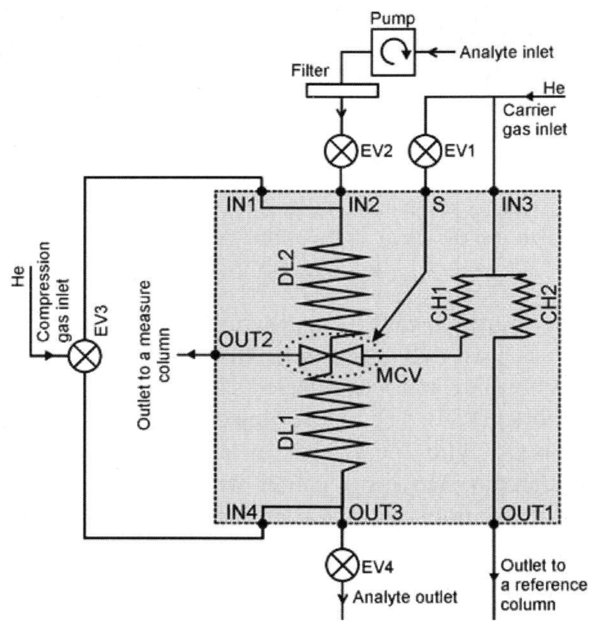

Fig. 4.143. The glass–silicon injector: a) pneumatic scheme of the device (gray area), additional valves (EV) are required for proper working of the device, b) chip layout, c) micro-cross-valve schematic views and cross sections (B-B) left: closed valve – two gas lines separated; right: opened valve – a gas flows from line B to A (injection).

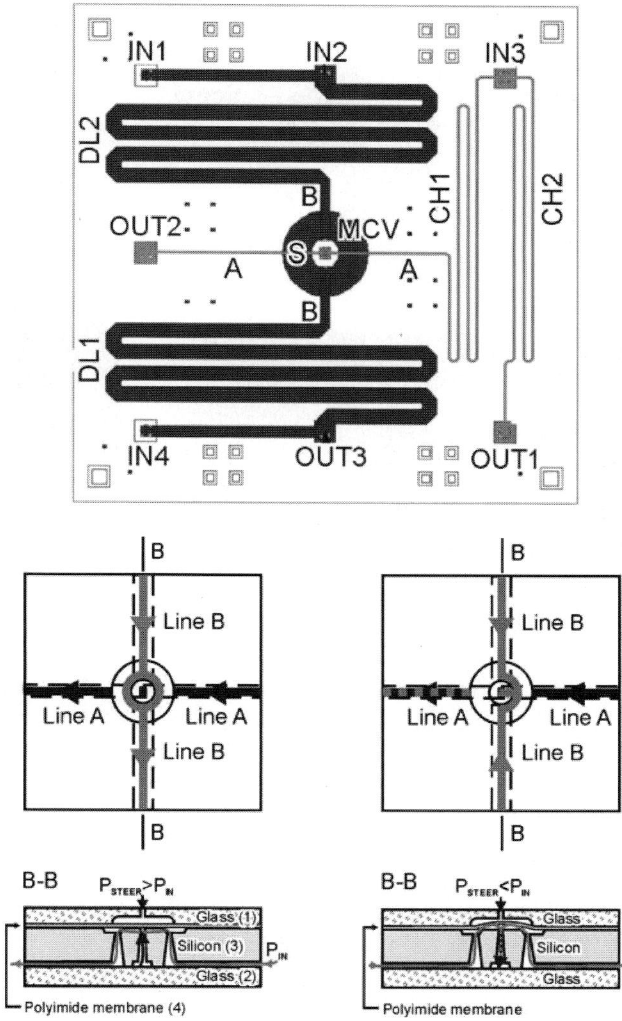

Fig. 4.143. *Continued.*

integrated micro cross-valve MCV and two meanders CH1 and CH2, characterized by a high value of the gas impedances.

The injector consists of two glass covers and a double-sided micromachined silicon body (Fig. 4.144) in which the cross-valve seat and all gas channels are made. The upper glass cover and the silicon body were foil-bonded through the Teflon® coated 50 μm-thick polyimide foil of Dupont (USA), working as the acting part of the valve. The back-side glass cover was anodic-bonded (450°C, 1200 V) directly to the oxidized silicon body. The cross-valve works similarly to the previously described shut-valve with Teflon® membrane. For steering pressure higher than actual pressure under the membrane, the membrane is

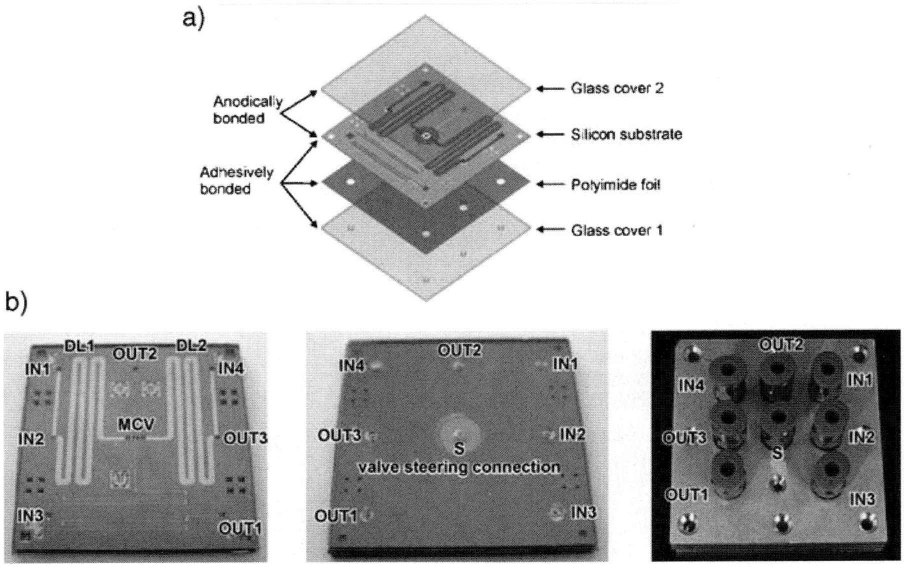

Fig. 4.144. The injector: a) expanded view of the chip, b) front/back-side views of the chip and assembled device in a metal case.

pressed down and seals the valve gasket. Lines A-A and B-B remain separated. For P_s lower than the actual pressure of the flowing gas valve is opened, A-A and B-B lines are cross-connected.

A description of a step-by-step work of the injector is shown in detail in paper [265] and will not be discussed here. Injections are fast and repeatable. The minimal injected volume is about 12 μL. the volume of injected gas portion may be adjusted in the range 12–14 μL by proper steering of the opening time of the cross-valve. The repeatability of the injection has been determined to be better than 0.5%.

4.4.4.3.2. Thermal-conductivity detectors

Detector with Pt-spirals

Bonding of glass to silicon covered with a thick SiO_2 layer has been applied in the fabrication process of the integrated thermal conductivity meter TCD, designed especially for the integrated gas chromatograph [265–267]. The main parts of TCD are two microspirals, made of a very thin platinum wire, which are positioned inside the cavity ending at either side with V-grooved gas channels wet deeply etched in a (100) silicon wafer. A glass cover is anodically bonded after assembly of the spirals. One of the spirals plays the role of electrical heater, the second one is used as a thermosensitive resistor (Fig. 4.145).

The fabrication process of a flowmeter with microspirals proceeds as follows. First, cavities are etched anisotropically in a silicon wafer in order to form the gas channels and ports. Then the silicon wafer is covered with a thick SiO_2

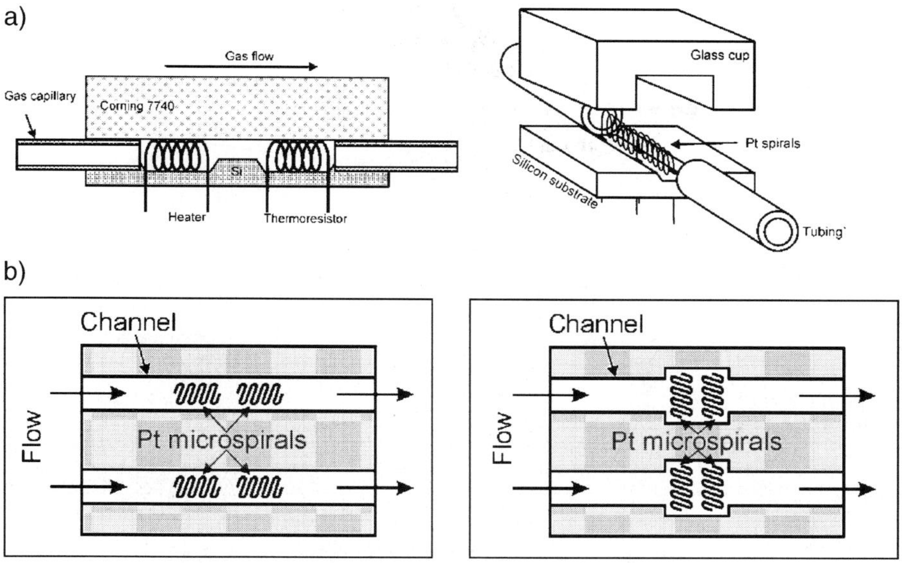

Fig. 4.145. TCD with microspirals; a) schematic view, b) two configurations of spirals.

layer (> 1 μm) and sewn into chips. Two spirals with external diameter $ED =$ 400 μm and length 1000 or 1200 μm, made of 15 μm Pt, are positioned inside etched cavities. Deep grooves, about 800 μm, are mechanically formed in the Borofloat 33 2 mm-thick wafer, then the wafer is sewn into small parts – covers of the TCD. Covers are anodically bonded (420°C, 600 V) to silicon chips. In the next step a silicon–glass chip is fixed to the epoxy laminate, jutting out Pt wires are soldered to conductive paths on laminate and metal or glass capillaries, connecting sensors with a gas system are assembled using hard epoxy glue with copper powder filling (Fig. 4.146).

Detector with thermo-components on membrane

The TCD consists of two symmetrical parts (Fig 4.147). Each part includes two devices: one heating resistor and one thermo-sensitive resistor made of the magnetron sputtered, 0.2 μm thick thin-film Pt layer patterned in the form of a long meander. The Pt layer is deposited onto a surface of a dielectric CVD Si_3N_4 layer, 0.15 μm thick, made on a (100), *n* type silicon wafer. This layer serves as the suspending membrane for thermo-components, as well as a mask for etching of the gas channels (TMAH 80°C, 3 hours). The glass cup, prepared as described earlier for the TCD with PT spirals, is anodically bonded to the silicon chip through a Si_3N_4 layer (450°C, 1200 V). The sensor die is then packaged in a standard metal case, electrical connections are thermosonically wire-bonded and short glass capillaries (ID 0.075 mm) are mounted.

The quality of the injector, the TCD with Pt spirals as well as of the TCD with thin-film thermo-components, is sufficiently good for their application in the model chromatograph [265]. In this device a successful separation and

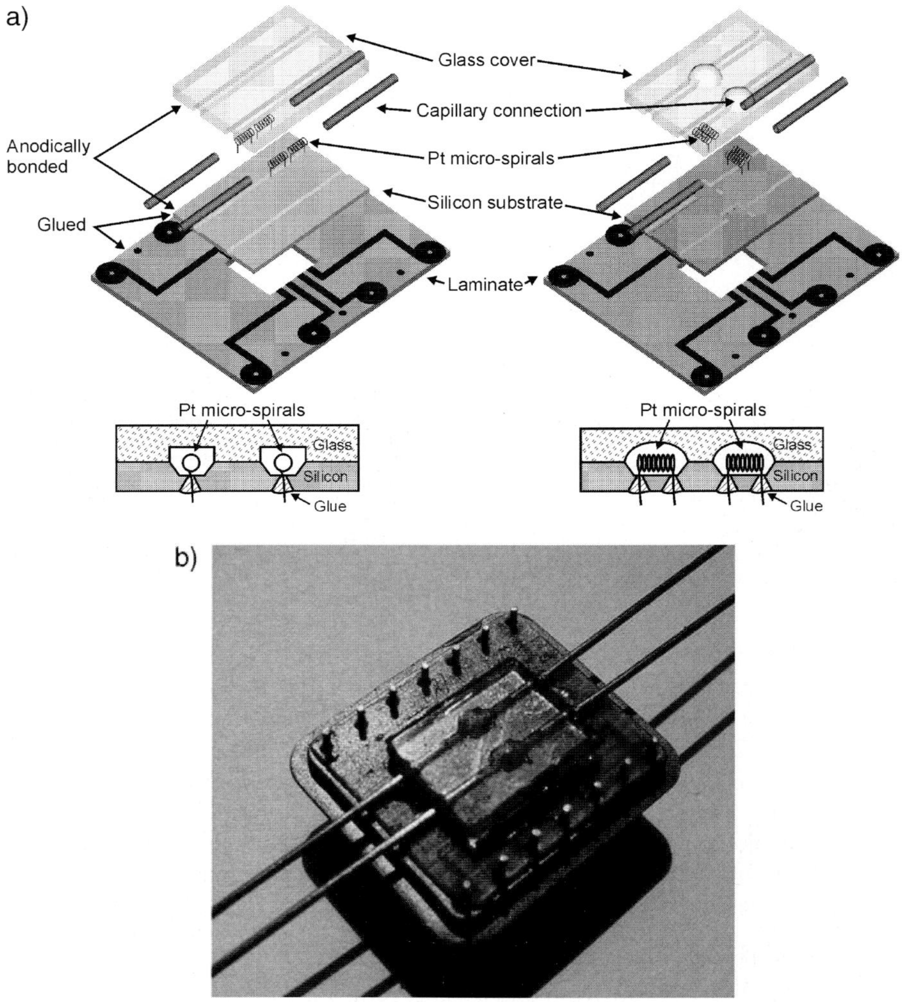

Fig. 4.146. TCD with spirals: a) expanded view and cross-section of the packaged sensor, b) the finished TCD.

detection of the components of gas mixture containing methane (CH_4) has been performed (Fig. 4.148).

4.4.4.4. Silicon–glass capillary devices

Single- or multi-layer silicon–glass, capillary, integrated liquid/gas analytical systems are the subject of scientific and application-oriented investigations [268–280] that are currently being developed on a large scale. For the present these systems are being introduced into the market as devices and analytical systems for microchemistry and combinatorial chemistry, and as components of biochips and lab-on-chips.

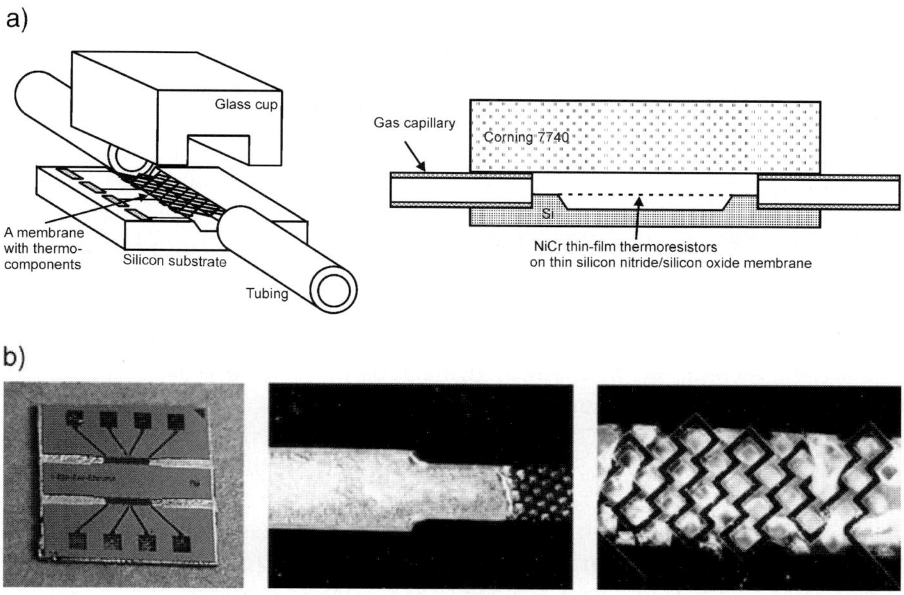

Fig. 4.147. TCD with thin-film thermo-components on the membrane: a) schematic view, b) view and details of the construction.

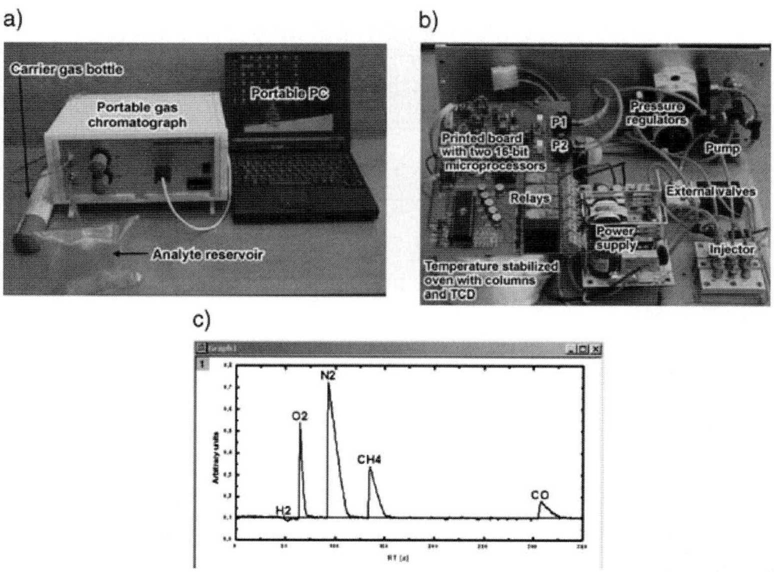

Fig. 4.148. The integrated gas chromatograph; a) a view of the instrument ready-to-work, b) internal parts, c) chromatogram of a deep coal mine industrial atmosphere.

4.4.4.4.1. *Microreactor with meandered capillary*

The fabrication process of a silicon–glass microreactor with meandered U-shaped capillary is presented schematically in Fig. 4.149. The capillary was etched isotropically in HF:HNO_3 by a thermal SiO_2 mask. After thick thermal oxidation (>0.8 μm) of the micromachined silicon, and proper washing and activation, a 2 mm-thick Corning 7740 glass substrate was anodically bonded ($t = 420°C$, $U_p = 900$ V) to the silicon wafer, from the capillary side. The bonded sandwich was annealed at 405°C for 30′ and slowly cooled (2 hours) down to the ambient temperature. Two versions of fluidic connections were formed: first, by back-side anisotropic etching of via-holes in KOH and anodic bonding ($t =$ 400°C, $U_p = 600$ V) of glass tubes and curing of bond at 405°C for 30′ (Fig. 4.150); second, by mechanical drilling of holes in the glass substrates followed by bonding of silicon–glass connections.

The fabrication procedure described above has been used in order to fabricate different microreactors with 30 μm wide, 11 μm deep capillaries, several meters long, integrated on chips of several square centimeters (Fig. 4.151).

The sophisticated solution of a multi-layer silicon–glass capillary made of three Pyrex glasses with horizontal connections (Fig. 4.152) was presented in paper [280]. Glass wafers were bonded (450°C, 700 V) through thin film layers of LPCVD deposited polysilicon.

4.4.4.4.2. *CE/MS/bio-chips, gas separation column*

In many microanalytical devices a liquid analyte is introduced (injected) into the stream of liquid buffer flowing in the capillary, driven by hydrostatic pressure

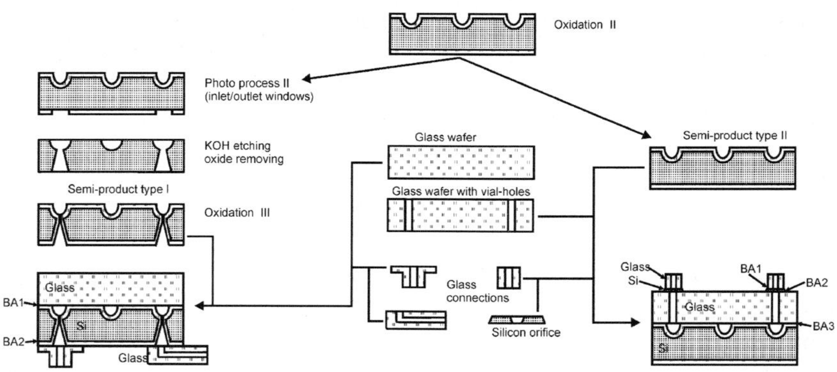

Fig. 4.149. Fabrication flow-chart of meandered microreactor in two constructional versions.

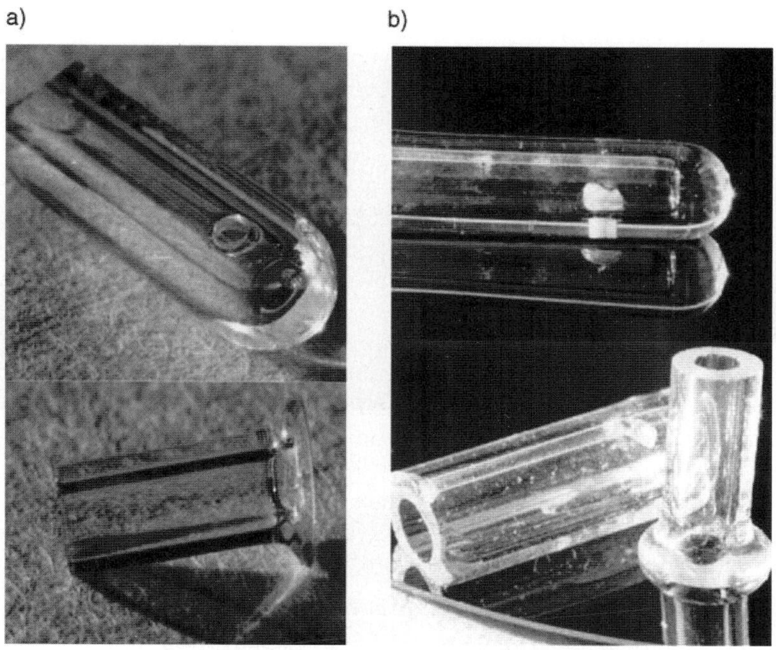

Fig. 4.150. Glass connections: a) glass tubes prepared for bonding, b) bonded connection.

or hydro-electro-pumping. Fractions of liquid analyte are fractionated by two principles: sorption-desorption or electrophoresis fractionation [268, 272, 273]. Separated fractions (eluents) reach the capillary outlet where ion-sensitive sensors (CHEM-FET or conductometric), or a suitable optical sensor (fluorimetric, photometric, etc.) are placed. Analysis of the particular eluent retention time and/or its charge allows qualitative and quantitative identification of the substance. In such analytical microsystems, simple as well as very complex capillary systems are used [241, 242, 245–247, 272, 273].

There are three main, important and positive characteristic features of the microanalysers discussed: very small dead-volumes (ranging from several femtoliters to a few dozen picoliters), short analysis time and very high resolution.

The chosen examples of fluidic capillary microanalyzers, developed in own studies [275–278] are shown in Fig. 4.153. Capillary channels in all of these devices were wet, isotropically etched in silicon. Before anodic bonding to glass, the silicon wafers were thick (> 1.5 μm) wet high-temperature oxidized, to obtain an electrically and chemically insulating SiO_2 layer. Multi-layer wafer-scale anodic and anodic–cathodic bonding of the microanalyser "body" ($t = 420°C$, $U_p = 900$ V, flat metallic-glass cathode, followed by curing at 405°C for 2 hours) and bonding of small details (fluidic interconnections) were used in the fabrication of these devices.

A similar fabrication procedure was used in order to fabricate the integrated, capillary, gas, chromatographic columns, covered with a continuous silicon

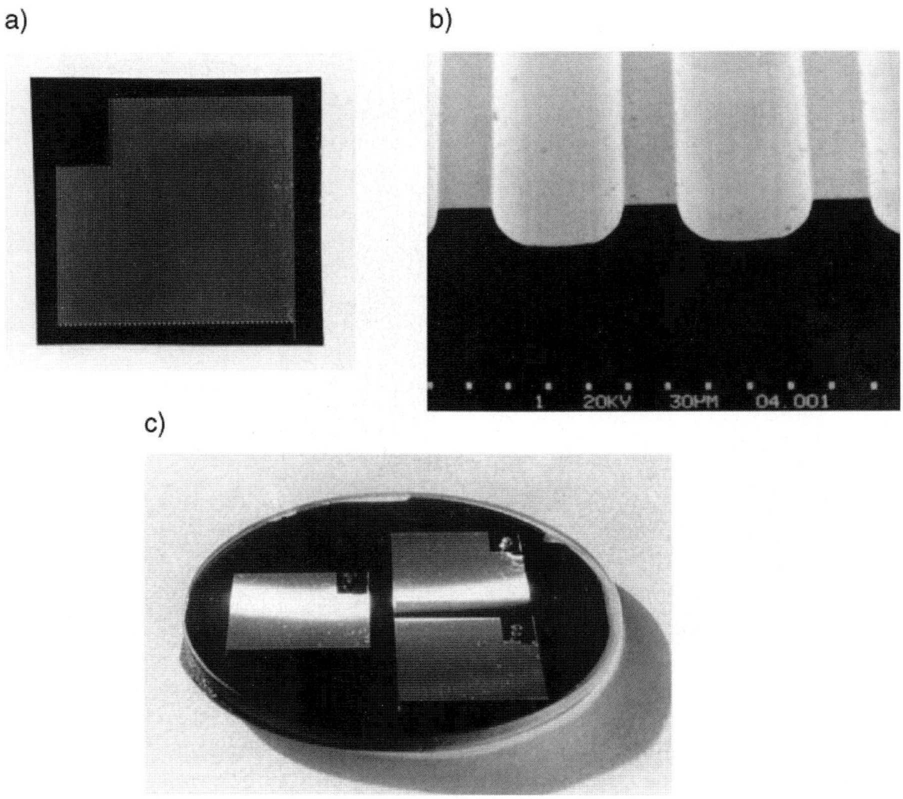

Fig. 4.151. The microreactor: a) top view of 20 mm × 20 mm chip, b) cross-section of the silicon structure, c) 3″ glass–silicon plate with three microreactors after anodic bonding, before sewing in chips.

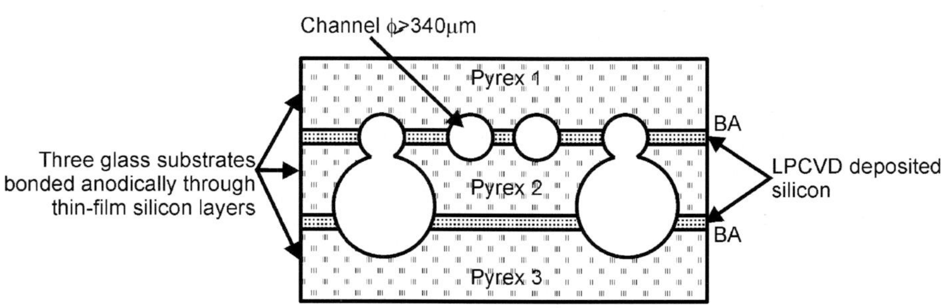

Fig. 4.152. Cross-section of triple-layer capillary column [280].

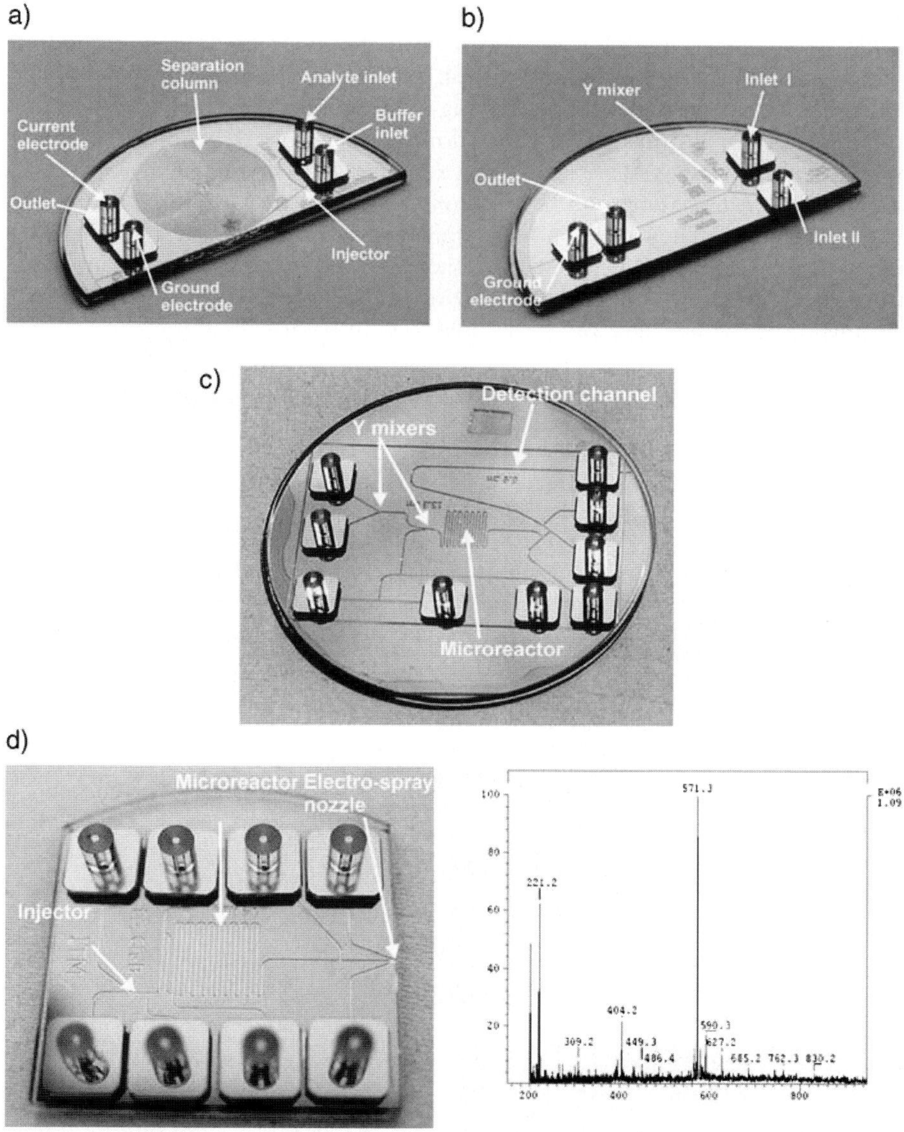

Fig. 4.153. Silicon–glass capillary microanalyzers: a) liquid, flow-injection capillary-electrophoresis (FI-CE) chromatograph with a spiraled separation microcolumn, b) Y mixer and CE separator, c) immunoassay bio-chip, d) electrospray chip microanalyzer for mass spectrometry and (MS) mass spectrogram of S Gramicidin obtain by use of this chip. Analysis was carried out at the Neurochemistry Group on the Faculty of Chemistry at the Jagiellonian University at Cracow, Poland, by the courtesy of prof. Jerzy Silberring.

oxide layer or porous silicon dioxide, the capillary length of which reached a dozen or so meters and ID equaled from a dozen or so to a few hundred micrometers [266, 277, 278, 281] (Fig. 4.154). The technology of the columns consisted of uniform wet isotropic etching of the spiraled channel, thick oxidization or selective porous silicon layer formation followed by oxidation, and bonding of a 3″ glass wafer (1 mm thick Corning 7740 or Borofloat 33) to a 3″ silicon wafer. After fabrication the columns were covered with liquid stationary phase (squalene). They have shown good separation of mixtures of aliphatic hydrocarbons (n-pentane, n-hexane, n-heptane, n-octane and n-decane) and aromatic hydrocarbons (benzene, toluene, ethyl-benzene, cumene, sec-bethyl-benzene, u-pentyl-benzene and naphthalene) (Fig. 4.155).

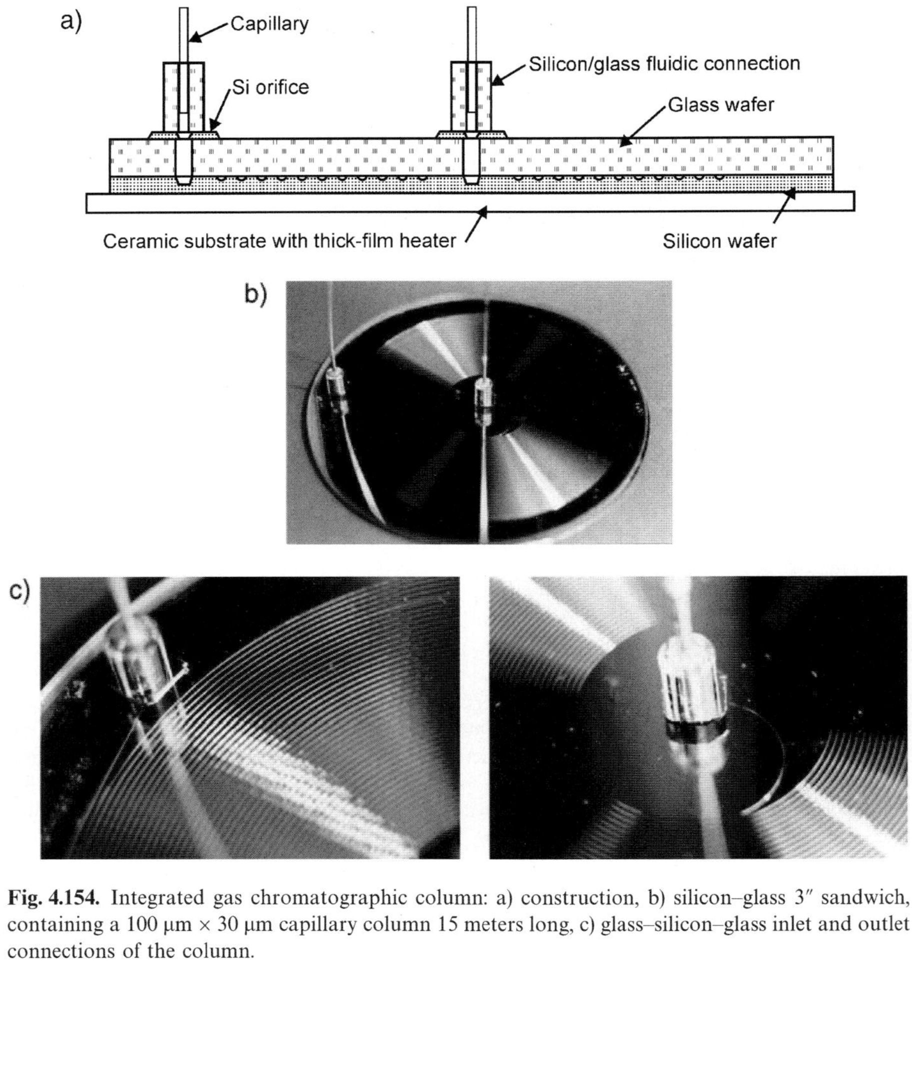

Fig. 4.154. Integrated gas chromatographic column: a) construction, b) silicon–glass 3″ sandwich, containing a 100 μm × 30 μm capillary column 15 meters long, c) glass–silicon–glass inlet and outlet connections of the column.

a)

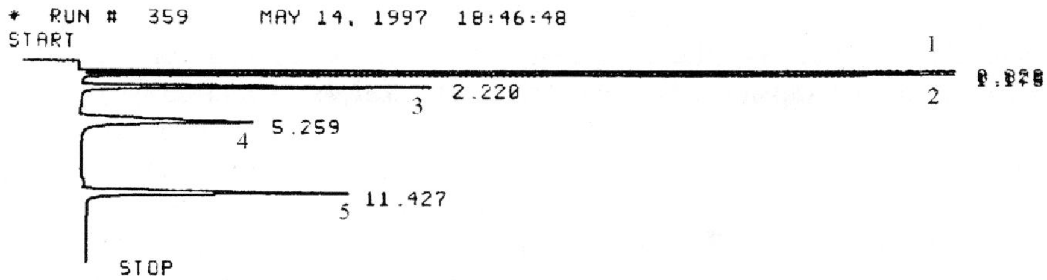

RUN# 359 MAY 14, 1997 18:46:48

AREA%

RT	AREA	TYPE	WIDTH	AREA%	
.828	728201	PB	.052	21.17636	1-n-pentane (n-C_5H_{12})
1.175	695865	BB	.084	20.23602	2-n-hexane (n-C_6H_{14})
2.220	589866	PB	.194	17.15353	3-n-heptane (n-C_7H_{16})
5.259	679350	BB	.455	19.75575	4-n-octane (n-C_8H_{18})
11.427	745463	PB	.324	21.67834	5-n-decane (n-$C_{10}H_{22}$)

TOTAL AREA=3438746
MUL FACTOR=1.0000E+00

b)

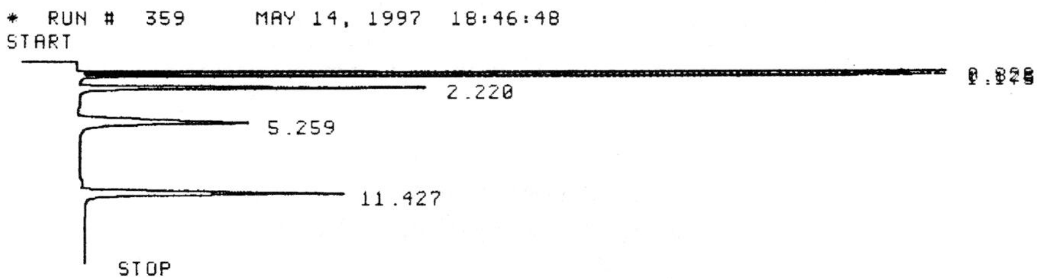

RUN# 359 MAY 14, 1997 18:46:48

AREA%

RT	AREA	TYPE	WIDTH	AREA%
.828	728201	PB	.052	21.17636
1.175	695865	BB	.084	20.23602
2.220	589866	PB	.194	17.15353
5.259	679350	BB	.455	19.75575
11.427	745463	PB	.324	21.67834

TOTAL AREA=3438746
MUL FACTOR=1.0000E+00

Fig. 4.155. Chromatograms of a mixture of hydrocarbons separated by the integrated column from Fig. 4.154: a) separation of aliphatic hydrocarbons, b) separation of aromatic hydrocarbons, carrier gas: helium, split 1:12 000, flame ionization detector. Study of the separation was carried out by M.Sc. eng. Jerzy Ruszczyński from the Faculty of Chemistry of the Warsaw University of Technology.

4.4.4.4.3. TFFF

Thermal Field-Flow Fractionation (TFFF) is a chromatographic-like elution technique of separating the components of solutions, colloids and particles by thermal gradients inside very "thin" channels with the one, usually top surface "hot" and the second, bottom "cold". A relatively small temperature difference applied between top and bottom walls of the channel produces the high thermal gradient. The principles of the separation process are shown in Fig. 4.156.

The macro-scale TFFF device is mechanically machined. Such devices pose good separation parameters, but need a large volume of samples and a high power supply. Only a few miniaturized (μTFFF) systems have been described [282, 283]. These are made of glass parts sealed by the use of SU8 bonding. The all-silicon–glass μTFFF [284] is designed as a 3″ chip. A 30 μm-deep separation channel (2 mm wide and 5 cm long) is etched in the silicon body and in the top glass plate (Fig. 4.157). Anodic bonding seals these parts. The heater is made of a deeply micromachined silicon piece; and a long, narrow chip, bonded anodically to the top glass plate. The channel of the cooler is deeply wet-etched on the back-side of the silicon body. The bottom glass plate is back-side bonded to the silicon body. The separation channel has one common input and two outputs connected to "hot" and "cool" sides of the separation channel. Heater and cooler are equipped with independent heating/cooling fluidic inlets and outlets, made by anodic bonding of glass tubes and silicon.

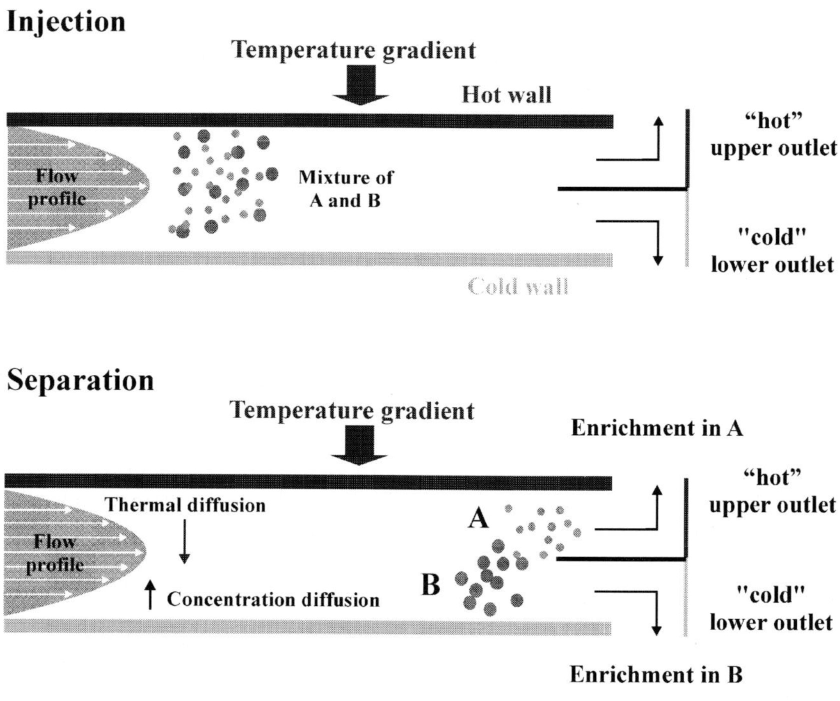

Fig. 4.156. The principles of separation; separated solution flows from inlet to two outlets.

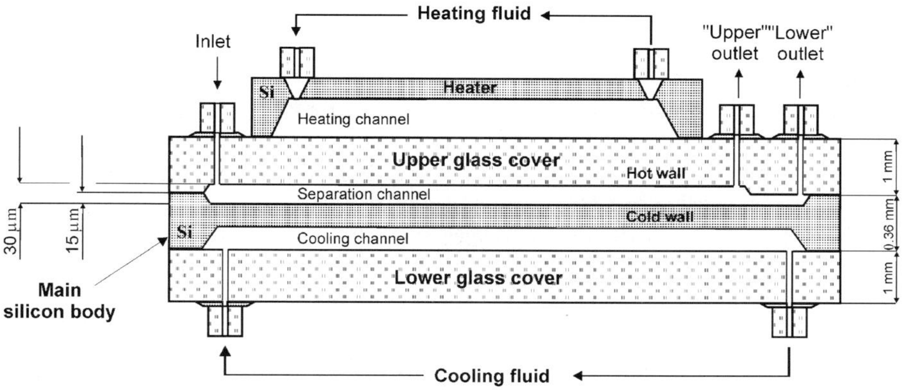

Fig. 4.157. Cross section of the μTFFF.

The complete device consisted of seven layers of silicon and glass, bonded in stack. The fabrication process was described in detail in reference [284]. Here we concentrate on assembly by anodic bonding.

As mentioned earlier, all parts of the μTFFF device (Fig. 4.158) are anodically bonded. Bigger parts of the device are bonded at 450°C for polarization at about 1.5 kV. Small details of fluidic connections are bonded at 400°C, 600 V and cured at 405°C for 30 minutes.

In the first step the silicon–glass outlets are bonded with the top and bottom glass plates. Next, the separation channel is formed between two bonded parts: the silicon body and the top glass plate. Sealed parts are post-bond cured at 450°C at about 600 V for 30 minutes. Next the heater is bonded to the top glass plate; a heater channel is formed. Finally, the silicon wafer and the bottom glass are bonded; the cooling channel is formed.

4.4.4.5. *Chemical sensors*

4.4.4.5.1. *Conductometric*

This detector consists of a silicon structure with channel etched 150 μm deep and 300 μm wide, glass cover plate with electrodes and silicon–glass outlets (Fig. 4.160).

The device was produced on the *n*-type, (111) oriented silicon wafer. The fluid channel is isotropically etched in HNO3:HF (9:1) solution. Three thin-layer electrodes (Ti–W–Au) are deposited in vacuum onto the Borofloat 33 glass cover. Two holes are mechanically drilled in the cover, then the cover is washed and activated (full procedure is used), fluidic connections are anodically bonded to glass, and finally the cover is bonded to silicon (450°C, 1000 V) The dead volume of the detector is about 96 nL. The sensor sensitivity is comparable to sensitivity of the macro conductometric sensors but it reacts better for time-dependant concentration of the measured solution (Fig. 161).

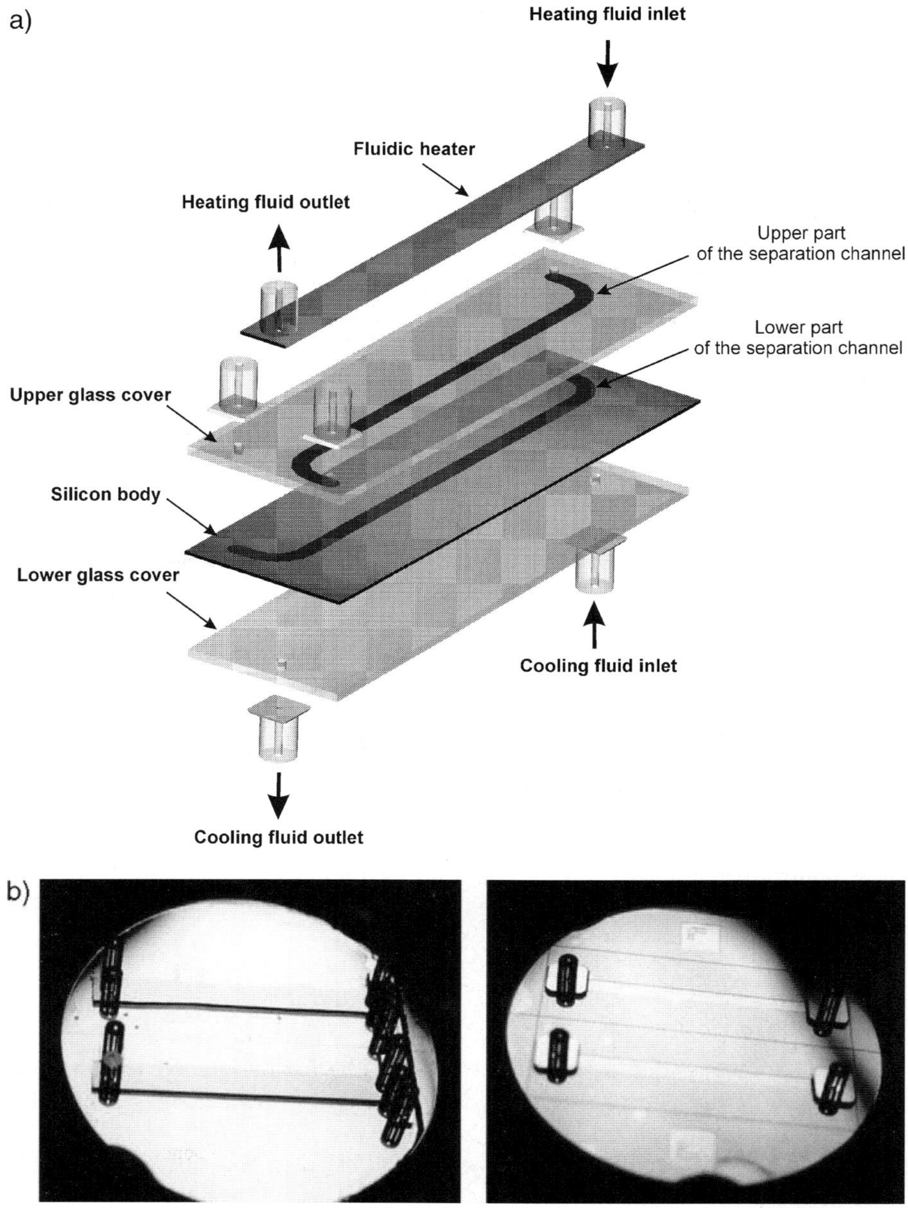

Fig. 4.158. The μTFFF device: a) expanded view, b) front- and back-side view of the chip with two devices.

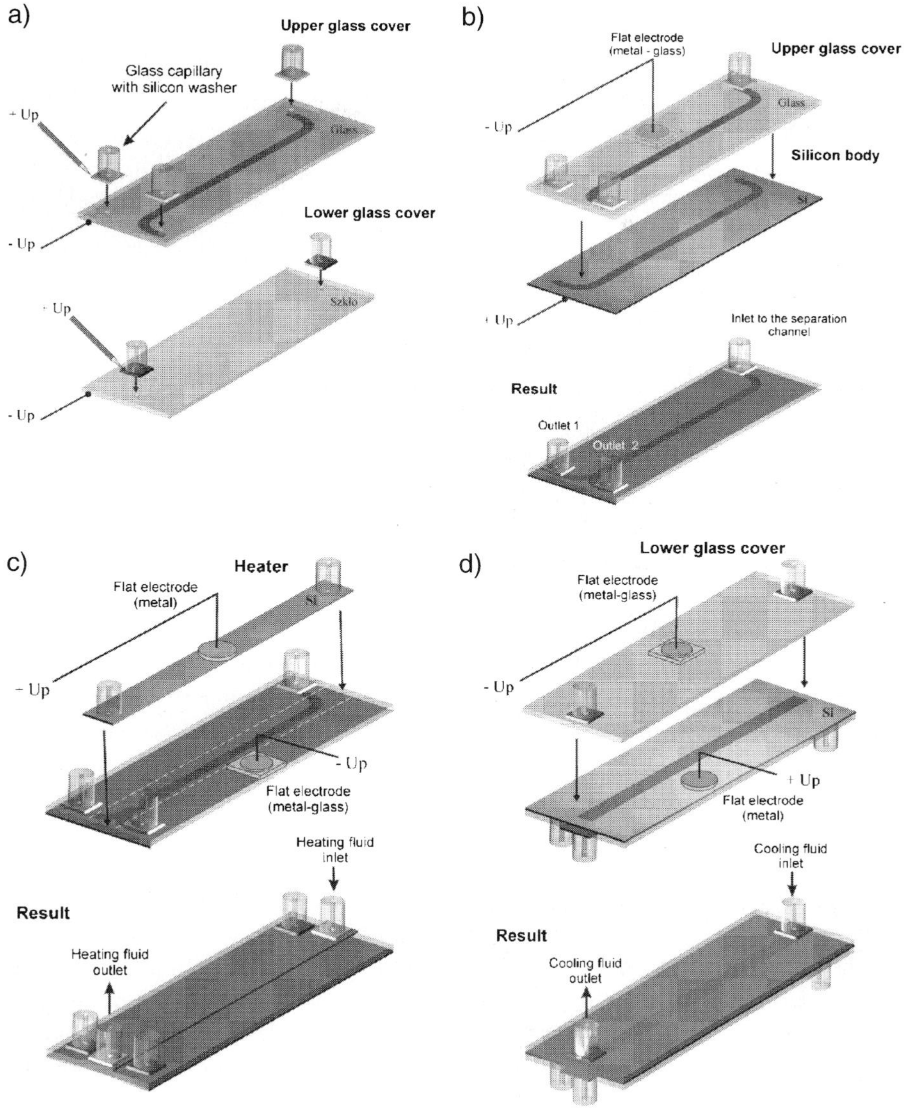

Fig. 4.159. Step-by-step assembly procedure of a μTFFF device: a) outlets bonding, b) separation channel forming, c) heater bonding, d) bottom glass bonding.

4.4.4.5.2. Spectrofluorimetric

The spectrofluorimetric method, based on the fluorescence effect, is widely used in analytical chemistry. In this method the small fluorimetric indicator (fluorescein, erythrosine) is added to an analyte which is illuminated by an excitation light. The fluorescent light is induced proportionally to the amount of the

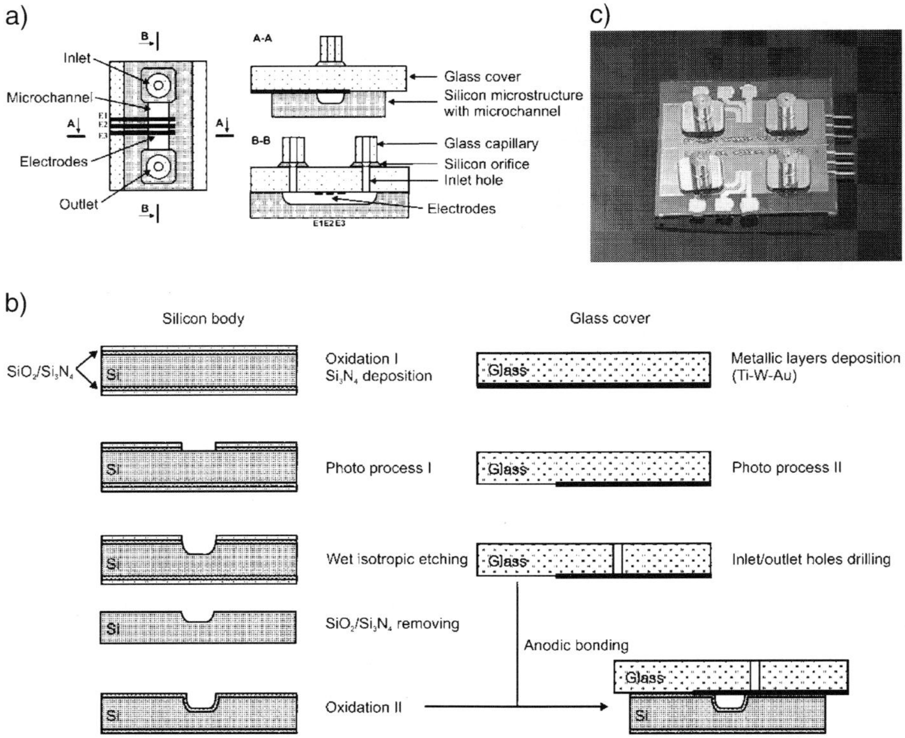

Fig. 4.160. Integrated conductometric detector: a) lay-out and cross-section, b) fabrication process, c) the chip with two detectors.

indicator complexed to the investigated eluent. The light intensity is a measurable parameter of a quantitative and/or qualitative analysis.

The detector discussed here [285] consists of a silicon chip with etched channels, a glass cover, two fibers and fluidic connections (Fig. 4.162). Sample flows through a 500 μm by 200 μm V-grooved channel. A standard telecommunication (ED125 μm, core size 62.5 μm) multimode glass fiber (fiber 1) is positioned in a V-groove etched perpendicularly to the sample channel. This fiber is used to illuminate 17 nL of a sample. PMMA (polymethyl methacrylate) plastic fiber (fiber 2) collects and guides the fluorescent light to an external or built-in light-sensitive diode.

The fabrication process of the device (Fig. 4.163) begins with channel formation in the *n*-type, (100) silicon wafer, in 10M KOH at 80°C followed by 0.3 μm-thick wet high-temperature oxidation. Next, in the 1 mm-thick Borofloat 33, 35 mm × 35 mm glass cover two holes were drilled. Afterwards, the glass substrate was anodically bonded to the silicon chip through a thermal SiO_2 layer (450°C, 1200 V, curing for 30 min). Fiber 1 is positioned in the V-groove and a small drop of UV curable glue (Epo-Tek, UVO-114) is added. Two fluid

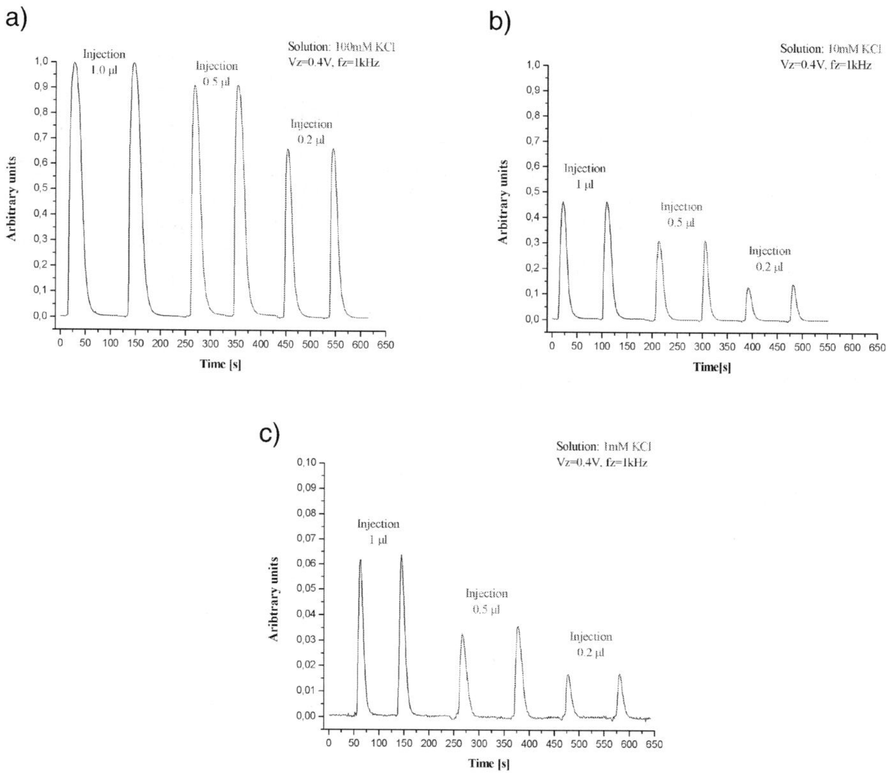

Fig. 4.161. Dynamic output signals obtained for various doses of KCl solution in water: a) 100 mM, b) 10 mM, c) 1 mM. Supply: voltage $U = 0.4$ V, frequency $f = 1$ kHz [317].

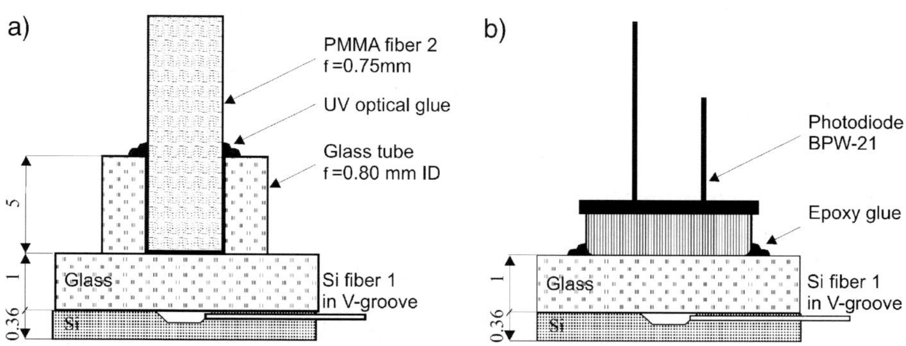

Fig. 4.162. Schematic view of the spectrofluorometric sensor: a) version with two fibers, b) version with photo-diode.

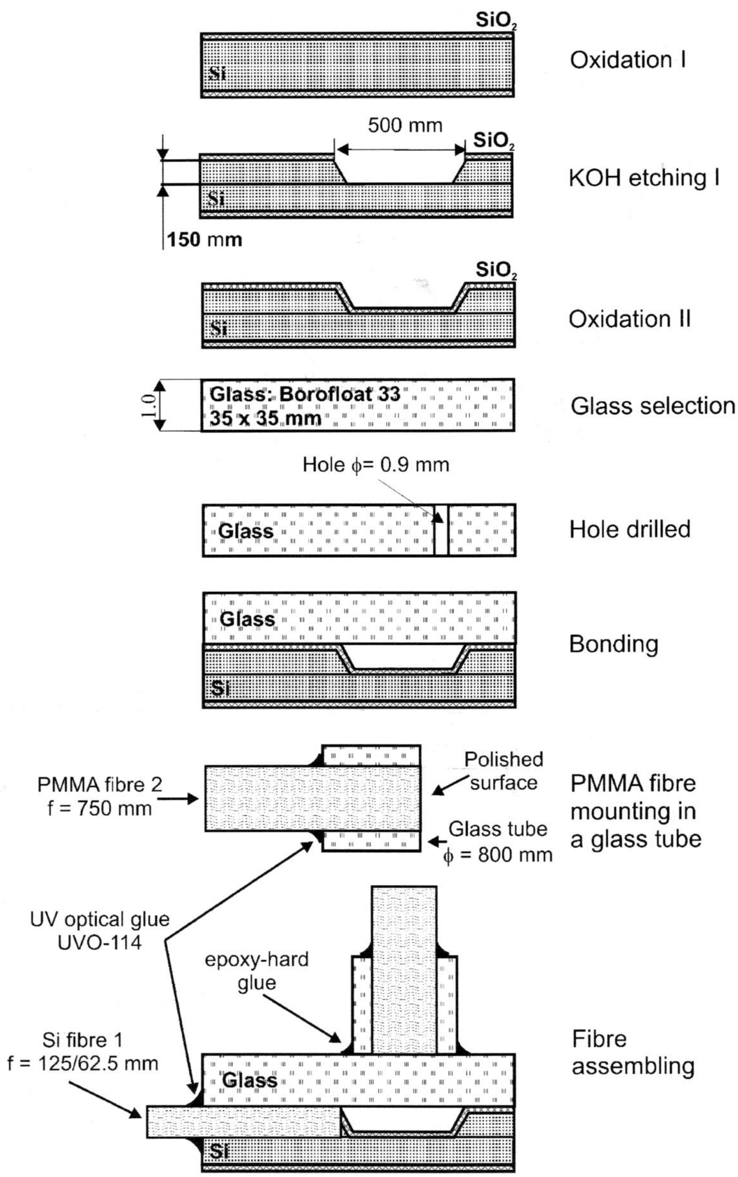

Fig. 4.163. Step-by-step fabrication process of spectrofluorimetric sensor in a two-fiber version.

inlet/outlet glass tubes and support of the plastic fiber 2 are epoxy-hard glued to the cover (Fig. 4.164).

The sensor has shown a linear response to the 0–0.2 mg/mL fluorescein and 0–0.14 mg/mL erythrosine concentration range. The sensitivity and peak wavelengths of emitted fluorescent light λ_{EM} (509 nm for fluorescein and 554 nm for erythrosine) is comparable with results obtained for a standard Fluka measurement set-up (414 nm and 556 nm respectively) [329].

4.4.4.5.3. Spectrophotometric

Spectrophotometric sensors are very popular in analytical chemistry because of their high reliability and sensitivity. In such sensors the absorption of light by a fluid flowing through a flow-through cell is measured. In microminiaturized versions of such sensors [285] the cell and fluidic channels are micromachined in silicon, and light sources and light collectors are made of optical fibers (Fig. 4.165). The fabrication process (Fig. 4.166) of the spectrophotometric detector includes three main steps: silicon micromachining, glass cover machining and bonding, assembly of the fluidic ports and optical fibers.

The fluidic cell and two V-grooves (150 μm-deep both) are wet anisotropically etched in a 3″, n-type, (100) oriented silicon wafer, the wafer is oxidized in steam, and a 0.3 μm-thick silicon dioxide layer is formed. The silicon wafer is divided into 25 mm × 25 mm × 0.36 mm chips. Two in/out holes are drilled in a 1 mm-thick, 35 mm × 35 mm, Borofloat 33 glass cover. The cover is anodically bonded (450°C, 1.5 kV) to a silicon chip. The optical fibers are positioned in V-grooves and immobilized by a small drop of UV glue.

The spectral characteristic of the sensor is comparable to or almost identical with the spectral characteristic of standard spectrophotometric equipment used in analytical chemistry (Fig. 4.167). The analytical usefulness of the spectrophotometric detector was confirmed by the determination of phosphate concentrations in water. The response of the detector was linear for concentrations of analyte from 0.2 to 1.6 μg P/mL.

4.4.4.6. New or unique solutions

4.4.4.6.1. MEMS cell of the atomic cesium clock

The atomic cesium clock, the most precise tool for time measurement, was developed in the 1950s. In this instrument the vapor of cesium is irradiated by frequency-modulated microwaves, cesium atoms become excited only when the actual microwave frequency precisely matches the frequency source dictated by the quantum condition of excitation of electrons orbiting the cesium atom. This is, for $f_r = 9192, 631\ 770$ Hz. Atoms excited by the microwave are separated by an internal magnetic field and reach the detector, giving a valuable output signal peak when the actual microwave frequency is f_r. The signal peak is used for correction and bringing the oscillator exactly to f_r. In the classical macroscale device, made of metal and glass, cesium source is heated to boiling point, the vapor of cesium fills a glass chamber, several centimeters long and wide. Such

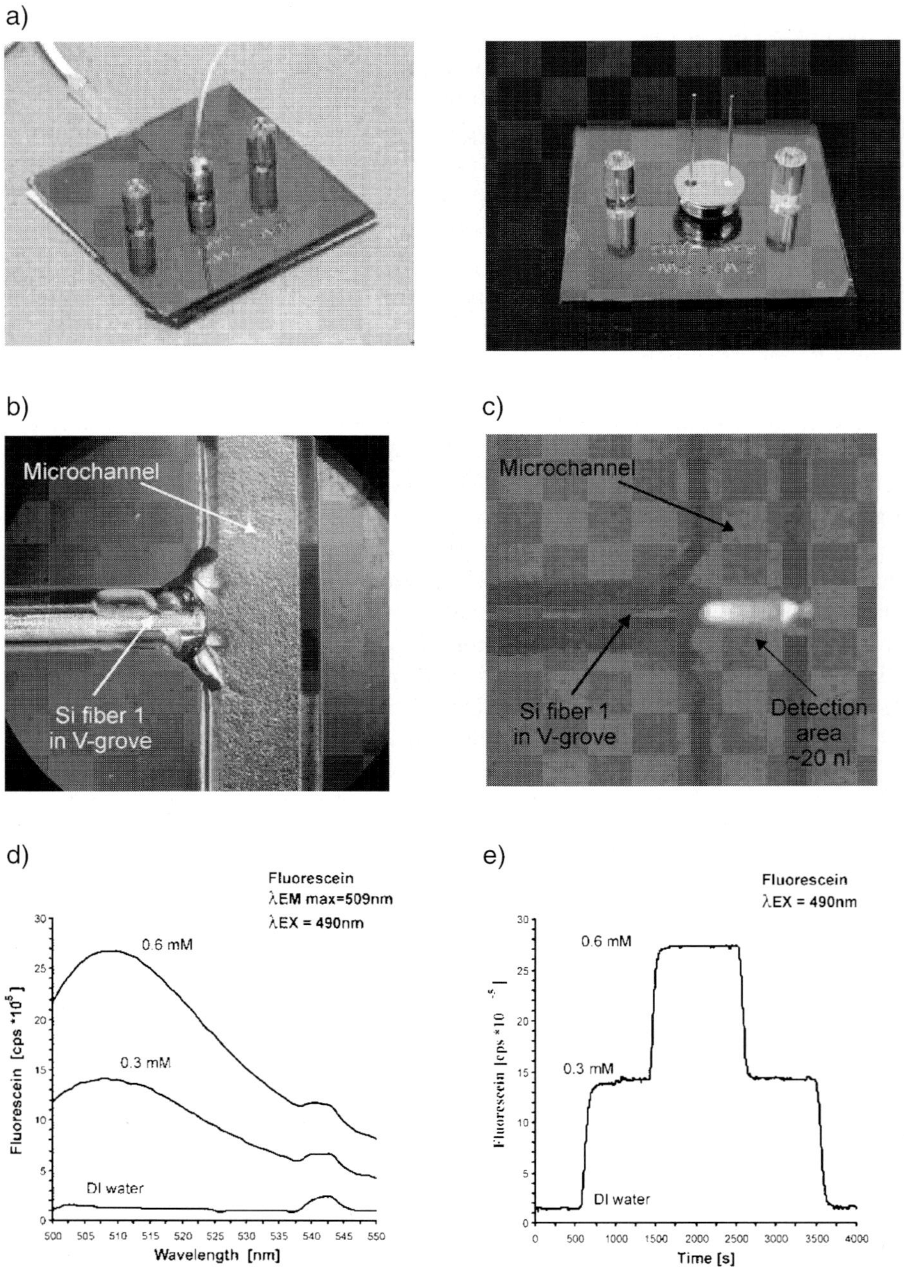

Fig. 4.164. The spectrofluorimetric micro sensor; a) the chip, b) detail of construction, c) light beam illuminating about 20 nL of the analyte, d) time-dependent output signal for 0.3 and 0.6 mM of fluorescein solution in DI water, e) dynamic output signals for 0.1 mM fluorescein solution in DI water and varying dose.

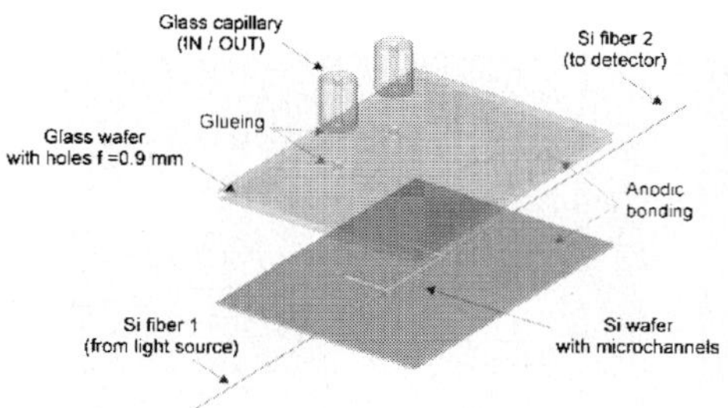

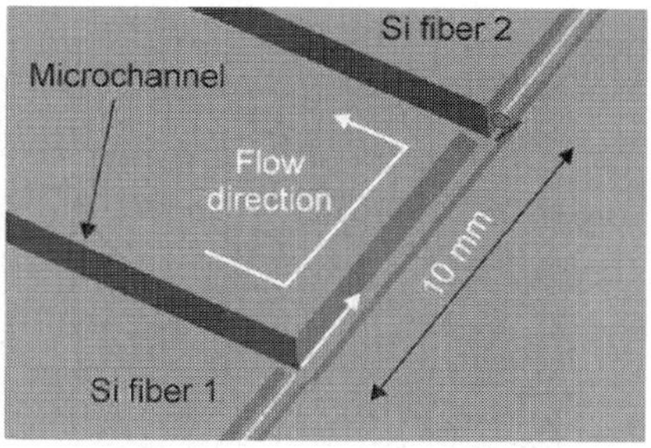

Fig. 4.165. Spectrophotometric microsensor – schematic view.

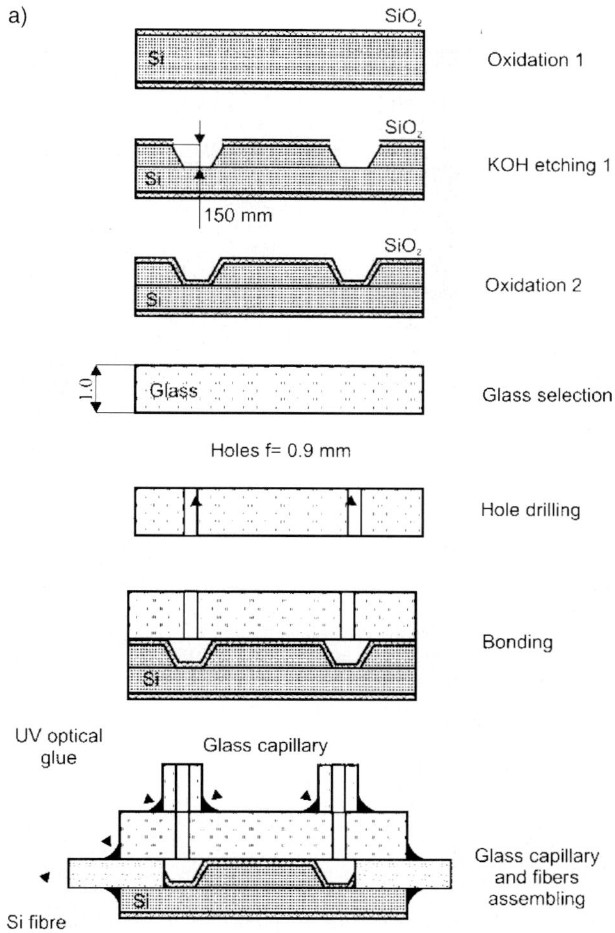

Fig. 4.166. Spectrophotometric sensor continued: a) step-by-step fabrication process of spectrophotometric sensor, b) detail of construction, c) assembled device, d) light beam inside the absorption cell.

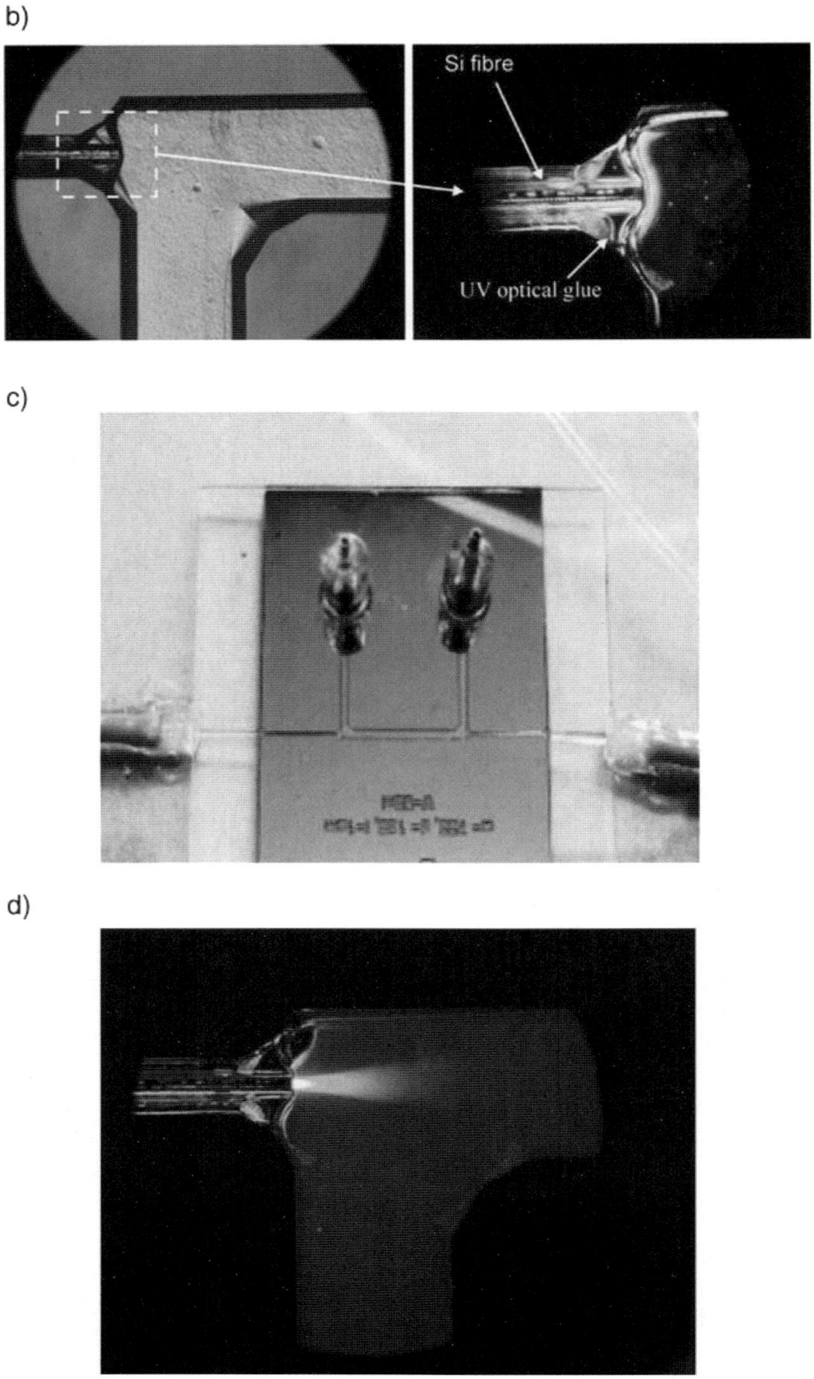

Fig. 4.166. *Continued.*

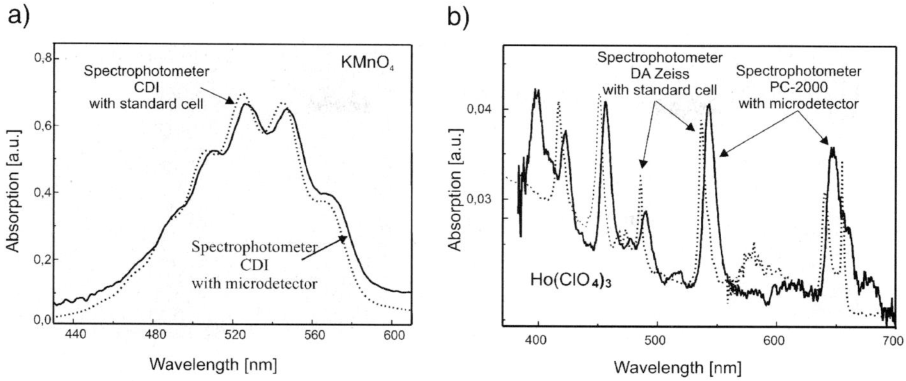

Fig. 4.167. Characterization of the microspectrophotometric sensor; a) test spectrogram of a $KMnO_4$ solution in water obtained by use of CDI spectrophotometer equipped with the standard, classical cuvette (cell) and microsensor, b) comparison of spectrograms obtained by use of DA Zeiss spectrometer with standard cell and PC-2000 spectrometer with microsensor.

a method cannot be applied in the fabrication procedure of a microscale device, made of silicon and glass.

The chip-scale atomic clock (CSAC) was designed and built in 2004 at NIST by Kitching and co-workers [286] under the DARPA project. The device work is based on the coherent population trapping (CPT) phenomenon. In the CSAC the light of the VCSEL laser, modulated at a frequency equal to one-half of the ground-state hyperfine splitting of the cesium atoms, is passed through the glass–silicon–glass cell filled with cesium vapor with a small addition of a buffer gas – nitrogen. The intensity of trespassing light decreases for f_r, which is observed as a small (contrast only 0.91%) signal, measurable by PIN diode.

The fabrication process of the glass–silicon cell of CSAC utilizes anodic bonding of silicon to glass (Fig. 4.168) The rice-grain-size cell consists of a square, 1 cm × 1 cm silicon chip with central via-hole DRIE etched in a 1 mm-thick double-side polished wafer. Next, the chip is anodically bonded (300°C, 1000 V) to a similarly sized, 1 mm-thick Corning 7740 cover 1. Following this, cesium chlorine is added to a solution of 15% barium azide (BaN_6) and water; a drop of these chemicals is introduced into the cavity formed by bonded glass and silicon and baked in order to evaporate water. Then the chip is placed inside the vacuum chamber of the bonding machine on a hot plate, the chamber is evacuated to about 10^{-5} Pa, and the chip is annealed at 120°C. In such conditions barium azide decomposed to solid-state barium and nitrogen, which is pumped away. In the next step, the vacuum chamber is filled with nitrogen under the desired pressure (about 20 kPa absolute); Corning 7740 glass cover 2 is bonded (slow temperature increase from 120°C to 200°C, 1 kV). At around

* http://www.boulder.nist.gov/timefreq/ofm/smallclock/Cellfab.htm.

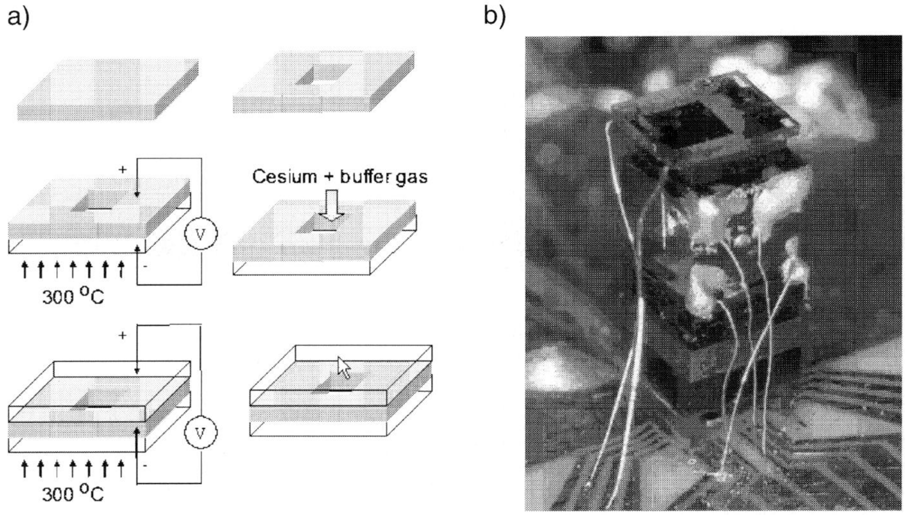

Fig. 4.168. Cesium cell; a) fabrication process flow-chart, b) the chip*.

200°C the barium reacts with cesium chloride according to the reaction:

$$Ba + CsCl = BaCl + Cs$$

4.4.4.6.2. Microlens and on-chip confocal microscope

The new concept of an on-chip integrated optical scanning confocal microscope was proposed by Gorecki [287]. The microscope is designed as a so-called "smart pixel" solution. The individual pixel is configured from the vertical cavity surface emitted laser (VCSEL) – flip-chip bonded to the microactuator moving up and down the integrated microlens, "flying" directly above the specimen. The laser light is focused by the microlens on the 3-D specimen. The light reflected from the specimen involves modulation of the laser light. The use of optical feedback of the laser cavity as an active detection system, integrated to the laser chip, simplifies microscope design because the light source and detector are unified parts of the VCSEL itself. The microscope can be fabricated in the form of a single device or as an array-type device, so called multi-probe architecture. The innovation – over other optical microscope functions – offers new capabilities of nanoscale measurements. In particular, such integrated architecture permits the performance of microscopy tasks in a lab-on-chip approach, improving the effective-space requirement and the deterioration of signal strength. The major challenge here is realization of the single miniaturized confocal microscope scheme or an array of the microscopes "on the top" of the fluidic microsystem (Fig. 4.169). The crucial part of the confocal microscope is a microlens. The lens should be made from glass, thus ensuring easy integration to a microactuator, by use of the deep micromachining technique.

There are many types of microlens and its arrays. Microlenses are made of

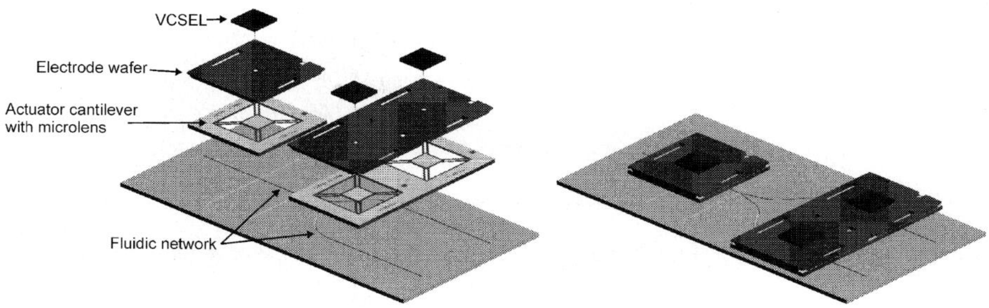

Fig. 4.169. The smart pixel confocal microscope integrated to fluidic chip – a vision.

plastic by varyous fabrication methods; polyimide small drops temperature curing, patterned SU-8 resin re-flow, hot embossing from PMMA foil using a mechanically machined matrix. Among others, one of the most promising possible methods of glass microlens fabrication utilizes deep mask-less etching of silicon matrix [288] followed by vacuum anodic bonding (450°C, 1200 V, 30 min) of 1 mm-thick Borofloat 33 glass plate. Next, the glass is re-flowed by thermal annealing, near the flow point (720°C), cooled down to ambient temperature and thinned by mechanical polishing. Following this, the silicon is selectively removed by DRIE etch, and the lens, suspended on a thin glass membrane, is completed (Fig. 4.170) [289].

Vertical actuation of the "smart pixel" for 3D imaging needs a specially developed actuator carrying a glass microlens in a central part. A moving silicon membrane (Fig. 4.171) can meet these requirements. The use of electrical or pneumatic actuation seems to be very natural here. The fabrication of the pneumatically operated actuator includes: fabrication of the glass cover with gas channels, deep wet micromachining of the body of the microscope with silicon bossed membrane, forming of the glass microlens at the membrane by mask-less etch/anodic bonding/re-flow and polishing as shown in Fig. 4.170, DRIE etch of central via-hole and releasing of the microlens, patterning of a thin glass layer for sealing of the glass cover and the body of the actuator, anodic bonding of cover and the body (250°C, 200 V), saw dicing of wafer in single chips, packaging of VCSEL laser, and packaging of the microscope to a ceramic or metallic case.

The boss of a membrane protects the microlens and stiffens the membrane while moving. The on-chip integrated microscopes are developed intensively, their application will cover many fields of bioscience (cell evaluation, 3-D transdermal anticancer scanning), chemistry and precise mechanics [336].

4.4.4.6.3. Array of SiC MOLD tips

The domain of vacuum microelectronics covers the research, designing and fabrication of vacuum devices with micrometer dimensions, in which free, ballistic movement of electrons occurs [290]. Devices for vacuum microelectronics

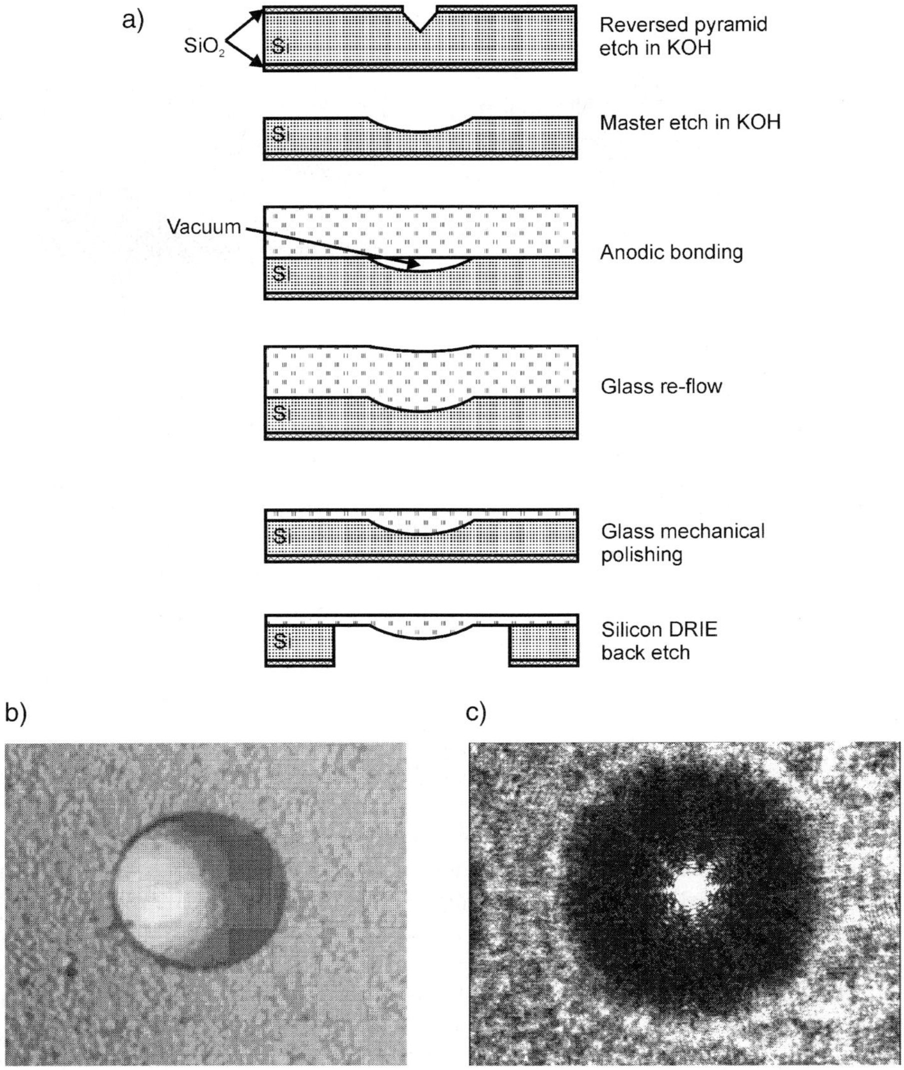

Fig. 4.170. Glass microlens: a) fabrication flow chart, b) the lens, c) true picture of a spot of the focused laser light beam.

are produced using the techniques of three-dimensional engineering, on the basis of microelectronic and micromechanical procedures [291]. Most often the devices are fabricated in the form of field emission arrays (FEAs) with submicrometer dimensions, which have already found many interesting applications, among others in microlamps (triodes, pentodes), logic devices with very high working frequency (300 GHz–3 THz), various micromechanical sensors and field emission displays (FEDs) that arouse the biggest interest on the market [292].

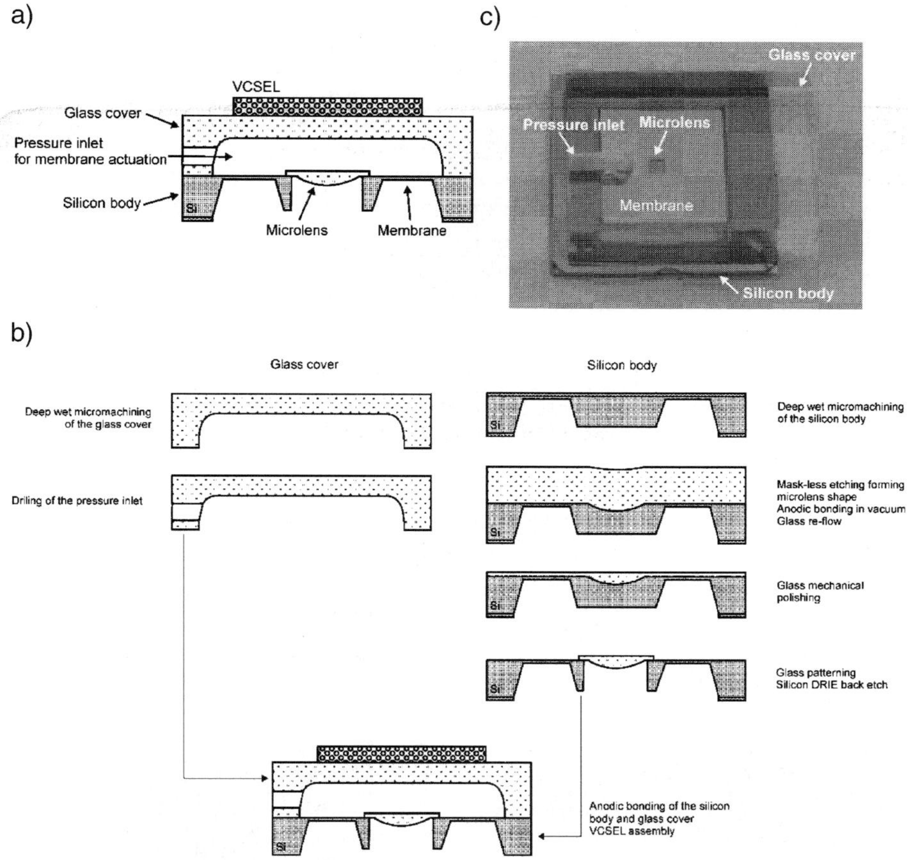

Fig. 4.171. The actuator of the confocal microscope chip with a glass lens: a) cross-section view, b) fabrication process, c) the chip.

Field emitters array

One of the most unique methods of fabrication of FEAs is the MOLD technique developed by Nakamoto [293, 294] and currently being intensively studied. The first step in the MOLD method consists in anisotropic etching in a (100) oriented silicon substrate of reversed pyramids – vials with walls limited by (111) planes. Summits of pyramids are sharpened by thermal wet oxidation (Grove-Deal's effect) and the micromachined substrate is covered with the type of material (W, Mo, α-diamond, TiN, GaN, SiC*), which is selected in order to attain high field emission and resistance to degradation processes (chemical

---

* SiC/MOLD technology was developed by dr inż. Anna Górecka-Drzazga from the Faculty of Microsystem Electronics and Photonics of Wrocław University of Technology with the collaboration of the author of this book.

stability, resistance to reverse ion bombardment, etc.). Next, the prepared substrate is anodically bonded to a Pyrex-like glass plate from the micromachined side and silicon is etched away in KOH. Etching stops on the oxide layer.

Silicon carbide (SiC) is characterized by very good emission properties and high chemical stability. It is very hard and mechanically resistant. Although the material is technologically difficult, it allows the fabrication of emitters with very advantageous properties. The application of the FEAs of SiC in switches of large power for electrical power engineering, subminiature X-ray sources, V-MOSFET (vacuum metal-oxide-semiconductor field-effect transistors) and logic circuits with extremely high switching rates arouses great expectations.

The process of fabrication of the MOLD SiC array [295, 296] proceeds as shown in Fig. 4.172. An array of reversed pyramids is etched in 10M KOH at 80°C, etch self-stop is utilized. Sharpening is performed at 1050°C in steam. A thin SiC layer is magnetron sputtered, silicon substrate is anodically bonded (450°C, 1.5 kV, 15 minutes) through the SiC layer to the 1 mm-thick Borofloat

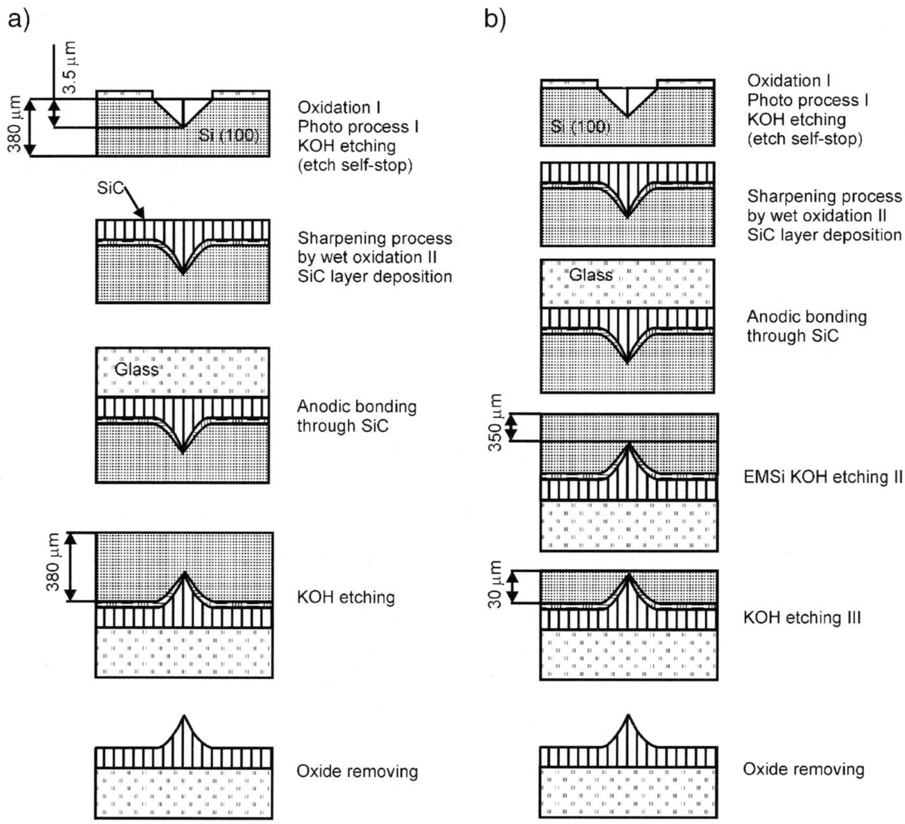

Fig. 4.172. MOLD-type FEAs fabrication process flow-chart, a) standard etching in KOH, b) EMSi method, a) [296], b) [297].

33 substrate glass substrate. Silicon is completely etched away in a 10M KOH solution (the EMSi process is particularly suitable for this procedure [297]) and silicon dioxide is also removed; an array of the sharp SiC field emitters on the glass substrate is obtained (Fig. 4.173). The array of gated SiC emitters (G-FEA) is formed by covering the emitters with an insulating thin-film layer followed by vacuum deposition of Ti–W–Au thin-film layers and photolithographic patterning of a Ti–W–Au and insulating layer, finally forming the G-FEA matrix (Fig. 4.174).

The process described above has been applied to produce for the first time an electron gun (cold cathode) with field emission, containing SiC microtips with the following dimensions: base 4 μm × 4 μm, height 2.8 μm, tip radius < 10 nm. Efficient and stable field emission from the microtip array has been obtained (Fig. 4.174).

The modified process of MOLD type G-FEA made of SiC was described in paper [298]. Gate is made of heavily boron-doped p^+ silicon, which is KOH etch resistant. The fabrication procedure starts from forming of p^+ doped regions, followed by reversed pyramid etching and sharpening by wet thermal oxidation. Next a SiC layer is deposited, the silicon wafer is anodically bonded to silicon and superfluous silicon is etched away in 7M KOH at 80°C. Etching stops at heavily doped areas (Fig. 4.175).

Volcano-like tips

The array of MOLD-type SiC tips with a nano-sized vial hole positioned precisely on their summits, with a shape similar to a volcano crater, is made by proper etching and oxidization of a silicon substrate and utilizing of the MOLD technique described above (Fig. 4.176) [299].

The oxidized silicon substrate is photo-patterned and a 15 μm-thick membrane is back-side etched in 10M KOH at 80°C. The micromachined wafer is oxidized again, photo-patterned from the front side and etched again in 10M

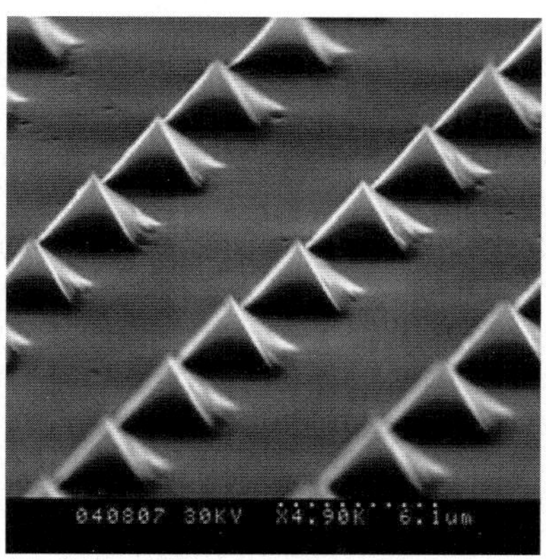

Fig. 4.173. MOLD-type FEA of SiC. (SEM pictures made by M.Sc. Stanisław Łasisz from the Faculty of Microsystem Electronics and Photonics of the Wrocław University of Technology).

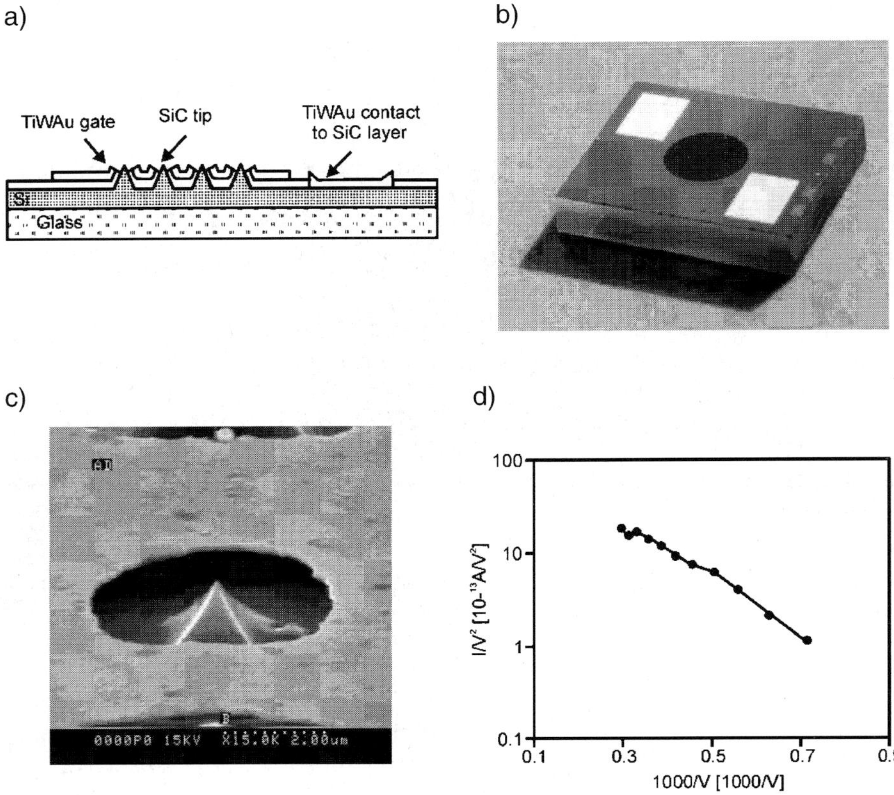

Fig. 4.174. MOLD-type G-FEA SiC cold cathode; a) cross-section of the device, b) the chip, black area ($\phi = 2$ mm) contains about 38 000 emitters, c) single gated emitting SiC tip, d) the Fowler-Nordheim's characteristic of emission. (Photo 4.174b was taken by dr B. Jankowski from the Faculty of Microsystem Electronics and Photonics of Wrocław University of Technology).

KOH, at 80°C, for 10 minutes, to form reversed pyramids the depth of which corresponds only to the patterns in the oxide layer etch-window. Following this, double-side micromachined wafer is wet oxidized at 1050°C for a period of time sufficient to obtain the controlled perforation of the membrane. For a typical square etch window 5 μm width, the ID of holes lies in the range of parts of a micrometer. Next, the back-side of the wafer is covered with the DC-pulsed magnetron-sputtered layers of SiC, each about 500 nm thick. The number of layers is adjusted to the dimensions of holes. Deposited material tends to build a regular 3-D pattern (construction) being a negative replica of the pyramid. Thus, the ID of holes becomes dependent on the deposited layer thickness, and usually six layers are sufficient to form the nano-sized holes. After deposition of SiC, the micromachined silicon wafer is front-side anodically bonded to the 1 mm-thick Borofloat 33 glass substrate, and the silicon is etched in 10M KOH at 80°C. This unique method of fabrication allows the manufacture

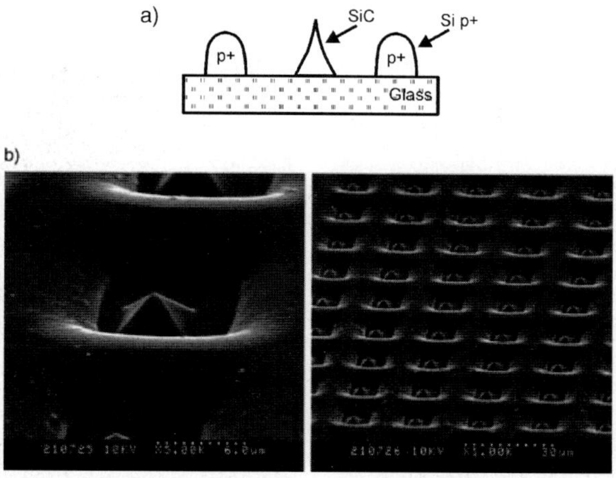

Fig. 4.175. MOLD-type G-FEA SiC with p^+ gate: a) cross-section of a single tip, b) SEM pictures of fabricated tips.

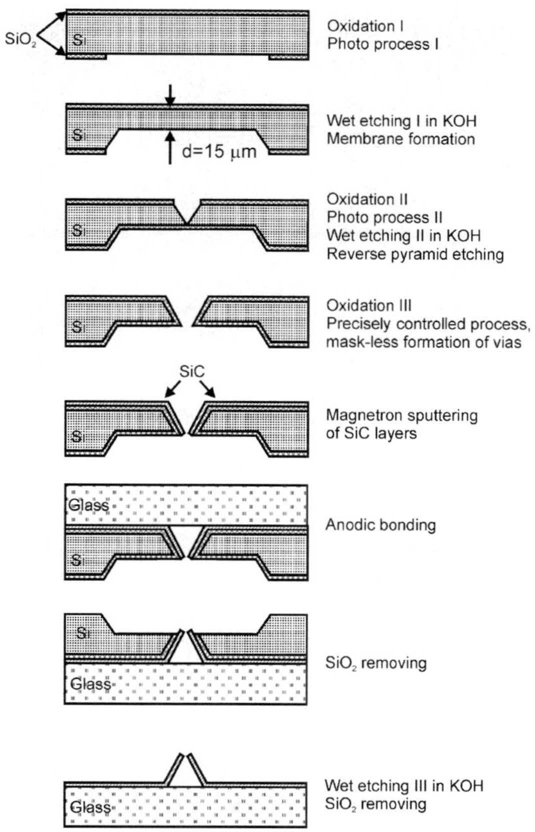

Fig. 4.176. SiC MOLD Volcano array of tips fabrication flow-chart.

Fig. 4.177. SiC MOLD-type volcano tips; a) array of pyramids, b) single volcano tip, c) white light passes through the matrix of nano-holes.

of arrays of almost identical SiC pyramids with – as shown in Fig. 4.177 – 80 nm by 100 nm holes, precisely at the summits of pyramids, without using submicron lithography. Such a construction can be utilized as an array of nozzles or a nano-size *camera-obscura*.

LITERATURE

[1] S. Schulze, Silicon bonding in microsystem technology. NEXUS Workshop IMSAS; 1995.

[2] S. Farrens, Low temperature wafer bonding. Proceedings of Electrochem. Soc. DV 97–36; and original materials of dr Shari Farrens from the University of California–Davis, Electrical and Computer Eng. Dept., Davis, Ca. 956/6, USA.

[3] S. Bengston, Semiconductor wafer bonding: A review of interfacial properties and applications. *Journal of Electronic Materials*, 21(8), 1992, 841–862.
[4] W. Kern, D.A. Puotinen, *RCA Review*, 31(187), 1970.
[5] Ch. Harendt, H.G. Graf, E. Penteker, B.H. Flinger, Wafer bonding: Investigation and in situ observation of the bond process. *Sensors and Actuators A*, 21–23, 1990, 927–930.
[6] M. Schimbo, K. Furukawa, K. Fukuda, K. Tanzawa, Silicon-to-silicon direct bonding method. *J. Appl. Phys.*, 60(8), 1986.
[7] J. W. Berenschot, J. Gardeniers, T. Lammerink, M. Elwenspoek, New applications of r.f.-sputtered glass film as protection and bonding layers in silicon micromachining. *Sensors and Actuators A*, 41–42, 1994, 338–343.
[8] R. Lagtenberg, S. Bouwstra, M. Elwenspoek, Low temperature glass bonding for sensors applications using boron oxide films. *J. Micromech. Microeng.*, 1, 1991 157–160.
[9] H. J. Ouenzer, W. Benecke, Low-temperature silicon wafer bonding. *Sensors and Actuators A*, 32, 1992, 340–344.
[10] B. Müller, A. Stoeffel, Tensile strength characterisation of low-temperature fusion bonded silicon wafers. *J. Micromech. Microeng.*, 1, 1991, 161–166.
[11] L. Tenerz, B. Hök, Silicon microcavities fabricated with a new technique: *Electronics Letters*, 22(11), 1986, 615–616.
[12] S. Sanchez, G. Gui, M. Elwenspoek, Spontaneous direct bonding of thick silicon nitride. *J. Micromech. Microeng.*, 7, 1997, 11–113.
[13] R. C. Frye, J.E. Griffith, Y.H. Wong, A field assisted bonding process for silicon dielectric isolation. *J. Electrochem. Soc.*, 133(8), 1986, 1673–1677.
[14] W. D. Maszara, G. Goetz, A. Caviglia, J.B. McKitterick, Bonding of silicon wafers for silicon-on insulator. *J. Appl. Phys.*, 64, 1988, 4943–4951.
[15] X. L. Xu, Q.Y. Tong, Novel two steps SDB technology for high performance thin film SOI/MOSFET applications. *Electron. Lett.*, 25, 1989, 394–395.
[16] M. Reiche, K. Gutjahr, D. Stolze, D. Burczyk, M. Petrold, The effect of plasma pretreatment on the Si–Si bonding behaviour. *Electroch. Soc. Proceed.*, PV 97–36, 473–444.
[17] K. Hermasson, U. Lindberg, G. Palmskog, Wetting properties of silicon surfaces. *Electrochem. Soc. Fall Mtg, Proceed.*, PV 91–2, 1991.
[18] N. W. Ashcroft, N.D. Mermin, *Solid State Physics*, Holt, Rixehart, Winston (eds.). New York–Sydney, 1976.
[19] W. Kissinger, G. Kissinger, Microstructures for perfect wafer bonding in different temperature ranges. *Electrochem. Soc. Fall Mtg, Proceed.*, PV 91–2, 1991.
[20] U. Bäckhund, K. Hermasson, L. Smith, U. Lindberg, *Electrochem. Soc. Fall Mtg, Proceed. PV*, 91–2, 1991.
[21] J. Dziuban, I. Halas, M. Depka, Washing and activation receipts – non published materials, in Polish. Reports of the R&D Programme PBZ 2705 on Silicon Micromechanical Sensors, financed by The Committee of Scientific Research, Poland, 1997.
[22] W.P. Maszara, Silicon-on isolator by wafer bonding – a review. *J. Electrochem. Soc.*, 138(1), 1991, 341–347.
[23] P.W. Barth, Silicon fusion bonding for fabrication of sensors, actuators and microstructures. *Sensors and Actuators A*, 21–23, 1990, 919–926.
[24] C. Harendt, C.E. Hundt, W. Appel, H.G. Grat, B. Höfflinger, E. Petenker, Silicon on insulator material by wafer bonding. *J. Electron. Meter.*, 20, 1991, 267–277.
[25] G. Kissinger, W. Kissinger, Hydrophilicity of silicon wafers for direct bonding. *Phys. St. Solidi A*, 123, 1991, 185–192.
[26] S. Bengtsson, O. Engström, IC Process compatible preparation of silicon interfaces using the silicon-to-silicon direct bonding method. Proc. ESSDERC 89, A. Henberger, H. Ryssel, P. Lange (eds.). Berlin, Springer 1985.
[27] S. Bengtsson, O. Engström, Low temperature preparation of silicon/silicon interfaces by the silicon to silicon direct bonding method. *J. Electrochem. Soc.*, 137, 1990, 2297–2303.
[28] T. Abeoc. M. Nakano, T. Itoh, Silicon wafer bonding technology for SOI structures. Proc.

4th Int. Symp. On Silicon-on-Insulator Techn. and Dev., Electrochem. Soc. Proc., 90–6, 1990, 61–71.
[29] T. Abe, T. Takei, A. Uchiyama, K. Yoshizawa, Y. Nakamoto, Silicon wafer bonding mechanism for silicon-on-insulator structures (Part III). *Jpn. J. Appl. Phys.*, 29, 1990, L2311–2314.
[30] M. Grundner, H. Jacob, Investigations on hydrophilic and hydrophobic silicon (100) wafer surfaces by X-ray photoelectron and high resolution electron energy loss spectroscopy. *Appl. Phys.*, A39, 1986, 73–82.
[31] K. Mitani, Silicon wafer bonding, an overview. *Electrochem. Soc. Proc.*, PV9736, 1998, 1–12.
[32] G. Gui, M. Elwenspoek, J.G.E. Gardeniers, P.V. Lambeck, Present and future role of CMP in wafer bonding. *Electrochem. Soc. Proc.*, PV9736, 1998, 114–123.
[33] C. Gui, H. Albers, J.G.E. Gardeniers, M. Elwenspoek, P.V. Lambeck, Fusion bonding of rough surfaces with polishing technique for silicon micromachining. *Microsyst. Technol.*, 1997, 122–128.
[34] D.L. Gay, P.T. Baine, B.M. Amstrong, H.S. Gamble, SOI production using polish stops formed by trench before bond technique. *Electrochem. Soc. Proc.*, PV9736, 1998, 536–543.
[35] J.B. Lasky, Wafer bonding for silicon-on insulator technologies. *Appl. Phys. Lett.*, 1(48), 1986, 78–80.
[36] R. Stengl, K.Y. Ahn, U. Gösele, Bubble-free silicon wafer bonding in a non-clean room environment. *Jpn. J. Appl. Phys.*, 27, 1998, L2364–2366.
[37] S.N. Farrens, C.E. Hunt, B.E. Roberts, J.K. Smith. A kinetics study on the strength of direct bonded wafers. *J. Electrochem. Soc.*, 141(11), 1994, 3225–3230.
[38] R. Stengl, T. Tan, U. Gösele, A model for the silicon wafer bonding. *Jpn. J. Appl. Phys.*, 28, 1989, 1735–1741.
[39] U.K. Weldon, V.E. Marisco, Y.J. Chabal, A. Agerwalda, D.J. Eglesham, Y. Candano, S.B. Christman, E.E. Chaban, A mechanistic study of silicon wafer bonding. *Electrochem. Soc. Proceed.*, PV 9736, 1998, 229–248.
[40] J. Harosma, G.A.C.M. Spierings, U.K.P. Bierman, Silicon-on-insulator wafer bonding – wafer thinning technological evaluation. *Jpn. J. Appl. Phys.*, 28, 1989 1426–1443.
[41] M. Horuchi, S. Aoki, A mechanism of silicon wafer bonding. *Electrochem. Soc. Fall Mtg.*, PV 91–2, 1991.
[42] K. Gutjahr, T. Martini, U. Gösele, Concepts of wafer debonding. *Electrochem. Soc. Proceed.*, PV 9736, 1998, 72–78.
[43] S. Mack, H. Baumann, U. Gösele, H. Werner, R. Slögly, Analysis of bonding relates gas enclosure in micromachined cavities sealed by silicon wafer bonding. *J. Electrochem. Soc.*, 144, 1997, 1106–1111.
[44] C. Maleville, O. Rayssac, H. Moriceau, B. Biasse, L. Baroux, B. Aspar, M. Bruel, Detailed characteristic of wafer mechanisms. *Electrochem. Soc. Proceed.*, PV 9736, 1998, 46–55.
[45] V. Lehman, U. Gösele, K. Mitani, Contamination protection of semiconductor surfaces by wafer bonding. *Solid St. Techn.*, 33, 1990, 91–92.
[46] F. Secco d'Aragona, T. Iwamoto, Hern-Der-Chiou, A. Mirza, A study of silicon direct wafers bonding for MEMS applications. *Electrochem. Soc. Proceed.*, PV 9736, 1998, 127–137.
[47] R.D. Black, E.L. Hall, N. Lewis, R.S. Gilmore, S.D. Arthur, R.D. Lillguist. *J. Electrochem. Soc.*, 107, 1988, 495–497.
[48] C. Parkes, E. Murray, H.S. Gamble, B.M. Amstrong, S.J.N. Mitchell, G.A. Amstrong, Characterisation of electronic devices employing silicon bonding technology. *Electrochem. Soc. Fall Mtg., Proceed.*, PV 91–2, 1991.
[49] R. Wilson, H.S. Gamble, S.J.N. Mitchell, Improvement of silicon power device characteristic using bonding technology. *Electrochem. Soc. Fall Mtg., Proceed.*, PV 91–2, 1991.
[50] T. Abe, J.H. Matlock, Wafer bonding technique for silicon-on insulator technology. *Solid St. Techn.*, 11(33), 1990, 39–40.
[51] K. Mitani, V. Lehmann, R. Stengl, D. Feijoo, U.M. Gösele, H.Z. Massond, Causes and prevention of temperature dependemt bubbles in silicon wafer bonding. *Jpn. J. Appl. Phys.*, 30, 1991, 615–622.

[52] U. Lehmann, U. Gösele, K. Mitani, Contamination protection of semiconductor. *Solid. St. Techn.*, 4(33), 1990, 91–92.

[53] C. Harendt, H.G. Graf, E. Penteker, B. Höfflinger, Wafer bonding: investigation and in situ observation of the bond process. *Sensors and Actuators A*, 21–23, 1990, 927–930.

[54] K. Furukawa, M. Shimbo, K. Fuknala, K. Tanzawa, Lattin configuration and electrical properties of the interface of direct bonded silicon. Ext. Abstr. 18th Int. Conf. on Solid St. Dev. and Mater., Tokyo, 1986 (The Jpn. Soc. of Appl. Phys. Tokyo), 533–536.

[55] H. Takagaki, R. Maeda, T.R. Chung, T. Suga, Low-temperature direct bonding of silicon and silicon dioxide by the surface activation method. *Sensors and Actuators A*, 70, 1998, 164–170.

[56] K.Y. Ahn, R. Stengl, T.Y. Tan, U. Gösele, Stability of interfacial oxide layers during silicon wafer bonding. *J. Appl. Phys.*, 65, 1989, 561–563.

[57] K.Y. Ahn, R. Stengl, T.Y. Tan, U. Gösele, P. Smith, Growth, shrinkage and stability of interfacial oxide layers between directly bonded silicon wafers. *J. Appl. Phys.*, A-50, 1990, 85–94.

[58] F.P. Widdershoven, J. Harisma, J.P.M. Naus, Boron contamination and antimony segregation at the interface of directly bonded silicon wafers. *J. Appl. Phys.*, 68, 1990, 6253–6258.

[59] L.J. Pakulti, P. Ling, P. Leonov, H. Kawayoshi, R. Ormond, J. Uan, Radiation response of CMOS SOI devices formed by wafer bond and etchback. *IEEE Trans. Nuclear Sci.*, 35, 1988, 1653–1656.

[60] C.A. Desmond, J.J. Olup, P. Obalgasem, J. Folta, G. Jernigan, Analysis of nitride bonding. *Electrochem. Soc. Proceed.*, PV 9736, 1988, 171–178.

[61] P. Rangsten, Ö. Vallin, K. Ljungberg, Y. Bäcklund, Quartz to quartz direct bonding. *Electrochem. Soc. Proceed.*, Abstract Meeting 2061, Vol. 97–2.

[62] T. H. Lee, Q.-Y. Tong, Y.-L. Chao, U. Gösele, Silicon and quartz by a smarter cut process. *Electrochem. Soc. Proceed.*, Abstract Meeting 2061, vol. 97–2.

[63] M. Bruel, B. Asper, C. Maleville, H. Moriceau, Unibond® SOI wafers achieved by smart-cut® process. *Electrochem. Soc. Proceed.*, Abstract Meeting 2061, vol. 97–2.

[64] *New bonding technology for SOI, Unibond®*. SOI substrates of SOI TEC S.A., 1 Place Firmin Gautier, 38000, Grenoble, France, Technical information.

[65] Q.Y. Tong, T.-H. Lee, L.-J. Huang, Y.L. Chao, W.J. Kim, R. Scholz, T.Y. Tan, U. Gösele, Design considerations for Si and SiC layer transfer by H^+ implantation. *Electrochem. Soc. Proceed.*, PV 9736, 1998, 521–528.

[66] P. Kopperschmidt, G. Kästner, D. Hesse, Recent development in semiconductor-on-sapphire wafer bonding. *Electrochem. Soc. Proceed.*, PV 97–36, 1998, 179–186.

[67] B.E. Roberts, K.D. Choquette, K.M. Geib, S.H. Kravitz, R.D. Twesten, S.N. Farreus, Wafer bonding of GaAs, InP and Si annealed without hydrogen for advanced device technologies. *Electrochem. Soc. Proceed.*, PV 97–36, 592–595.

[68] E. Jalaquier, B. Aspar, S. Pocas, J.F. Michand, A.M. Papon, M. Bruel, Transfer of thin InP films onto silicon substrate by proton implementation process. IRPM 99, Proceed. 11th Int. Conference on InP and Rel. Materials, Piscataway, USA, 1999, 26–27.

[69] E. Jalaquier, B. Aspar, J.F. Michand, M. Zussy, A.M. Papon, M. Bruel, Transfer of 3″ GaAs film on silicon substrate by proton implementation process. *El. Lett.*, 35(6), 1999, 477–478.

[70] K. Petersen, P.W. Barth, J. Poyolock, J. Brown, J. Mallon, J. Bryzek, Silicon fusion bonding for pressure sensors. *IEEE Solid St. Sensors and Actuators Workshop*, 1988, 114–147.

[71] K. Petersen, J. Brown, T. Veurmellen, P. Barth, J. Mallon Jr., J. Bryzek, Ultrastable high temperature pressure sensors using silicon fusion bonding. *Sensors and Actuators A*, 21–23, 1990, 96–101.

[72] Product information, NovaSensor/Lucas Nova Sensor, 1055, Mission Court, Fremont, CA 94539, USA.

[73] Y. Wang, X. Zheng, L. Liu, Z. Li, A novel structure of pressure sensors. *IEEE Tran. El. Dev.*, ED-38(8), 1991, 1797–1801.

[74] L. Parameswaran, V.M. McNeil, M.A. Huff, M.A. Schmidt, Sealed cavity microstructure

using wafer bonding technology. Techn. Digest, 7th Int. Conf. Solid St. Sensor and Actuator, Transducers 93, Yokohama, Japan, 7–10 June, 1993, 274–277.

[75] C.H. Hsu, M.A. Schmidt, Micromachined structures fabricated using a wafer-sealed cavity process. Techn. Dig. IEEE Solid St. Sensors and Actuators Workshop, 1994, 151–155.

[76] L. Parameswaran, A. Mirza, W.K. Chan, M.A. Schmidt, Silicon pressure sensors using a wafer bonded sealed cavity process. Tech. Dig. The 8th Int. Conf. on Solid St. Sensors and Actuators Transducers 95, Eurosensors IX, Stockholm, Sweden, 25–29 June, 1995, 582–585.

[77] J.M. Noworolski, E. Klaasen, J. Logan, K. Petersen, N. Maluf, Fabrication of SOI wafers with buried cavities using silicon fusion bonding and electrochemical etchback. Tech. Dig. the 8th Int. Conf. Solid St. Sensors and Actuators Transducers 95, Eurosensors IX Stockholm, Sweden, 25–29 June, 1995, 71–74.

[78] M.A. Samber, T.P.L. Ho-Tran, Evaluation of the fabrication of pressure sensors using bulk micromachining before IC processing. *Sensors and Actuators A*, 46–47, 1995, 147–150.

[79] J. Mandla, D. Lefort, A. Migeon, A new micromachining silicon high-accuracy pressure sensor. *Sensors and Actuators A*, 46–47, 1995, 129–132.

[80] J. Hermann, C. Bourgeois, F. Porret, B. Kloeck, Capacitive differential pressure sensor. Tech. Dig. the 8th Int. Conf. Solid St. Sensors and Actuators Transducers 95, Eurosensors IX, Stockholm, Sweden, 25–29 June, 1995, 620–623.

[81] S. Chatzandroulis, D. Goustouridis, P. Normand, D. Tsoukalas, A solid-state pressure-sensing microsystem for biomedical applications. *Sensors and Actuators A*, 31, 1997, 651–657.

[82] Y. Matsumoto, M. Iwahiri, H. Tanaka, M. Ishiala, T. Nakamura, A capacitive accelerometer using SDB-SOI structure. Tech. Dig. the 8th Int. Conf. Solid St. Sensors and Actuators Transducers 95, Eurosensors IX, Stockholm, Sweden, 25–29 June, 1995, 550–553.

[83] E. Klaasen, K. Petersen, J.M. Noworolski, J. Logan, N.I. Maluf, J. Brown, C. Storment, W. Culley, G.T.A. Kovacs, Silicon fusion bonding and deep reactive ion etching: a new technology for microinstrumentation. Tech. Dig. the 8th Int. Conf. Solid St. Sensors and Actuators Transducers 95, Eurosensors IX, Sweden, 25–29 June, 1995, 556–559.

[84] Y.T. Lee, H.D. Seo, A. Kawamura, T. Yamada, Y. Matsumoto, M. Ishida, T. Nakamura, Compensation method of offset and its temperature drift in silicon piezoresistive pressure sensor using double-bridge configuration. Tech. Dig. the 8th Int. Conf. Solid St. Sensors and Actuators Transducers 95, Eurosensors IX, Stockholm, Sweden, 25–29 June, 1995, 570–573.

[85] N. Schweinger, J. Burgold, A. Achermann, Piezoelectric micropumps based on a new deposition technology for ZnO films. Microsystem Technologies, 4th Int. Conf. on Micro-Opto-Mech. System and Components, Potsdam, 1994, 1035–1044.

[86] H. Jerman, Electrically activated micromachined diaphragm valves. Techn. Dig. IEEE Solid St. Sensors and Actuators Workshop, 1990, 65–69.

[87] J. Drake, H. Jerman, A precision flow restrictor for medical infusion therapy. Tech. Dig. the 8th Int. Conf. on Solid St. Sensors and Actuators Transducers 95, Eurosensors IX, Stockholm, Sweden, 25–29 June, 1995, 373–376.

[88] H. Jerman, Electrically-activated normally-closed diaphragm valves. Tech. Dig., 6th Int. Conf. Solid St. Sensors and Actuators, 1991, 1045–1048.

[89] R. Zengerle, A. Richter, H. Sandmeier, A micromembrane pump with electrostatic actuation. Proceed. IEEE Micro Electro Mechanical Systems Workshop, 1992, Travemünde, 19–24.

[90] R. Pinker, D. Cammack, B. Khan, Novel fabrication of miniature discharge lamps by wafer bonding. *Electrochem. Soc. Proceed.*, PV 97–36, 193–199.

[91] Y. Bäckhund, K. Hjort, S. Johansson, L. Stenmark, Micro Sculpturing: Somewhat new materials and micromachining methods to meet new applications. Proceed. Eurosensors XIII, 13th European Conf. On Solid-St. Transducers, den Haag, Holland, September 12–15, 1999, 527–530.

[92] London, A. Epstain, A. Ayon, S.M. Spearing, T. Harrsion, J. Kerrebrock, A microfabricated high pressure bipropellant rocket engine. Book of Abstracts Eurosensors XIV, 14th European Conf. On Solid-St. Transducers, Kopenhagen, Denmark, August 28–30, 2000.

[93] P.-F. Indermühle, N.F. de Rooij, Integration of a large tip with high aspect ratio on an X-Y

microstage for AFM imagining. Tech. Dig. 8th Int. Conf. Solid St. Sensors and Actuators, 1996, 636–639.

[94] P.F. Indermühle, V.P. Jachlin, J. Brugger, C. Linder, N.F. de Rooij, N. Binggeli, AFM imaging with X-Y micropositioner with integrated tip. *Sensors and Actuators A*, 46–47, 1994, 562–565.

[95] J. Welham, J. Greenwood, M. Bartoli, A lateral resonant pressure sensor fabricated via fusion bonding, wafer thinning and reactive-ion-etching. Eurosensors XII, Proceed. 12th Europ. Conf. on Solid St. Transducers Southampton, 13–16 September 1998, N.M. White (ed.). IOP Publishing 1998, Vol. 1, 353–356.

[96] J. Kräenert, C. Deter, T. Gessner, W. Dötzel, Laser display technology. Proceed. 11th IEEE Workshop on Microelectromechanical Systems, January 25–29, Heidelberg, 1998, 110–115.

[97] P. Menotti, A. Ferreira, *Les micromachines*, ed. Hermes, Paris, 1998.

[98] Fe.F. Regnier, Microfabricated liquid chromatography columns based on collocated monolith support structures. Proceed. μTAS 98 Workshop, 1998, Kluwer, Ac. Publ. ed: D.J. Harrison, A. van den Berg, 451–456.

[99] E. Kälvesten, T. Cornan, H. Huiku, K. Weckström, P. Meriläninen, G. Stemma, A silicon IR source and CO_2 chamber for CO_2 measurements. Proceed. 11th Workshop on Microelectromechanical Systems, January 25–29 Heilderberg, 1998, 69–74.

[100] Z. Lin, D.L. De Voe, Micromechanism fabrication using silicon fusion technology. *Rob. Comp. Inetgrat. Manufac.*, 17(1–2), 2001, 131–137.

[101] K. Schjølberg-Henriksen, M.M. Visser Taklo, A. Hanneborg, G.U. Jensen, Oxide charges induced by plasma activation for wafer bonding. *Sensors and Actuators A*, 102(1–2), 2002, 99–105.

[102] S. Bengtsson, P. Amirez, Room temperature wafer bonding of silicon, oxidized silicon, crystalline quartz. *J. Electr. Mat.*, 29(17), 2000, 909–915.

[103] www.suss.com

[104] O. Zucker, W. Iangerheinrich, M. Kulozik, H. Goebel, Application of oxygen plasma processing for silicon direct bonding. *Sensors and Actuators A*, 36, 1993, 227–231.

[105] W. Kissinger, G. Kissinger, Void-free silicon-wafer-bond strengthening in the 200–400°C temperature range. *Sensors and Actuators A*, 36, 1993, 1–3.

[106] D.-Y. Tong, G. Cha, R. Gafitenau, U. Gösele, Low temperature wafer direct bonding. *J. Microelectromech. System*, 3(1), 1994.

[107] M. Horiuchi, S. Aoki, Characteristics of silicon wafer bond strengthening by annealing. *J. Electrochem. Soc.*, 139, 1992, 2589–2594.

[108] Q.Y. Tang, Q. Gan, G. Hudson, G. Fountain, P. Enquist, P. Scholtz, V. Gosele, Low temperature hydrophobic bonding. *Phys. Lett.*, 83(23), 2003, 767–769.

[109] Q.Y. Tang, Q. Gan, G. Fountain, P. Enquist, Fluorine enhanced low-temperature-bonding of native-oxide covered Si wafers. *Appl. Phys. Lett.*, 85(17), 3731–3733.

[110] W.H. Ko, J.T. Sumito, G.J. Yell, *Bonding techniques for microsensors; in: Micromachining and micropackaging of transducers*, C.D. Fung (ed.). Elsevier: New York, 1985, 41–61.

[111] S.S. Deng, J. Wei, C.M. Tan, M.L. Nai, W.B. Yu, H. Xie, Low temperature silicon wafer bonding by sol-gel processing. *Int. J. Comp. Eng. Sc.*, 4(3), 2003, 655–658.

[112] S. Schulze, I. Albrecht, M. Noffke, W. Benecke, Low temperature wafer bonding by sol-gel processing. Proceed. 188 Spring Electrochem. Soc. Meet., Chicago, 1995.

[113] W. Kern, G.L. Schnable, Chemically vapour deposited boro-phosphoro-silicate glasses for silicon device applications. *RCA Review*, 43, 1993, 423–447.

[114] L.A. Fields, R.S. Muller, Fusing silicon wafers with low melting temperature glass. *Sensors and Actuators A*, 21–23, 1990, 919–926.

[115] S. Li, C.B. Freindhof, R.M. Young, R. Ghodssi, Fabrication of micronozzles using low temperature wafe-level bonding with SU 8. *J. Microeng. Micromechanics*, 13, 2003, 732–738.

[116] F. Niklaus, Low temperature BCB bonding. *Sensors and Actuators A*, 92(1–3), 2001, 235–24.

[117] K.W. Oh, A. Han, S. Bhansali, Ch. A. Ahn, A low temperatures bonding technique using spin-on fluorocarbons polymers to assemble microsystems. *J. Micromech. Microeng.*, 2, 2003, 187–191.

[118] H. Nakanishi, T. Nishimoto, R. Nokamura, A. Yotsumoto, S. Shoji, Studies on SiO_2–SiO_2 bonding with hydrofluoric acid – room temperature and low stress bonding technique for MEMS. Proc. 11th IEEE MEMS Workshop, 1998, 609–614.

[119] H. Nakanishi, T. Niskimoto, N. Nakamura, S. Nagamachi, A. Arai, Y. Iwata, Y. Mito, Fabrication of electrophoresis devices on quartz. Proceed. 10th IEEE MEMS Workshop, 1997, 299–304.

[120] H. Becker, K. Lowachi, A. Manz, Planar quartz chip with submicron channels for two-dimensional capillary electrophoresis applications. *J. Micromech. Microeng.*, 8(1), 1999, 24–28.

[121] D.Y. Sim, T. Kurabayashi, M. Esashi, Bakable silicon pneumatic valve. Tech. Dig. 8th Int. Conf. Solid St. Sensors and Actuators Transducers '95, Eurosensors IX, Stockholm, Sweden, 25–29 June, 1995, 280–283.

[122] D.F. Wolfenbuttel, Low-temperature intermediate Au–Si wafer bonding: electric or silicide bond. *Sensors and Actuators A*, 62 1997, 680–686.

[123] A.-L. Tiensuu, M. Bexel, J.-A. Schweiz, L. Smith, S. Johannson, Assembling three-dimensional microstructures using Au–Si eutectic bonding. *Sensors and Actuators A*, 45, 1994, 227–236.

[124] U.M. Mescheder, M. Alavi, K. Hiltmann, Ch. Lietzau, Ch. Nachtigall and H. Sandmaier, Local laser bonding for low temperature budget. *Sensors and Acuators A*, 97–98, 2002, 422–427.

[125] M. Waelti, N. Schnecberger, O. Paul, H. Baltes, Low temperature packaging of CMOS infrared microsystems by Si–Al–Au bonding. *Electrochem. Soc. Proceed.*, PV 97–36, 1998, 147–145.

[126] P.M. Zawracky, B. Vu, Patterned eutectic bonding with Al/Ge thin films for MEMS. Proceed. SPIE, vol. 2639, 1995, 46–52.

[127] M.S. Ismail, R.W. Bower, Platinum silicide fusion bonding. *Electron. Lett.*, L7, 1992, 1153–1155.

[128] D.J. Yao, G.C Hen, C-J. Kim, Low-temperature eutectic bonding for in-plane type micro thermoelectric cooler. 2001 ASME Int. Mech. Congress and Exposition, Nov. 11–26, 2001, NY, USA.

[129] M.A. Schmidt, Wafer to wafer bonding for microstructure formation. *Proceed. IEEE*, 68(8), 1998, 1575–1585.

[130] H.Y Wang, R.S. Foote, S.C.Jacobson, J.H Schneibel, J.M. Ramsey, Low temperature bonding for microfabrication of chemical analysis devices. *Sensors and Actuators B*, 45(3), 1997, 199–207.

[131] R. Zengerle, A. Richter, H. Sandmeier, A micromembrane pump with electrostatic actuation. MEMS92 Proceed., Travemünde 4–7 February, 1992, 19–24.

[132] J.H. Correia, M. Bartek, R.F. Wolffenbutel, Bulk-micromachined tunable Fabry-Perrot microinterferometer for the visible spectral range. Eurosensors XII, Proceed. 12th Europ. Conf. on Solid St. Transducers Southampton, 13–16 September 1998, N.M. White (ed.). IOP Publishing 1998, Vol. 1, 287–290.

[133] J.M. Ruano-Lopez, M. Aguirregabiria, Marroyo, J. Berganzo, F.J. Blanco, P. de la Fuente, E. Castano, K. Mayora, An optical microfluidic fluidic platform based on the combination of a novel SU 8 multilayer technology, waveguides and photodiodes on silicon. *MicroTAS*, 2004, 26–20 September, Malmő, Sweden.

[134] C.C. Lee, C.Y. Wang, G. Matijesievic, Au–In bonding below eutectic point or silicode bond. *IEEE Trans. Comp. Hybrids and Manufac. Techn.*, 16, 1999, 311–316.

[135] B. Lee, D.R. Ciarlo, P.A. Krulevitch, S. Lehew, J. Trevino, M.A. Northrup, A practical microgripper by fine alignment, eutectic bonding and SMA actuator. Tech. Dig. the 8th Int. Conf. on Solid St. Sensors and Actuators Transducers 95, Eurosensors IX, Stockholm, Sweden, 25–29 June, 1995, 368–369.

[136] Wallis, D. Pommerantz, Field assisted glass–metal sealing. *J. Appl. Phys.*, 40(10), 1969, 3346–3949.

[137] H. Iback, Thermal expansion of silicon and zinc oxide. *Phys. St. Solidi*, vol. 31, 625–634.

[138] G. Slack, S.F. Bartraum, Thermal expansion of some diamond like crystals. *J. Appl. Phys.*, 46, 1975, 89–98.
[139] G.K.J. White, Thermal expansion of reference materials: copper, silica and silicon. *J. Phys.*, D-6, 1973, SF 2070–2078.
[140] N. Dutta, Lattice constants and thermal expansion of silicon up to 900°C by X-Ray Method. *Phys. St. Solidi*, 2, 1962, 984–986.
[141] M.A. Norton, J.N. Berthold, S.F. Jacobs, W.A. Plummer, Precise measurements of the thermal expansion of silicon near 40°C. *J. Appl. Phys.*, 47, 1976, 1683–1685.
[142] Product specification:
SD-2 glass for anodic bonding, Hoya Japan
Borofloat Flat Glasses: Schott Glaswerke Nr 03SPQ-05/04
Alkali free and alkali low thin glasses: AF45-D263, Desag TKT
Pyrex 7740, 7070 Corning Co.
[143] Y. Hachitani, H. Sagara, Glass substrates for silicon sensors. Abstract: Tech. Dig. Int. Conf. on Solid St. Sensors and Actuators Transducers 93, Yokohama, Japan, 7–10 June, 1993.
[144] T. Rogers, J. Kowal, Selection of anodic bonding conditions and materials compatibility for silicon–glass capacitive sensors. *Sensors and Actuators A*, 46–47, 1995, 113–120.
[145] T. Rogers, Consideration of anodic bonding for capacitive type silicon/glass sensor fabrication. *J. Micromech. Microeng.*, 2, 1992, 164–166.
[146] For details of material properties of glass for anodic bonding look at: SD-2: www.hoyacandeo.co.jp; Pyrex: www.corning.com; Borofloat 33: www.schott.com/borofloat.
[147] K. Hildendorf, P. Krause, E. Obermeier, Reduction of the influence of the anodic bonding process on the behaviour of pressure sensors by using new glass substrates. Microsystem Technologies 96, Potsdam 1996, 331–336.
[148] P.R. Younger, Hermetic glass sealing by electrostatic bonding. *J. Noncrystall. Solids*, 38–39, 1980, 909–914.
[149] M. Haerz, H. Engelke, Curvature changing or flattening of anodically bonded silicon and borosilicate glass. *Sensors and Actuators A*, 55, 1996, 201–209.
[150] T.R. Anthony, Anodic bonding of imperfect surfaces. *J. Appl. Phys.*, 54(5), 1983, 2419–2428.
[151] G. Wallis, Direct current polarisation during field-assisted glass–metal sealing. *J. Am. Ceramic. Soc.*, 53(10), 1970, 563–567.
[152] E. Carlson, K.W. Hang, G.F. Stockdale, Electrode "polarisation" in alkali containing glasses. *J. Am. Ceramic. Soc.*, 55(7), 1972, 337–341.
[153] M.P. Borom, Electron-microprobe study of field assisted bonding of glasses to metals. *J. Am. Ceramic. Soc.*, 56(5), 1973, 254–257.
[154] E. Carlson, K.W. Hang, G.F. Stockdale, *J. Am. Ceramic. Soc.*, 57(7), 1974, 295–300.
[155] K.B. Albaugh, Electrode phenomena during anodic bonding of silicon to sodium borosilicate glass. *J. Electrochem. Soc.*, 138, 1991, 3089–3094.
[156] M. Aizpurha, J.M. Artola, E. Castario, F.J. Gracia, A proposed model for the transferred charge in an anodic bonding process. Proceed. Eurosensors XI, 11th European Conf. On Solid-St. Transducers, Warsaw, Poland, September 21–24, 1997, 1253–1256.
[157] B. Schmidt, P. Nitzche, K. Lange, S. Grigull, U. Kreissig, In-situ investigation of ion drift processes in glass during anodic bonding. *Sensors and Actuators A*, 67, 1998, 191–198.
[158] K.B. Albaugh, D.H. Rassmussen, Rate processes during anodic bonding. *J. Am. Ceramic. Soc.*, 75(10), 1992, 2644–2648.
[159] M. Despont, H. Gross, F. Arrouy, C. Stebler, U. Staufer, Fabrication of a silicon–Pyrex–silicon stack by a.c. anodic bonding. *Sensors and Actuators A*, 55, 1996, 219–224.
[160] C.G. Wilson, A.C. Carter, The self diffusion of sodium ions in a borosilicate glass and a soda-lima glass. *Phys. Chem. Glasses*, 5, 1964,.
[161] P. Jorgessen, Effect on electric field on silicon oxidation. *J. Chem. Phys.*, 37(1), 1962, 874–877.
[162] R. G. Gossile, SIMS Analysis of a field-assisted glass-to-metal seal. *J. Am. Ceram. Soc.*, 61(11–12), 1978, 539–540.
[163] K. Lange, S. Grigull, M. Harz, U. Kreissig, B. Schmidt, Ion drift behaviour in borosilicate

glass during anodic bonding to silicon or metals: Semiconductor Wafer Bonding. The Electrochemical Society Proceed. Series PV-95-7, 1995, Pennington, NJ, 371–378.
[164] Cozma, B. Puers, Characterisation of the electrostatic bonding of silicon and Pyrex glasses. *J. Micromech. Microeng.*, 5, 1995, 98–102.
[165] J. Dziuban, A. Górecka-Drzazga, J. Koszur, P. Kowalski, An influence of anodic bonding on silicon pressure sensors quality. Proceed. 21st Conference Int. Soc. Hyb. Micr. Poland Chapter, Ustroń, 5–8 October, 1997, 111–120.
[166] H. Baumann, S. Mack, H. Minneli, Bonding of structured wafers, Semiconductor wafer bonding: physics and application. *Electrochem. Soc. Proceed.*, PV 95–7, 1995.
[167] S. Go, Y.H. Cho, Experimental evaluation of anodic bonding process using Taguchi method for maximum interfacial fracture toughness. IEEE Workshop MEMS 98, Heidelberg, 1998, 318–321.
[168] Y. Kanda, K. Matsunde, Ch. Murayama, J. Sugaya, The mechanism of field assisted silicon glass bonding. *Sensors and Actuators A*, 21–23, 1990, 939–943.
[169] J. Dziuban, Activation energy of silicon to glass anodic bonding process. Proceed. Eurosensors XIII, 13th European Conf. On Solid-St. Transducers, den Haag, Holland, September 12–15, 1999, 192–193.
[170] H. Nese, A. Hanneborg, Anodic bonding of silicon to silicon wafers coated with aluminum, silicon oxide, polysilicon or silicon nitride. *Sensors and Actuators A*, 37–38, 1993, 61–67.
[171] Hanneborg, M. Nese, P. Ölckers, Silicon-to-silicon anodic bonding with borosilicate glass layer. *J. Micromech. Microeng.*, 1, 1991, 139–144.
[172] P. Krause, M. Sporys, E. Obermeier, K. Lange, S. Grigull, Silicon to silicon anodic bonding using evaporated glass. Tech. Dig. 8th Int. Conf. Solid St. Sensors and Actuators Transducers '95, Stockholm, Sweden 25–29 June, 1995, 228–231.
[173] J. Mack, H. Baumann, U. Gösele, H. Werzer, R. Shögl, Analysis of bonding-related gas enclosure in micromachined cavities sealed by silicon wafer bonding. *J. Electrochem. Soc.*, 144, 1997, 1106–1111.
[174] W.B. Choi, B.K. Ju, Y.H. Lee, S.J. Jeong, N.Y. Lee, M.Y. Sung, M.H. Oh, Glass-to-glass bonding for vacuum packaging of field emission display in an ultra-high-vacuum chamber using silicon thin film. *J. Electrochem. Soc.*, 146(1), 1999, 400–404.
[175] H. Heuni, S. Shoji, Y. Shoji, K. Yoshini, M. Esashi, Vacuum packaging for microsensors by glass–silicon anodic bonding. *Sensors and Actuators A*, 43, 1994, 243–248.
[176] S. Shoji, M. Esashi, Bonding and assembling methods for realizing at μTAS. Proceed. μTAS 94, Twente, 1995 167–179.
[177] J.B. Sanders, A.Q. Tool, Effect of heat treatment on the expansive of a Pyrex glass. *Burr. Stand. J. Res.*, 11, 1933, 795, cited after pos. [149].
[178] A. Harz, Anodic bonding for the third dimension. *J. Micromech. Microeng.*, 2, 1992, 161–162.
[179] S. Nehlsen, V. Relling, F. Kraus, J. Lübke, J. Müller, Lateral Pyrex thin film anodic bonding and KOH deep etching of silicon substrates for microfluidic applications. *Microsyst. Technol.*, 96, 1996, 217–222.
[180] J.A. Plaza, J. Estere, E. Lora-Tamayo, Effect of the silicon oxide, silicon nitride and polysilicon layers on the electrostatic pressure during anodic bonding. *Sensors and Actuators A*, 67, 1998, 181–184.
[181] J.A. Plaza, J. Estere, E. Lora-Tamayo, Non destructive in-situ test for anodic bonding. *Sensors and Actuators A*, 60, 1997, 176–180.
[182] J. Dziuban, Ł. Nieradko, Multilayer anodic bonding for microTAS. Proc. 21st Conf. Int. Microel. Pack. Soc. IMAPS Polish Chapter, Ustroń, 5–8 October 1997, 121–123.
[183] J. Dziuban, S. Patela, Glass on insulator on silicon planar waveguides. Proceed. 12th Europ. Conf. on Solid St. Transducers Southampton, 13–16 September 1998, N.M. White (ed.). IOP Publishing, vol. 1, 728–729.
[184] J. Dziuban, S. Patela, I. Hałas, GIS-type integrated optical waveguide. Proc. 22nd Conf. Int. Microel. Pack. Soc. IMAPS Polish Chapter, Zakopane, 1–3 October 1998, 135–138.
[185] J. Dziuban, A. Górecka Drzazga, I. Hałas, M. Kramkowska, T. Ohly, E. Prociów, Anodic

bonding of silicon with unconventional surface layers. Proc. 22nd Conf. Int. Microelec. Pack. Soc. IMAPS Polish Chapter, Zakopane, 1–3 October 1998, 127–130.

[186] P.T. Baire, L.J. Quinn, B. Lee, S.J.N. Mitchell, H.S. Gamble, B.M. Amstrong, Electrostatic bonding for silicon on glass applications. *Electrochem. Soc. Proceedings*, PV 97–36, 1997, 214–221.

[187] W.-B. Choi, B. Ewon, S. Jeong, N.-Y. Lee, K.-H. Koh, M.R. Hiskard, M.-Y. Sung, M.-H. Oh, Anodic bonding technique under low-temperature and low voltage using evaporated glass. Proceed. at 9th International Vacuum Microelectronic Conference, St. Petersburg 1996, 427–430.

[188] D. Brooks, R.P. Donovan, C.A. Hardesty, Low temperature electrostatic silicon-to-silicon seals using sputtered borosilicate glass. *J. Electrochem. Soc.*, 119, 1972, 545–546.

[189] S. Weichel, R. de Rues, M. Lindahl, Silicon-to-silicon wafer bonding using evaporated glass. *Sensors and Actuators A*, 70, 1998, 179–184.

[190] W.Y. Lee, F. Sequeda, J. Salem, Field assisted bonding below 200°C using metal and glass thin film foil-layers. *Appl. Phys. Lett.*, 50, 1987, 522–524.

[191] Esashi, A. Nakano, S. Shoji, H. Hebiguchi, Low-temperature silicon to silicon anodic bonding with intermediate low melting glass. *Sensors and Actuators A*, 21–23, 1990, 931–934.

[192] Hameborg, M. Nesse, H. Jakobsen, R. Holsn, Silicon-to-thin film anodic bonding. *J. Micromech. Microeng.*, 2, 1992, 317–321.

[193] W.-B. Choi, B.-K. Ju, Y.-H. Lee, J.-W. Jeong, M.P. Raskard, N.-Y. Lee, M.-Y. Sung, M.-H. Oh, Experimental analysis of the anodic bonding with an evaporated glass layer. *J. Micromech. Microeng.*, 7, 1997, 312–316.

[194] R. Puers, A. Cozma, Bonding wafers with sodium silicate solution. *J. Micromech. Microeng.*, 7, 1997, 114–117.

[195] R.C. Frye, J.E. Griffith, Y.H. Wong, A field-assisted anodic bonding process for silicon dielectric isolation. *J. Electrochem. Soc.*, 133, 1986, 1673–1677.

[196] G. Wallis, J. Dorsey, J. Beckett, Field assisted seals of glass to Fe–Ni–Co alloy. *Ceramic. Bulletin*, 50(12), 1971, 958–961.

[197] Y. Sim, T. Karabayashi, M. Esashi, A backable microvalve with KOWAR–glass–silicon–glass structure. *J. Micromech. Microeng.*, 6, 1996, 266–271.

[198] Polish norma PN FeNiCo 2N29PR.

[199] J. Bryzek, K. Petersen, J. Mallon Jr, L. Christel, F. Pourohmeadi, *Silicon sensors and microstructures*. Nova Sensor, Fremont, 1991, CA, USA.

[200] *NEXUS analysis of microsystems 1996–2002*. NEXUS office, Fraunhofer ISIT, Billenburger Strasse 53, D/4/99, Berlin, (patrz również MST News, 3, 1998, 37–40).

[201] E. Peeters, S. Vergote, B. Puers, W. Sansen, A combined silicon fusion and glass/silicon anodic bonding process for an uniaxial capacitive accelerometer. *J. Micromech. Microeng.*, 2, 1992, 167–169.

[202] G. Hashimoto, C. Cabuz, K. Minami, M. Esashi, Silicon resonant angular rate sensor using electromagnetic excitation and capacitive detection. *J. Micromech. Microeng.*, 5, 1995, 219–225.

[203] L.M. Roylance, A batch fabricated accelerometer. *IEEE Tran. El. Dev.*, ED-26(12), 1979, 1611–1917.

[204] K. Kwon, S. Park, A bulk micromachined three-axis accelerometer using silicon direct bonding technology and polysilicon layer. *Sensors and Actuators A*, 66, 1992, 250–255.

[205] J.A. Plaza, J. Esteve, E. Lora-Tamago, Simple technology for bulk accelerometer based on bond and etch-back silicon on insulator wafers. *Sensors and Actuators A*, 68, 1998, 299–302.

[206] R. Puers, S. Reyntjens, Design and processing experiments of a new miniaturised capacitive triaxial accelerometer. *Sensors and Actuators A*, 68, 1998, 324–328.

[207] H. Jakobsen, Sensor foundries and production of sensors of Sensonor Nov. A. S. *J. Micromech. Microeng.*, 6, 1996, 193–196

[208] G. Schröpfer, W. Elfein, M. de Labachelerie, H. Porte, S. Bellenoras, Lateral optical accelerometer micromachined in (100) oriented silicon with remote readout based on coherence modulation. *Sensors and Actuators A*, 68, 1998, 344–345.

[209] S. Roy, R.G. de Auna, A. Izad, M. Mehregany, Miniature ice detection sensor systems for aerospace applications. IEEE Workshop MME Systems, Heidelberg, 1998, 75–80.

[210] J. Schimkat, L. Kiesewatter, H.J. Geratter, F. Arndt, A. Steckenborm, H.-F. Schlack, Moving edge actuator: an electrostatic actuator for use in microrelay. *Microsyst. Technol.*, Berlin, 1994, 989–996.

[211] J. Bergvist, F. Rudolf, A new condenser microphone in silicon. Abstract, Eurosensors III, Montreux, Switzerland, 1989, 94–95.

[212] T. Cormann, P. Enakson, G. Stemme, Deep wet etching of borosilicate glass using an anodically bonded silicon substrate or mask. *J. Micromech. Microeng.*, 8, 1998, 84–87.

[213] R. Varlan, W. Sansen, Micromachined conductometric p (CO_2) sensor. *Sensors and Actuators B*, 44, 1997, 309–315.

[214] B. Chévrier, K. Baert, T. Slater, A. Verbist, Micromachined infrared pneumatic detector for gas sensor. *Microsyst. Technol.*, Berlin, 1994, 445–453.

[215] K. Petersen, Silicon as a mechanical material. *Proceed. IEEE, El.Dev.*, 70(5), 1982, 420–457.

[216] for examples: www.kulite.com, www.us.sbt.siemens.com, www.sssinternational.com.

[217] W.S. Czarnocki, Media isolated sensor. *Sensors and Actuators A*, 67, 1998, 142–145R.

[218] W.S. Czarnocki, J.P. Schuster, Automotive sensors. *Konf.*, COE 94, 1994, 171–187.

[219] B. Hufenbach, S. Habinc, P. Vuillenmier, Space applications for smart sensors. Keynote lecture, Proceed. Eurosensors XIII, 13th European Conf. On Solid-St. Transducers, den Haag, Holland, September 12–15, 1999, 3–6.

[220] R. Puers, D. de Bruyker, A. Cozma, A novel combined redundant pressure sensor with self test function. *Sensors and Actuators A*, 60, 1997, 6871.

[221] Z. Pruszkowski, J. Dziuban, A. Górecka-Drzazga, U. Lipowicz, Epoxy glue for positioned packaging of piezoresistive silicon pressure sensors (in Polish). *Elektronika*, 11, 1993, 19–20.

[222] Pressure sensors for 100 kPa, catalogue card of the Institute of Electron Technology, Warsaw, Poland.

[223] J. Dziuban, K.P. Friedel, The semi flip chip attachment of single or twin chips silicon piezoresistive pressure sensors. Proceed. 12th Europ. Microelectronics and Packaging Conf., 7–9 June, 1999, Harrogate, England, 61–67.

[224] J. Dziuban, K. Friedel, The silicon piezoresistive pressure sensor in the twin-chips configuration. Proceed. 4th Int. Workshop Elect. Cont. Measur. Signals, ECMS '99 31st May–1st June, 1999, Liberec, Czech Rep., 31–34.

[225] J. Dziuban, K. Friedel, A. Chotomski, The thermo-mechanical modelling of piezoresistive pressure sensors in the single or twin-chips configuration. Proceed. XXIII Conf. Int. Microel. and Packaging Soc. IMAPS, 21–23 Sept., 1999, Kołobrzeg, Poland, 257–262.

[226] R. Puers, E. van den Bossche, W. Sansen, A capacitive pressure sensor with low impedance and active suppression of parasitic effect. *Sensors and Actuators A*, 21–23, 1990, 108–114.

[227] H. Sander, J. Knutti, J. Meindl, A monolithic capacitive pressure sensor with pulse period output. *IEEE Tran. El. Dev.*, ED-17, 1990, 927–930.

[228] R. Puers, S. Vergote, A subminature capacitive moment detector using a composite membrane suspension. *Sensors and Actuators A*, 31, 1992, 90–96.

[229] R. Puers, G. Blasquez, Low cost high-sensitivity integrated pressure and temperature sensor. *Sensors and Actuators A*, 41–42, 1994, 338–401.

[230] X. Chauffer, G. Blasquez, P. Pous, Influence of the bonding conditions on the response of capacitive pressure sensors. *Sensors and Actuators A*, 46–47, 1995, 121–124.

[231] J. Dziuban, R. Walczak, Etching microwave silicon [EMSi] – microwave enhanced fast anisotropic etching of silicon for electromechanical systems [MEMS]. *Sensors and Materials*, 1, 2001, 041–045.

[232] J. Dziuban, R. Walczak, A silicon capacitive pressure sensor micromachined by EMSi method. Proceed. of the 12th Micromechanics Europe Workshop MME 2001, 16–18 Sept., 2001, Cork, Ireland, 123–127.

[233] Górecka-Drzazga, J. Dziuban, S. Bargiel, Fibre-optic micromechanical pressure sensor (in Polish). Proceed. VI Conf. Optoelectronics and Electronics Sensors, COE 2000, 2, 144–149.

[234] M. Mehregany, *Silicon microactuators in advanced actuators*, A.P. Dorey, J.H. Moore (eds.). Institute of Physics Publishing, Bristol-Philadelphia.
[235] H. Suzuki, Shell body fabrication for micromachines. *J. Micromech. Microeng.*, 5, 1995, 36–40.
[236] M. Mehregany, K. Gabriel, W. Trimmer, Microgears and turbines etched from silicon. *Sensors and Actuators A*, 12(4), 1987, 341–348.
[237] L. Fau, S. Woodman, Batch fabrication of mechanical platforms for high density data storage. Eurosensors IX, Stockholm, Sweden, 25–29 June, 1995, 434–437.
[238] J. Dziuban, A. Górecka-Drzazga, Silicon microturbines for liquids flow sensors. Eurosensors X, 10th European Conf. on Solid-St. Transd. 8–11 September, Leuven, Belgium, 1996, 457–460.
[239] Shoji. M. Esashi, Photoetching and electrochemical discharge drilling of Pyrex glass. *Tech. Dig. 9th Sensor Symp.*, 1990, 27–30.
[240] A. Manz, N. Graber, H.M. Widmer, Miniaturized total chemical analysis systems, a novel concept for chemical sensing. *Sensors and Actuators B*, 1, 1990, 244–248.
[241] P. Bergveld, The challenge of developing μTAS. μTAS 94, Twente 1994, 1–4.
[242] A. Manz, S. Verpoorte, D.E. Raymond, C.S. Effenhauser, N. Burrgrat, H.M. Widmer, μTAS: miniaturised total chemical analysis systems. μTAS 94, Twente 1994, 5–23.
[243] A. Okumura, Y. Miyahara, M. Sakairi, Integrated chemical analysis systems as a step toward ionoelectronics. μTAS 96, Basel 1996, 22–23
[244] J.M. Ramsey, Miniature chemical measurement systems. μTAS 96, Basel 1996, 24–27.
[245] Proceedings of μTAS Conferences in years 1994–2004
[246] A. Götz, C. Cané, I. Grâcia, E. Lora-Tamayo, A sensor chip for biomedical analysis. μTAS 96. Bazel 1996, 211–213.
[247] S. Shoji, H. Huruya, T. Ohori, Micromachined micro flow devices for a medical μTAS. μTAS 96, Bazel 1996, 61–64.
[248] H.T.G. van Linten, F.C.M. van den Pol, S. Bouwstra, A piezoelectric micropump based on micromachining of silicon. *Sensors and Actuators A*, 15, 1988, 153–167.
[249] J.G. Smits, Piezoelectric pump with three valves working peristatically. *Sensors and Actuators A*, 21–23, 1990, 203–206.
[250] T. Gerlach, A simple micropump employing dynamic passive valves made in silicon. MST 94, Berlin 1994, 1025–1034.
[251] S.W. Lee, O.C. Jong, S.S. Yong, The fabrication of microinjector actuated by boiling and (or) electrolysis. IEEE Work. MME Systems, Heidelberg 1998, Proceed., 51–56.
[252] K. Miura, S. Shoji, Fabrication of injection and switching valve for whole blood control. Proceed. μTAS98 Workshop, Bauff, Can., Kluwer Ac. Publishers, ed: D.J. Harrison, A. van den Berg, 1998, 85–88.
[253] S. Shoji, S. Nakagawa, M. Esashi, Micropump and sample-injector for integrated chemical analysis systems. *Sensors and Actuators A*, 21–23, 1990, 185–192.
[254] S. Kamisuki, T. Hagata, C. Tezuka, Y. Nase, M. Fuji, M. Atabe, A low power small electrostaticall-driven commercial ink-jet head. IEEE Work. MME Systems, Heidelberg 1998, Proceed. 63–68.
[255] G.J. Schaubmueller, M. Koch, A.G.R. Evans, A. Braunnschweiler, Micromachined chemical reaction system realised on a microfluidic circuit board. Eurosensors XII. Proceed. 12th Europ. Conf. on Solid St. Transducers Southampton, 13–16 Sept., 1998, N.M. White (ed.). IOP Publishing 1998, Vol. 2, 571–574.
[256] C. Terry, A gas chromatographic air analyser fabricated on silicon wafer. *IEEE Tran. El. Dev.*, ED-26(2), 1978, 1880–1886.
[257] www.mtigc.com.
[258] Catalogue of MTI Analytical Instruments, MTI's P200 Gas Chromatograph 1995/1996, Microsensors Technology Inc. 41762 Christy Street Fremont, CA 94538, USA.
[259] J. van Kuijk, T.S.J. Lammerinck, H.E. de Bree, M. Elwenspoek, J.H.J. Fluitman, Multiparameter in fluid flows. *Sensors and Actuators A*, 46–47, 1995, 365–372. .

[260] T.S.J. Lammerinck, F. Dijkstra, Z. Hoker, J. van Kujik, Intelligent gas-mixture flow sensor. *Sensors and Actuators A*, 46–47, 1995, 380–386.

[261] A. Olsson, P. Enohsonn, G. Stemme, E. Stemme, A valve-less planar pump isotropically etched in silicon. *J. Micromech. Microeng.*, 6, 1996, 87–91.

[262] A. Olson, G. Stemme, E. Stemme, A valve-less planar fluid pump chambers. *Sensors and Actuators A*, 46–47, 1995, 549–556.

[263] K. Małecki, J. Dziuban, A. Górecka-Drzazga, A silicon valve-less pump IC compatible technology. *MST News Poland*, 3, 1997, 5–8.

[264] J. Dziuban, A. Górecka-Drzazga, Ł. Nieradko, K. Małecki, Silicon components for gas chromatography. Proceed. Eurosensors XIV, 14th Europ.Conf.on Solid-St. Trancducers. Copenhagen, Denmark, August 27–30, 2000, 55–58.

[265] J.A. Dziuban, J. Mróz, M. Szczygielska, M. Małachowski, A. Górecka-Drzazga, R. Walczak, W. Buła, D. Zalewski, Ł. Nieradko, J. Łysko, J. Koszur, P. Kowalski, Portable gas chromatograph with integrated components. *Sensors and Actuators A*, 115, 2004, 318–330.

[266] J. Dziuban, A. Górecka-Drzazga, K. Małecki, Ł. Nieradko, J. Mróz, M. Szczygielska, Silicon components for gas chromatograph. *Mat. SPIE*, 4516, 2001, 247–257.

[267] J. Mróz, M. Sczygielska, J. Dziuban, A. Górecka-Drzazga, Laboratory model of gas microchromatograph – construction and testing. *Mat. SPIE*, 4516, 2001, 258–266.

[268] S.C Jakeway. A.J. de Mello, E.I Russel, Miniaturized total analysis systems for biological analysis. *J. Anal. Chem.*, 366, 2000, 525–539.

[269] P. Dario, M.C. Carozza, A. benvenuto, A. Menciassi, Micro-systems in biomedical applications. *J. Micromech. Microeng.*, 10, 2000, 235–244

[270] A. Manz, Micromachining of noncrystalline silicon and glass for chemical analysis systems: A look into the next century's technology or just a fashionable erase? *Tren. Anal. Chemistry*, 10, 1991, 144–148.

[271] J. Harrison, A. Manz, Z. Fan, H. Lüdi, H.M. Widmer, Capillary electrophoresis: and sample injection systems integrated on a planar glass chip. *Anal. Chemistry*, 64, 1992, 1926–1932.

[272] J. Harrison, K. Fluri, Z. Fan, K. Seiler, Integration of analytical systems incorporating chemical reactions and electrophoretic separation. μTAS 94, Twente 1994, 105–115.

[273] K. Seiler, D.J. Harrison, A. Manz, Planar glass chips for capillary electrophoresis: repetitive sample injection, quantitation and separation efficiency. *Anal. Chemistry*, 65, 1993, 1481–1488.

[274] J. Dziuban, A. Górecka-Drzazga, Ł. Nieradko, J. Mróz, Silicon microcolumns for FIA chromatography. μTAS 96, Basel 1996, 946–953.

[275] J. Dziuban, A. Górecka-Drzazga, Ł. Nieradko, J. Mróz, Silicon integrated capillary for μTAS. *MST News*, 1(5), 1997, 8–11.

[276] J. Dziuban, A. Górecka-Drzazga, Ł. Nieradko, J. Mróz, Micromachined silicon microcolumns for FIA chromatography with anodic bonding inlets and outlets. The 192nd Electrochemical Meeting Paris 1997. *ECH Proceed.*, PV 97–19, 946–953.

[277] J. Dziuban, A. Górecka-Drzazga, Ł. Nieradko, J. Mróz, IC process compatible technology of silicon integrated capillary column fabrication. Eurosensors XI Proceed. 11th Europ. Conf. on Solid St. Transducers, Warsaw, 21–24 September 1997, Vol. 3, 3PI2, 1291–1295.

[278] J. Dziuban, Z. Najzarek, Technological aspects of integrated chemical throughput synthesing and analysing microsystems development. *Int. Meet. Chem. Eng., Env. Prot. and Biotech.*, ACHEMA 2000, Frankfurt am Main, 22–27 May, 2000, 164–165.

[279] P. Salomon, Bio-MEMS and microfluidic demonstrate their potential at the Transducer 99. *MST News*, 3/99, 27.

[280] A. Gretillat, F. Paoletti, P. Thiébaud, S. Roth, M. Koudelka-Hep, N.F. de Roji, A new fabrication method for borosilicate glass capillary tubes with lateral inlets and outlets. *Sensors and Actuators A*, 60,1997, 219–222.

[281] L. Nieradko, *Integrated microcolumn for gas chromatograph.* Thesis, The Faculty of Microsystem Electronics and Photonics of the Wroclaw University of Technology, 2000.

[282] J. Janca, Micro-channel thermal Field-Flow-Fractionation; new challenge in analysis of macromolecules and particles. *J. Liq. Chromatogr. & Rel. Technol.*, 25(5), 2002, 683–704.

[283] T.I. Edwards, B.K. Gale, A.B. Frazier, A micromachined thermal field-flow fractionation system, proc. Transducers '99, Intern. Conf. Solid.-St. Sensors and Actuators, Sendai, Japan, June 7–11, 1999.

[284] S. Bargiel, A. Górecka-Drzazga, J.A. Dziuban, A micromachined system for the separation of molecules using thermal field-flow fractionation method. *Sensors and Actuators A*, 110, 2004, 328–325.

[285] S. Bargiel, A. Górecka-Drzazga, J.A. Dziuban, P. Prokaryn, M. Chudy, A. Dybko, Z. Brzózka, Nanoliter detectors for flow systems. *Sensors and Actuators A*, 115, 2004, 245–251.

[286] Li-A. Liew, S. Knappe, J. Moreland, H. Robinson, L. Hollberg, J. Kitching, Microfabricated alkali atom vapor cells. *Appl. Physics. Lett.*, 84(14), 2004, 2694 – 2696.

[287] C. Gorecki, S. Khalfallah, H. Kawakatsu, Y. Arakawa, New SNOM sensor using optical feedback in a VCSEL-based compound-cavity. *Sensors and Actuators A*, 87, 2001, 113–123.

[288] D.W. de Lima,o.Akhzar-Mehr, P.M. Sarro, G.Vdovin, Single mask microfabrication of aspherical optics using KOH anisotropic etching of Si. *Optics Express*, Sept., 2003.

[289] R. Carrasco, J.A. Dziuban, I. Moreno, C. Gorecki, R. Walczak, M. Kopytko, L. Nieradko, M. Jozwik, *Optical microlenses for MEMS*. Proceed. Microtechn. New. Millenium, 9–11 May, Seville, Spain, 2005.

[290] H. H. Busta, Vacuum microelectronics 1992. *Review, J. Micromech. Microeng.*, 2, 1992, 43–74.

[291] J. Dziuban, A. Górecka-Drzazga, On process silicon microemitters with sharp tips. *MST New Poland*, 1996, 6–11A.

[292] Spindt, FEA's state of art and applications. Proceed. 2nd IWVM of 12th IVMC, Wrocław-Darmstadt, invited lecture.

[293] M. Nakamoto, United States Patent, no 5499938, March 19, 1996.

[294] M. Nakamoto, K. Fukuala, Versatile field emitter arrays fabricated by transfer mold technique. Proceed. 12th Int. Vacuum Microel. Conf., 10–13 July 1999, 17–20.

[295] J. Dziuban, A. Górecka-Drzazga, E. Prociów, SiC field emitters array fabricated by transfer MOLD technique. Proceed. 12th Int. Vacuum Microel. Conf., 6–9 July 1999, Darmstadt, Germany, 31–37.

[296] Górecka-Drzazga, J. Dziuban, E. Prociów, SiC field emitters array fabricated by transfer MOLD technique. *J. Vac. Sc. Techn. B*, 18(2), 2000, 1–4.

[297] J. Dziuban, A. Górecka-Drzazga, MOLD type FEA's fabrication by use of fast silicon etching. *J. Vac. Sc. Techn. B*, 19(3), 2001, 897–899.

[298] A. Gorecka-Drzazga, J. Dziuban, W. Drzazga, Mold-type SiC field emitters with heavily boron-doped gates. *J. Micromech. Microeng.*, 14, 2004, 907–913

[299] A. Gorecka-Drzazga, J. Dziuban, S. Bargiel, Mold-type SiC emitters with nano-holes at the apex. *J. Meas. Sci. Technol.*, 17, 2006, 45–49.

Chapter 5

CLASSIFICATION OF BONDING AND CLOSING REMARKS

CLASSIFICATION OF BONDING

The classification of bonding techniques applicable in microsystem technology has been done according to Schulze's [1] (Table 5.1), Shoji's [2] (Table 5.2) and the author's methods.

The classifications of Schulze and Shoji are incomparable. What is more, they do not include many methods of bonding that are technologically used. In the author's classification, bonding is divided into two main groups: non-electric bonding (without the use of polarization) (Table 5.3) and anodic bonding (Table 5.4). Types of non-electric bonding are listed in the order corresponding to the degree of application of particular techniques of this method in microsystem technology (from DFB, which is applied most often, to eutectic bonding). Types of anodic bonding are listed in the order corresponding to the description presented in this book, which at the same time reflects the degree of application of this method in microsystem technology (from the most popular bonding of silicon to glass, to bonding of silicon to silicon).

CLOSING REMARKS

Apart from deep micromachining based on wet/dry etching techniques and assembly of microsystems by bonding of materials, several other technological procedures are applied for the three-dimensional precise machining of microsystems. They are of microelectronic origin: surface silicon micromachining, LIGA (***L**ithography – **G**alvanotechnik – **A**bformung*), stereolithographic surface deposition (*stereo CVD*) or coming from mechanical machining: precise micromachining, electro erosion, powder blasting, laser machining, molding etc.

Surface silicon micromachining sometimes described as "the sacrificial layer technique", is compatible to the CMOS technique. The process utilizes different etching characteristics of a thin-film layer sandwich. Most often, the sandwich

Table 5.1. Methods of bonding according to Schulze [1]

Method	Material	Interlayer	Temperature [°C]	Preparation of surface or additional cond.	Selective bonding
Anodic	Glass–Si Si–Si Si metal/glass	Pyrex Al, W, Ti, Cr	>250 >300 300–500	Voltage 50–1000 V	Photolithography etching, lift-off
Direct Si	Si–Si SiO_2–SiO_2	SiO_2	700–1000	Standard washing	Photolithography etching
Low-temperature to glass	Si–Si SiO_2–SiO_2	Na_2O–SiO_2 and other materials, e.g. sol-gel, boron glass	200–400 >450	Covering by spinning CVD, implantation	Photolithography etching
Low-temperature direct	Si–Si SiO_2–SiO_2		200–400	Plasma machining, wet activation of surface (immersion)	
Eutectic	Si–Si	Au, Al.	379, 580	Sputtering, electrocladding	Lift-off, etching
Soldering	Si–Si	Au, Pb–Sn	300	Evaporation, sputtering	Lift-off, etching
Adhesive	Si–Si Si–glass SiO_2–SiO_2 Si_3N_4–Si_3N_4	Binders, Photoresist	Room temperature – 200	Covering by spinning	Photolithography

Table 5.2. Methods of bonding according to Shoji [176]

Method	Material	Interlayer	Temperature [°C]	Applied voltage [V]	Applied pressure [kg/cm^2]	Selective bonding
Glueing	Si–Si glass–Si glass-glass	Spin-on binder	Room temperature	—	>0.2	×
		Photosensitive dry layer	>150	—	>0.5	Photolithography
Low-temperature to glass	Si–Si glass–Si glass-glass	Glass frit	>415	—	—	Screen process
		Liquid glass	>200	—	—	×
	Si–Si	Boron glass	>450	—	—	×
Eutectic	Si–Si	Au	370	—	95	Photolithography
Fusion	Si–Si	—	>1000	—	—	×
Anodic	glass–Si	—	>250	>200	—	SiO_2(>1µm) TiW/Au
	Si–Si	SiO_2	>850	>30	—	Photolithography
	Si–Si	Sputtered Pyrex glass	>300	>100	—	Lift-off
	Si–Si Si–Al/glass Si–ITO/glass	Sputtered fusible glass	Room temperature	>40	>0.8	Lift-off

Table 5.3. Non-electrical bonding

Method	Substrates	Interlayer	Temperature [°C]	U [V]	Comments
Direct fusion High temperature	Si–Si	SiO_2 Si_xN_y α-C SiC	>850 >1110	—	Cool down slowly
	Si–sapphire	—	400/800		Samples crack, strong bonding
	Si–quartz	—	150/270		
	GaAs–sapphire	—	500		Annealing in hydrogen
	GaAs–In	—	650		Standard method
Direct fusion Low temperature **Hydrophilic**	Si–Si	—	<300	—	Activation of surface by O_2 plasma
Direct fusion Low temperature **Hydrophobic**	Si–Si		<450		Activation by B_2H_6 or Ar plasma
Indirect, Through fusible glass Low-temperature	Si–Si	glass frit	>415	—	Screen process, 5570, good bonding.
		sodium silicate	200/400	—	Aqueous solutions, spin-on, very good bond cure for 30′
		aluminum phosphates	350	—	Very strong bond

continued

Table 5.3. *Continued*

Method	Substrates	Interlayer	Temperature [°C]	U [V]	Comments
Indirect	Si–Si	boron glass	450	—	Clamp
Through sputtered glass		phosphorus–silicon glass	900	—	Oxidized Si
Low-temperature	Glass-glass	NaOH	500	–	Anneal in vacuum
	Si–Si			-	
	Si–glass				
	quartz-quartz				
Room temperature	Glass–glass	SU-8, foil	20–180		Spin-on or laminated
	Si–Si	PDMS	20		Spin-on, O_2 plasma
	Si–glass	BCB	20/70/180		
		HF	20		1% HF, clamp strongly for many hours
Eutectic	Si–Si	Au	365		Cure for a short time
		Al	650	—	Cure for a short time
		Al–Au,	350	—	Clamp strongly 40 MPa
		Au–In–Au	424	—	—
		Al–Ge	350/700	—	Anneal for 2 h
		Pt–Si	455/520	—	Anneal in vacuum
	Si/SiO_2–Si	Cr–Au	520		—
		Ti–Au			

Table 5.4. Bonding with polarization – anodic bonding

Substrates	Interlayer	Temperature [°C]	U [V]	Comments
Silicon–glass	—	450	600–1500	SD2 glass
		350–450	500–1000	Corning 7740 glass
		350–400 (450)	< 1000	Borofloat 3.3 glass
	Thermal SiO_2	450	500–1500	Use flat cathodes
	Sputtered SiO_2	450	1000	Clamp, heat for a long time at given temperature, especially thick SiO_2 layers (≤ 1.5 μm), cool down slowly
	Plasma SiO_2	450	1000	Do not apply small voltage for SOG layers thicker than 3000 Å
	SOG	450	1300	
Oxidized silicon–glass	Si(V, B)	450	1800–2300	Apply elevated polarization and complete hydrophilizing washing
	Si(Pd)	—	not bondable	Bonds well, layers can be broken and cut
	SiC	450	2200	
Silicon–glass	Al	< 577	Select	Strength of bonding is limited by adhesion of Al to substrate, on which it was deposited.
	Al_2O_3	450	Select	During the bonding a thin Al_2O_3 layer evolves, apply excess oxygen in layer before bonding
	Si_3N_4	450–550	1000	Cure before bonding at 1050°C in steam for 1 h, voids, bubbles

Continued

Table 5.4. *Continued*

Substrates	Interlayer	Temperature [°C]	U [V]	Comments
Silicon–silicon	Pyrex-like glasses sputtering	20–450	30–200	Conditions are strongly dependent on the method of fabrication of interlayer
Silicon–FeNiCo	—	360	1200–1500	Kovar alloy needs to be oxidized, 450°C, air, 5 min
Silicon–silicon	SiO_2	>1000°C	<15	Type of direct bonding

is formed from silicon nitride/silicon dioxide and polysilicon layers (Fig. 5.1). First, the silicon nitride buffer layer is deposited and photo-processed to form the access windows to the silicon surface. Next, the sacrificial layer of silicon dioxide is deposited and photo-processed in a similar way to the buffer layer. This process is followed by polysilicon layer deposition and micromechanical detail formation in the layer. Finally, the sacrificial layer is etched away in a buffered hydrofluoric acid to free micromechanical details. Sacrificial layer technology is applied in the production of sensors and actuators. Well-known devices fabricated using this method are, for instance: gyroscopes, inclination and angle sensors, yaw sensors, accelerometers, linear and rotational microengines, microtools, microactuators and many other micromachines [3–6].

The LIGA technique utilizes X-ray lithography, electroplating and plastic molding. Two key technological steps can be specified here (Fig. 5.2). In the first step a thick layer of light-sensitive material, a resin, mainly PMMA, is deposited on the conductive substrate (e.g. silicon) [6].

Next, resin is irradiated with synchrotron radiation and developed; patterns of very high aspect-ratio coefficient (500:1) with very smooth walls (rms < 20 nm) are fabricated. Then, the produced pattern is electrochemically filled with metal. In the second step a resin layer is removed. The formed metal detail is used as a final product or as an insert mold for plastic details replication. The LIGA technique is quite often applied in micromachine technology for the production of microengines, gearboxes, systems of planetary gears and many other micromechanisms [7–13], as well as micromechanical sensors, vibrating structures for artificial ears [14], turbine microengines, including a microdrilling machine for machining (cleaning) of choked blood vessels, etc. [15].

Stereolithography was developed in the mid-1990s [16]. In this method a liquid monomer material is exposed by a scanned focused laser light beam. The

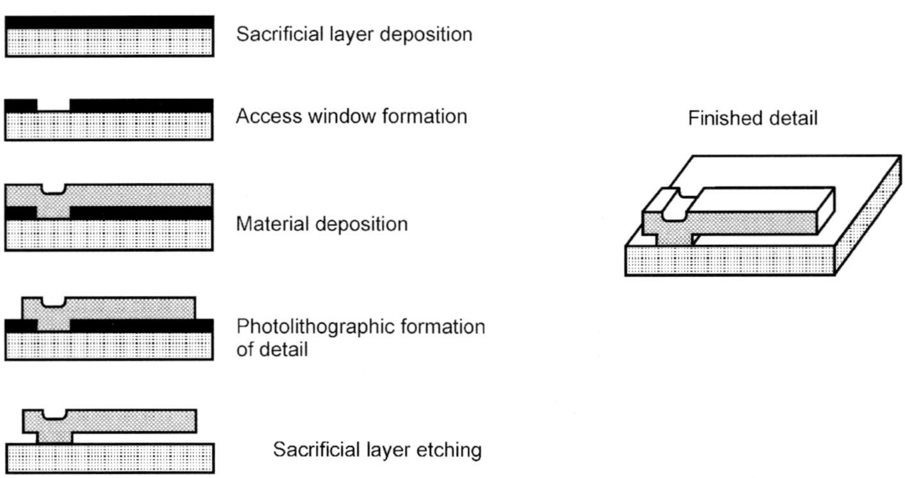

Fig. 5.1. Surface micromachining, the principle of the process.

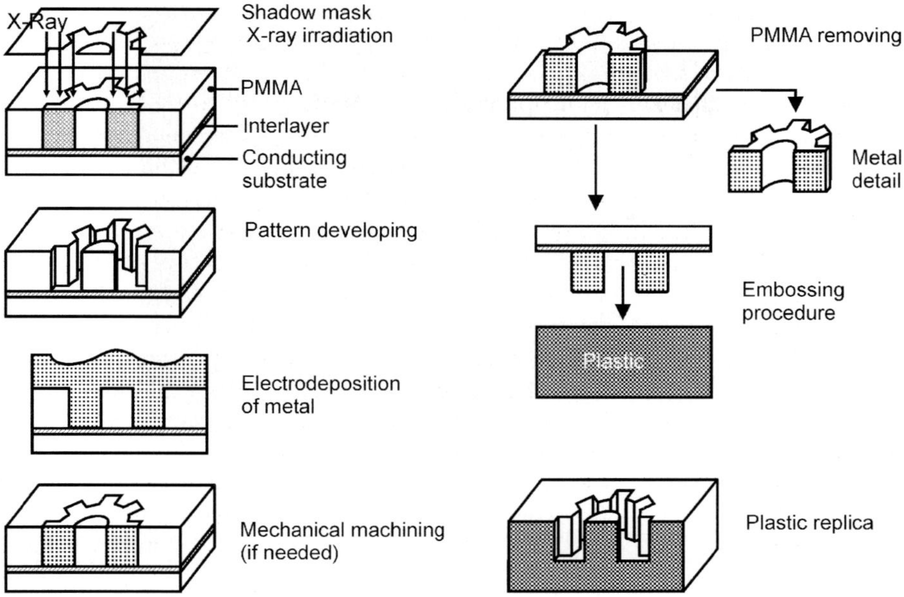

Fig. 5.2. LIGA: the principle of the process.

intensity of a spot of focused light is sufficient for the polymerization and hardening of the material. As a result a three-dimensional mechanical detail may be obtained (Fig. 5.3).

The stereolithography technique has become an innovative method for fabrication of various micromechanical details [17, 18]. Using this method, applied on the production scale, a subminiature submarine for the fluid system (cardiovascular system) diagnostics (Fig. 5.4) has been fabricated, as well as elements of inner-ear prosthesis, gas flow meters with rotational turbine [19] and analytical microsystems μTAS [20].

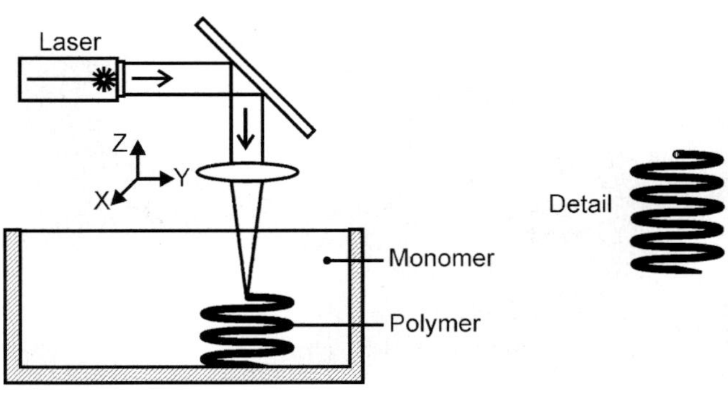

Fig. 5.3. Stereolithography: the principle of the process.

Fig. 5.4. Miniature submarine with the dimensions of about 4 mm × 0.8 mm sailing in a blood vessel. The rotating screw powered by an electric microengine drives the hulk; it can be equipped with a TV camera and different sensors (by courtesy of the Micro-TECH Co., Duisburg, Germany).

In the stereo deposition technique a focused laser scanned light beam activates deposition of solid-state structures from the gas phase [21]. Other methods of micromachining with the use of laser light are optoerosion and laser microstructurization. Electroerosion, precise drilling and milling [22–26] are non-electric micromachining techniques, rated among precision mechanics. The techniques enumerated above are being applied more and more often in non-silicon microsystem technologies. By means of these methods extremely precise metal micromachines have been produced. For instance, the model of TOYOTA from 1936 (the firm Nippon Denso) has been fabricated in a scale of 1:1000 in Japan [27] (Fig. 5.5). The car is driven by an electromagnetic engine, which is $\phi = 0.7$ mm in diameter, and consists of 18 different parts.

Fig.5.5. The Nippon-Denso microautomobile.

It is difficult to estimate how technologies such as deep silicon machining and bonding, surface silicon micromachining, the LIGA technique and other non-photolithographic methods of three-dimensional microengineering, will be used in future mass-scale microsystem manufacturing. But it is common knowledge that development of silicon and non-silicon microsystems is ongoing and intensive (Fig.5.6).

Dr. Janusz Bryzek, a great researcher and the founder of Nova-Sensor, said: "on the basis of the expected growth index, we will be able to call the next decade (2000–2010) the decade of MEMS [...] The social impacts of the exploding MEMS technology can be viewed from two angles. On the positive side, it will create a lot of a new job places [...]. On the problematic side, MEMS technology [...] contributes significantly to more efficient products and manufacturing systems [...]. Increase of productivity reduces the number of people required in the other market segments using MEMS technology. The author expects that, within the next several decades, several-fold reduction of employment in the manufacturing sector of world society could occur. To manage properly a transition of society to such a new social order will require a well-orchestrated action [...], reduced manufacturing employment may bring happiness via the increase of free leisure time." [28, p. 8].

Prof. Kurt Petersen, the creator of microsystems in the USA, admitted: "We are on the verge of revolution in all types of chemical analysis processes [...] The age of a chemical -analysis laboratory in a toolbox may not be far from reality" [29, p. 148].

The Federal Minister of Education and Research of the Federal Republic of Germany, during the conference on 21st-century innovations (Bonn, June 1999 [30]) stated that: "Since the eighties (of the 20th century) microsystem technology has been a testimony of the dynamic development of new technologies of the future. Microsystems offer the opportunity to expand the revolution, started by microelectronics, in other fields of techniques and technology. If we are able to fabricate miniaturized mechanical, optical, chemical and biochemical systems

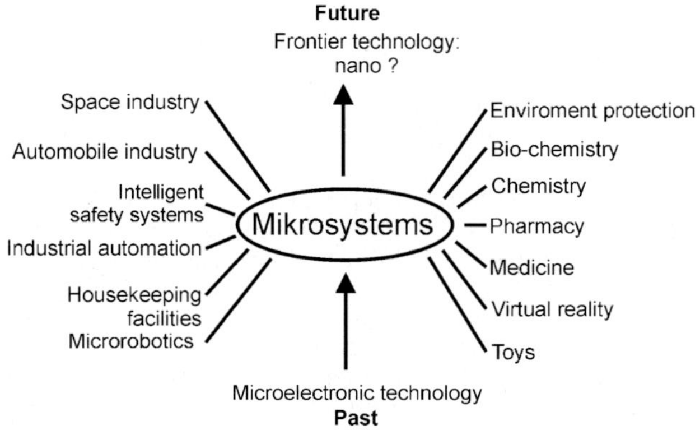

Fig. 5.6. Back to the future – from microelectronics to microsystem technology.

with the same integration scale as in microelectronics, obtaining the capacity of data processing and presenting similar to the capacity obtained in microelectronics, we will provide a significant assistance to the earlier successes (market successes) of other fields of technology".

LITERATURE

[1] S. Schulze, Silicon bonding in microsystem technology; *NEXUS Workshop IMSAS*, 1995.

[2] S. Shoji, M. Esashi, Bonding and assembling methods for realizing of μTAS. Proceed. μTAS 94, Twente, 1995 167–179.

[3] M. Mehregany, Y.C. Tai, Surface micromachined mechanisms and micromotors. *J. Micromech. Microeng.*, 1, 1991, str. 73–86.

[3] S.D. Rapoport, M.L. Reed, L.E. Weiss, Fabrication and testing of microdynamic rotor for blood flow measurement. *J. Micromech. Microeng.*, 1, 1991, str. 60–66.

[4] M. Mehregany, Silicon microactuators. In: *Advances in actuators*, A.P. Dorey, J.H. Moore (eds.). IOP Publishing, 1995.

[5] E. W. Becker, W. Ehrfeld, P. Hagmann, A. Maner, D. Münchmeyer, Fabrication of microstructures with high aspect ratio and great structural heights by synchrotron radiation Lithography, Electroforming and Plastic Moulding (LIGA process). *Microelectronic Eng.*, 4, 1986, str. 35–56.

[6] U. Gebhard, R. Günther, E. Just, P. Ruther, Fluidic driven linear motor fabricated by the LIGA process. Microsystem Technologies 96, 5th Int. Conf. Micro, Electro, Opto, Mechanical Systems and Components, Potsdam, Deutschland, Sept., 17–19, 1996.

[7] www, imm-mainz.de.

[8] H. Guckel, T.R. Christenson, K.J. Skrobis, Metal micromechanism via deep X-ray lithography, electroplating and assembly. *J. Micromech. Microeng.*, 2, 1992, str. 225–228.

[9] W.K. Schomburg, D. Mass, W. Bacher, B. Büstengs, J. Fahrenberg, W. Menz, D. Seidel, Assembly for micromechanics and LIGA. *J. Micromech. Microeng.*, 5, 1995, str. 57–63.

[10] W. Ehrfeld, Microtools at work. *MST News Poland*, 4, 1997, 3.

[11] W. Menz, Physikalische Mikrokomponenten für die Mikrosystemtechnik. 2. Statuskolloquium des Projektes Mikrosystemtechnik, 28/29 Nov., 1995, Kalsruhe, Deutschland, 11–14.

[12] T. Bieger, W. Wallrobe, Tribological investigations of LIGA – microstructure. *Microsyst. Techn.*, 2, 1996, 63–70.

[13] F. Wauro, F. Bartels, New implants for middle-ear surgery. *MST News* 19, 1997, 14–15, see catalogue Bartels Mikrotechnik GmbH, Emil-Figge Strasse 76, D-44287 Dortmund.

[14] P. Bley, Microsystem technology in medical engineering – activities at Forschungszentrum Karrlsruhe. *MST News*, 19, 1997, 10–11.

[15] U. Wallrabe, J. Fahrenberg, U. Gebhard, J. Mohr, P. Ruther, A. Ruzzu, Möglichkeiten der Mikrosystetechnik zur Herstellung von Mikrokomponenten für einen Herzkaheter. 2. Statuskolloquium des Projektes Mikrosystemtechnik, 28/29 Nov., 1995, Kalsruhe, Deutschland, 123–127.

[16] S. Zissi, A. Dertsch, J.Y. Jesequel, S.C. Corbel, D.J. Lougnat, J.C. Andre, Stereolithography and microtechniques. *Microsyst. Techn.*, 2, 1996, 97–102.

[17] T. Nakamoto, K. Yamagushi, P.A. Abraha, K. Mishima, Manufacturing of three dimensional microports by UV laser inducted polymerization. *J. Micromech. Microeng.*, 6, 1996, 240–253.

[18] A. Bertsch, H. Lorentz, P. Renaud, Combining microstereolithography and thick resist UV Lithography for 3D microfabrication. Proceed. IEEE MEMS 98, the 11th Annual Workshop on MEMMS, January 25–29, Heidelberg, Germany, 1998, 18–23.

[19] Catalogue MicroTECH, Bismarckstrasse 1426, 45057 Dusburg, Niemcy, www.microTEC-D.com.

[20] K. Ikuta, S. Marno, Y. Fukaya, T. Fujisawa, Biochemical integrated chip toward cell free

DNA protein synthesis. Proceed. IEEE MEMS 98, the 11th Annual Workshop on MEMMS, January 25–29, Heidelberg, Germany, 1998, 131–136.

[21] Optics in microstructure technology. *Microstructure bulletin, Newsletter for Swedish Microstructure Technology*, 3, 1994.

[22] D. Raynerts, P.H. Heeren, H. van Brussel, Microstructuring of silicon by electrodischarge machining EDM, Part I and II. Proceed. 10th Europ. Conf. on Solid-State Transducers, Eurosensors X, Leuven, Belgium, Sept., 8–11, 1996, 251–258.

[23] W. Ehrfeld, H. Lehr, F. Michel, A. Wolf, Microelectrodischarge machining as a technology in micromachining. Proceed. SPIE, Vol. 2879, 1996, 332–337.

[24] T. Hirata, T. Akashi, A. Bertholds, H.P. Gruber, A. Schmid, M.A. Gretillet, O.T. Guenat, N.F. de Rooij, A novel pneumatic actuator system realized by microelectrodischarge machining. Proceed. IEEE MEMS 98, the 11th Annual Workshop on MEMMS, January 25–29, Heidelberg, Germany, 1998, 160–165.

[25] P. Dario, M.C. Carozza, N. Croce, M.C. Montesi, M. Cocco, Non-traditional technologies for microfabrication. *J. Micromech. Microeng.*, 5, 1995, 64–71.

[26] N. Rizvi, Microstructuring with excimer laser. *MST News*, 1, 1999, 18–21.

[27] H. Suzuki, N. Ohya, N. Kawahara, M. Yokoi, S. Ohyanagi, T. Kurashi, T. Hattori, Shell body fabrication for micromachines. *J. Micromech. Microeng.*, 5, 1995, 36–40.

[28] J. Bryzek, Impact of MEMS technology on society. *Sensors and Actuators A*, 56, 1996, 1–9.

[29] K. Petersen, From sensors to microinstruments. *Sensors and Actuators A*, 56, 1996, 143–149.

[30] Opening Session Lecture (invated) of the Federal Minister of Education and Research of The Federal Republic of Germany. World Congress of Microsystems, MicroTECH 2000, September 25–27, Hannover Expo 2000, Germany.

Springer Series in

ADVANCED MICROELECTRONICS

1 **Cellular Neural Networks**
Chaos, Complexity
and VLSI Processing
By G. Manganaro, P. Arena,
and L. Fortuna

2 **Technology of Integrated Circuits**
By D. Widmann, H. Mader,
and H. Friedrich

3 **Ferroelectric Memories**
By J.F. Scott

4 **Microwave Resonators and Filters for Wireless Communication**
Theory, Design and Application
By M. Makimoto and S. Yamashita

5 **VLSI Memory Chip Design**
By K. Itoh

6 **Smart Power ICs**
Technologies and Applications
Ed. by B. Murari, R. Bertotti,
and G.A. Vignola

7 **Noise in Semiconductor Devices**
Modeling and Simulation
By F. Bonani and G. Ghione

8 **Logic Synthesis for Asynchronous Controllers and Interfaces**
By J. Cortadella, M. Kishinevsky,
A. Kondratyev, L. Lavagno,
and A. Yakovlev

9 **Low Dielectric Constant Materials for IC Applications**
Editors: P.S. Ho, J. Leu, W.W. Lee

10 **Lock-in Thermography**
Basics and Use
for Functional Diagnostics
of Electronic Components
By O. Breitenstein
and M. Langenkamp

11 **High-Frequency Bipolar Transistors**
Physics, Modelling, Applications
By M. Reisch

12 **Current Sense Amplifiers**
for Embedded SRAM
in High-Performance
System-on-a-Chip Designs
By B. Wicht

13 **Silicon Optoelectronic Integrated Circuits**
By H. Zimmermann

14 **Integrated CMOS Circuits for Optical Communications**
By M. Ingels and M. Steyaert

15 **Gettering Defects in Semiconductors**
By V.A. Perevostchikov
and V.D. Skoupov

16 **High Dielectric Constant Materials**
VLSI MOSFET Applications
Editors: H.R. Huff and D.C. Gilmer

17 **System-level Test and Validation of Hardware/Software Systems**
By M. Sonza Reorda, Z. Peng,
and M. Violante